Technology
and Global Environmental Issues

Editors

William J. Makofske

Eric F. Karlin

Ramapo College of New Jersey

Associate Editors

Howard Horowitz

Patricia McConnell

 HarperCollins*CollegePublishers*

Acquisitions Editor: Marshall Sanderford
Developmental Editor: Juliana Nocker
Publishing Services: Ruttle, Shaw & Wetherill, Inc.
Project Editor: Janet Nuciforo
Cover Design and Illustration: Louis Fuiano
Photo Researcher: Juliana Nocker
Electronic Production Manager: Angel Gonzalez Jr.
Manufacturing Manager: Angel Gonzalez Jr.
Printer and Binder: RR Donnelley & Sons
Cover Printer: The Lehigh Press, Inc.

Technology and Global Environmental Issues

Copyright © 1995 by HarperCollins College Publishers

Library of Congress Cataloging-in-Publication Data

Technology and global environmental issues / editors William J.
Makofske and Eric F. Karlin.
      p.   cm.
Includes bibliographical references and index.
ISBN 0-673-99181-4
1. Environmental sciences—Congresses.   2. Sustainable
development—Congresses.   3. Environmental protection—Congresses.
4. Man—Influence on nature—Congress.   I. Makofske, William J.
II. Karlin, Eric F.
GE5.T43   1995
363.7—dc20   94-12437
                                                                          CIP

96 97 98   9 8 7 6 5 4 3 2 1

Contents

Preface   v

Contributing Authors   vii

1   **Technology and Global Environmental Issues: An Introduction and Overview   1**
William J. Makofske, Eric F. Karlin, Howard Horowitz, and Patricia McConnell

2   **Population Growth and the Global Environment: An Ecological Perspective   19**
Eric F. Karlin

3   **Agriculture, Technology, and Natural Resources   38**
David Pimentel

4   **Sustainable Food Production   51**
Terry Gips

5   **Appropriate Technology for Sustainable Development   68**
Eric L. Hyman and John D. Ferchak

6   **World Energy: Sustainable Strategies   86**
Christopher Flavin

7   **Hydro-Development   107**
Eric F. Karlin, Howard Horowitz, and William J. Makofske

8   **The Loss of Biodiversity   126**
Howard Horowitz and Eric F. Karlin

9   **Tropical Forests: Problems and Prospects   151**
Howard Horowitz, Eric F. Karlin, and Barbara Bramble

10   **Desertification   174**
Howard Horowitz, Eric F. Karlin, and William J. Makofske

11 **Coastal Ocean Resources and Pollution: A Tragedy of the Commons**   191
John B. Pearce and Angela Cristini

12 **World Air Pollution: Sources, Impacts, and Solutions**   215
Eric F. Karlin and William J. Makofske

13 **Stratospheric Ozone Depletion**   245
William J. Makofske and Eric F. Karlin

14 **The Challenge of Global Climate Change**   269
William J. Makofske

15 **Predicting Climate Change: The State of the Art**   288
Kerry H. Cook

16 **Disaster Revisited: Bhopal and Chernobyl—What Are the Lessons?**   305
Michael R. Edelstein

17 **Gaia, a New Look at the Earth's Surface**   336
Lynn Margulis and Eric F. Karlin

Glossary   G.1

Index   I.1

Preface

The need for interdisciplinary approaches to solving environmental problems has been increasingly recognized, although the implementation of such approaches in undergraduate and graduate education is just beginning. This trend toward inter-disciplinary analysis has been encouraged by the recent emphasis given to research on changes in the global environment by the United States and other countries. With increasing frequency, these issues involve international political negotiations at the highest level. It is generally agreed that global environmental problems are inherently complex and require a broad spectrum of scientific, social, and policy analyses for their understanding and solution. Our objective has been to provide such an integrative and interdisciplinary approach.

Technology and Global Environmental Issues provides a unique, comprehensive, and up-to-date interdisciplinary perspective on today's and tomorrow's major global environmental issues. In keeping with the highly interdisciplinary nature of environmental problems, each chapter was written and/or edited by a team of environmental scholars with a wide range of disciplinary expertise. Scientific, social, and policy concerns have been integrated to provide the reader with a clear and concise overview of each issue. In addition, a broad framework for understanding many of the issues from a historical and developmental perspective is presented. Considerable detail on both existing and potential technologies, activities, and policies that would alleviate and reverse global environmental deterioration is provided. Heavily grounded in the current scientific literature, the book provides insight into the kinds of changes that may ultimately be necessary to provide a sustainable human society in balance with global ecosystems. Extensive references are another strength. Each of the chapters can stand alone and can therefore be read in any sequence. At the same time, cross-referencing and integration among the chapters has been made so that major linkages are highlighted. Chapter 1 provides an integrative synthesis of the book and Chapter 2 develops the central issues underlying global environmental problems; we therefore recommend that they be read first.

Depending on the course design, *Technology and Global Environmental Issues* may stand alone or supplement many areas of study including environmental science, environmental studies, natural resources, pollution, global issues, international business, and environmentally focused science and society courses. Practically all of the topics covered have received considerable media attention in the last few years. In spite of the large number of popularized articles that have appeared in newspapers and magazines, however, public understanding of these complex issues today appears to be limited at best. This volume provides an in-depth overview of each issue that is accurate, current, and based on all available scientific evidence. We have attempted to achieve a balanced and impartial presentation as well as discuss the tradeoffs between various alternatives and policies. At the same time, very real concerns about the problems addressed and their implications for humans is evident. Where it is critical, particularly in the rapidly evolving areas such as global climate change, stratospheric ozone depletion, acid deposition, and biodiversity, the very latest research findings are provided.

Our actions over the next few decades will significantly affect the lives of all who live after us. There is a strong likelihood that the cumulative impacts of human activities will result in a radically altered world, one which may be far less capable of supporting the human species. But we also have the ingenuity and potential to balance our population, technological prowess, and resource consumption with local and global environmental constraints. Such a balance must be achieved if the present global environment is to be maintained on a sustainable basis. It is our hope that this volume, by providing a detailed overview of the forces and interactions that are shaping the critical global environmental issues today and well into the next century, will stimulate the reader to research these issues further and to become actively involved in the creation of environmentally sound (sustainable) human societies.

Acknowledgments

A large contributing factor in the creation of this book has been the dynamic faculty of the environmental programs at Ramapo College, 5 of whom contributed chapters to this volume. Having worked together for roughly 15 years, this group has developed a powerful and integrative approach to the study of environmental issues. Indeed, we have made the interdisciplinary approach the foundation of this book, and have strived to incorporate our respective disciplines to provide comprehensive and detailed coverage of the subject matter. The Institute for Environmental Studies (Ramapo College of New Jersey), as a prime facilitator of our interdisciplinary collaborations, has also been critical to the success of this project. Indeed, this volume is an outgrowth of a proceedings titled *Technology, Development and the Global Environment* which was published in 1991 by the Institute for Environmental Studies .

Although the majority of the chapters have been written by Ramapo environmental faculty, many other people deserve credit for helping in this endeavor. Foremost among them are the other contributing authors who have generously shared their expertise and knowledge. Special thanks goes to our associate editors, Howard Horowitz and Patricia McConnell. We also appreciate the many thoughtful suggestions made by those who reviewed the book; their comments helped to improve clarity and cohesiveness. We especially thank Melissa Lerner of HarperCollins for her help in bringing this book to the attention of her publisher. It has been a pleasure to work with HarperCollins and we thank the editors and editorial staff for their support and encouragement.

January 1995   *William J. Makofske*

Eric F. Karlin

Contributing Authors

Barbara Bramble
Director of the International Program
World Wildlife Federation
1400 16th Street, NW
Washington, DC 20036

Kerry H. Cook
Assistant Professor
Atmospheric Sciences Program
Cornell University
Ithaca, NY 14853

Angela Cristini
Professor of Biology
School of Theoretical and Applied
    Science
Ramapo College of New Jersey
Mahwah, NJ 07430-1680

Michael R. Edelstein
Professor of Psychology
School of Social Sciences and Human
    Services
Ramapo College of New Jersey
Mahwah, NJ 07430-1680

John D. Ferchak
Agricultural Scientist
Appropriate Technology International*
1828 L Street, NW
Washington, DC 20036

Christopher Flavin
Vice-President of Research & Senior
    Researcher
Worldwatch Institute
1776 Massachusetts Ave., NW
Washington, DC 20036

Terry Gips
Executive Director
International Alliance for Sustainable
    Agriculture
The Newman Center
University of Minnesota
1701 University Ave., SE
Minneapolis, MN 55414

Howard Horowitz
Associate Professor of Geography
School of Theoretical and Applied
    Science
Ramapo College of New Jersey
Mahwah, NJ 07430-1680

Eric L. Hyman
Program Economist
Appropriate Technology International*
1828 L Street, NW
Washington, DC 20036

*Appropriate Technology International (ATI) is a not-for-profit development assistance organization created by the U.S. Congress and supported by USAID and international agencies.

Eric F. Karlin

Professor of Plant Ecology
School of Theoretical and Applied
  Science
Ramapo College of New Jersey
Mahwah, NJ 07430-1680

William J. Makofske

Professor of Physics
School of Theoretical and Applied
  Science
Ramapo College of New Jersey
Mahwah, NJ 07430-1680

Lynn Margulis

Distinguished University Professor
Department of Biology
University of Massachusetts
Amherst, MA 01003

Patricia McConnell

Research Associate
Institute for Environmental Studies
Ramapo College of New Jersey
Mahwah, NJ 07430-1680

John B. Pearce

Deputy Director
Northeast Fisheries Center
National Marine Fisheries Service
Woods Hole, MA 02543

David Pimentel

Professor of Insect Ecology &
  Agricultural Sciences
College of Agriculture and Life
  Sciences
Cornell University
Ithaca, NY 14853

Chapter 1

Technology and Global Environmental Issues: An Introduction and Overview

William J. Makofske, Eric F. Karlin, Howard Horowitz, and
Patricia McConnell

In this volume, we examine relationships between development, technology, and environmental dilemmas in both industrialized and developing countries. Ultimately, our goal is to understand these relationships and to identify options that would allow ecosystems and human societies to become sustainable. The environmental issues addressed in this book require a global perspective for their understanding and global cooperation for their ultimate solution. This was not always so and is still not always recognized. There are several reasons why this global perspective has come to be necessary in the latter half of the twentieth century.

First and foremost, the rapidly expanding world population, now over 5.7 billion people, and the widespread use of ever more powerful technologies, have an increasing impact on the planet. We do not know what the present carrying capacity is for the human population, but there is ample evidence that we are currently overstepping the boundaries of life-support systems throughout the world.

Second, human activities supplying resources for human needs are now at a scale unprecedented in human history. Monocultural agriculture has perhaps had the greatest impact on natural systems. Accelerated forest destruction, overgrazing, and poor soil management, resulting in soil erosion, can reduce the productivity of the land for several millennia. We have overfished the oceans so that yields have leveled off or decreased: many fisheries have been depleted. It is clear that resource consumption is affecting ecosystems on a global level.

Third, the extensive pollution created by the activities of over 5.7 billion people has a multitude of impacts on ecosystems and life-support systems. Recycling is a major process in natural ecosystems and we can create some wastes without destroying natural systems. However, the twentieth century, often called the exponential century, is taxing the very life-support systems that humans depend on. Acid deposition is severely degrading many thousands of lakes and large tracts of forests. Water pollution is overwhelming estuaries and providing ever more frequent algae

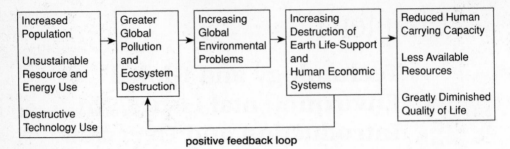

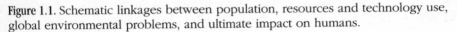

positive feedback loop

Figure 1.1. Schematic linkages between population, resources and technology use, global environmental problems, and ultimate impact on humans.

blooms and fish kills. The depletion of the ozone layer by chlorofluorocarbons (CFCs) and other gases threatens the viability of the planet for many forms of life. Fossil energy production and use are seriously damaging ecosystems and air quality. The greenhouse effect threatens to change the climate in ways that will affect agriculture, ecosystems, and many other societal activities adversely. The trends that we see in population, resource use, and destructive technologies will ultimately affect the livability of planet Earth, for ourselves and for future generations. Figure 1.1 shows schematically the relationships between these factors and the final result—reduced carrying capacity for humans and many other life forms.

In a very real sense, we are carrying out an unprecedented global experiment with our environment. The change is rapid and ominous, and we are unable to fully comprehend its ultimate effects. Ecology tells us about the range of carrying capacities of ecological systems: overshooting that capacity can disrupt the system. To what extent are our global life-support systems threatened? What will happen as the developing countries increase their consumption of resources and energy to meet the needs of their increasing populations and to raise their standard of living? Can developing countries integrate environmental and developmental needs to achieve sustainable growth? Can the industrial countries change their consumptive habits and reduce pollution and use of resources? How quickly do we have to become sustainable in order to avoid catastrophic damage to the environment? These and related issues were the major focus of the Earth Summit held in Brazil in 1992. They will undoubtedly remain as critical agenda items for the world in the twenty-first century.

Beyond the impact of population, resource use, and pollution, there are contributing factors that help explain why the global environment emerged as a focus of our attention in the latter half of the twentieth century. Part of our global concern comes from the increasing ability of technology to monitor the environment and to rapidly analyze and transmit information. We are now more aware of changes in the biosphere than in the past. For example, ozone depletion and the greenhouse effect would not be detectable in their early stages without the sophisticated equipment presently employed by scientists. Satellites showing the burning of tropical forests bring home the extent of change occurring in remote regions. Today, telecommunications and satellites are making Earth a global village. Even so, our ability to understand the causes and effects of our activities, and to predict Earth system behavior are still at a very primitive level.

Other changes in the way we live our lives have modified our perspective from a parochial to a more global one. Improved transportation systems and affluence in more developed countries (MDCs)[1] allows much more first-hand observation of various parts of the world. Conversely, it has allowed the traditional and indigenous peoples of less developed countries (LDCs)[2] more first-hand observation of western ways. Indeed, travel, coop placements, and study abroad increasingly allow our college students to gain an international perspective.

Newspaper and magazine coverage of international business and economic interdependence have contributed to this increasing "world consciousness" developing in the United States. The rise of international corporations and increasing world trade has necessitated a global perspective for many industries if they are to survive and prosper. The emergence of the North American Free Trade Agreement is just one outcome of this growing interdependence.

The focus on international news goes well beyond business. Socially, there are concerns about equity and human rights around the globe. Environmentally, we see increasing coverage about a range of global issues, from depletion of the ozone layer, acid deposition, and the greenhouse effect to issues surrounding ocean pollution, tropical rain forests, and endangered species. There is no doubt that the latter part of the twentieth century will be remembered as the time when both ecological and global consciousness first blossomed in a meaningful way on the planet. The real need, however, is to translate concern about what is happening into understanding and then into meaningful change.

There are alternatives to the dismal process described in Figure 1.1. Population stabilization, sustainable resource and energy use, and appropriate technology, as shown schematically in Figure 1.2, can be implemented to allow adequate resources for an improved quality of life for the world's people and to protect fragile life-support systems, now and in the future. It is our hope that this collection of papers provides insight into how these alternative approaches may be applied to specific problem areas.

Although we have subdivided the chapters into sections with general thematic headings, *Population, Natural Resources and Energy, Pollution,* and *Reexamining Technology and Our Relationship to Earth,* we recognize that the complexity and interconnectedness of the global problems that we face often defy such simple categorization schemes. Throughout the text, we have provided cross-referencing to other chapters in this book when appropriate. Each chapter itself also recognizes the complexity and interactions among the set of issues that we face. As an example, Figure 1.3 provides a schematic linking of one topic, hydro-development, to other problem areas, showing that there are possibilities for both positive and neg-

---

[1] Throughout this volume, the terms *more developed countries, developed countries,* and *First World* are used interchangeably.

[2] Throughout this volume, the terms *less developed countries, developing countries,* and *Third World* are used interchangeably.

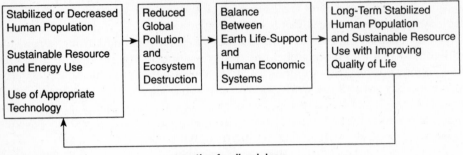

negative feedback loop

Figure 1.2. Alternative relationships between population, resources, and technology that could lead to a sustainable world system.

ative outcomes, depending upon how the hydro-development is approached and implemented.

In the remainder of this introductory chapter, we provide the reader with a brief discussion and overview of the issues that are presented later in the text. The aim is to provide some background and a broader perspective for understanding individual articles that follow and the linkages among them. We encourage readers to further develop their own linkages as they read the chapters in this book.

Overview

Population Growth

There has been an ongoing debate for many years about the relative importance of population size in contributing to environmental problems. Some argue that it is not the numbers per se that are important; it is technology and resource use, with its resulting impact that is paramount. Others, recognizing that rising populations will make every problem worse, have focused on population growth as the primary concern. We do not pretend to have the ultimate answer to this question. Perhaps viewing population growth, resource use, and technology as interacting synergistically, however, may be closer to the truth. In addition, in many developing countries rapid population growth has created problems that seem beyond that country's ability—political, technical, economic, or institutional—to solve.

More generally, the study of population growth and demography provides a useful perspective on the twentieth century and beyond, both for the industrialized countries and for the developing countries. The gradual decline in birth rates due to increasing standards of living, the use of contraceptive technologies, and the changing role of women in modern industrial societies need to be juxtaposed with the situation in many developing countries today. In most of these nations there has been a rapid decline of death rates due to modern medicine, disease control, and sanitation advances. According to conventional wisdom, the so-called demographic transition to a more stable population based on lowering birth rates is sup-

HYDRO-DEVELOPMENT	
Positive	Negative
Renewable and sustainable form of energy.	Accelerated erosion could lead to short useful lifetime for project.
Reduction in poverty from development could reduce pressure for population growth.	Development may encourage greater regional population growth and associated environmental impacts.
Reduction of air pollution and climate change impacts.	Associated development may increase air and water pollution.
Provides water reservoir for agriculture, human consumption, and recreation.	Changes ecosystem from flowing to standing water with potential human health impacts. Increases deforestation & biodiversity loss.
Provides electricity for sustainable development.	Possibly inappropriate technology depending on scale, location, and use of electricity.

Figure 1.3. Schematic linking of hydro-development to other problem areas, indicating several potential positive and negative effects on other problem areas.

posed to occur along with development, but what we see today is quite different. The transfer of modern technologies has led not to modernization, development, and the demographic transition, but to what Lester Brown of Worldwatch Institute has characterized as a "demographic trap." The rapid increase of population in many LDCs prevents economic gains from being made: this then feeds back into the continuation of high population growth rates while resources and life-support systems are consumed. With the melting of the cold war between the United States and the former Soviet Union, this may well be the greatest threat to world peace and stability as well as to the global environment that the world now faces. The impact of the human population growing to some 12 billion by the end of the next century will provide the main drama for the immediate future. The bulk of this population growth will occur in the LDCs and will enhance the differences between the haves and have-nots. Pressure for human survival may overstress natural systems, ultimately leading to decreased human carrying capacity in many countries.

Strongly coupled with population growth and economic development, the state of health of the world's human population is of utmost concern. Although a strong concern, it does not seem to be a top priority for many of the developing countries today. This is partly because health, defined broadly as physical, mental, and social well-being, is directly connected to a set of other conditions such as steady supply of food and adequate nutrition, good sanitation and water supply, literacy and education, family planning, and adequate housing and transportation. Although poverty is relative, there is agreement that much of the world's population is unable to afford the conditions leading to good health. The rapid population growth that we see today makes it difficult, if not impossible, to provide for basic human needs for a significant portion of the world's population. In Chapter 2, "Population Growth and the Global Environment: An Ecological Perspective," Eric Karlin argues that population stabilization is of the highest priority if humans are to improve the present situation.

Natural Resources and Energy: Sustainable Alternatives to Ecosystem Destruction

Agriculture

Agriculture is still the dominant activity in the world in terms of number of people involved. For vast numbers of people, the daily struggle is to produce enough food for survival. This is difficult for contemporary Americans to understand, because we are increasingly isolated from food production: a relatively small and decreasing number of farmers provide a surplus of food for this country. A comparison of subsistence and modern agriculture is a beginning point for understanding the current situation.

Today our modern food system is totally dependent on petroleum inputs. The use of large machinery, irrigation, pesticides and fertilizers, transportation, processing, packaging, and storing require huge energy inputs. Many analysts speak of "eating oil," since we put in about ten times the energy inputs into the food system than we get back as food energy. So-called primitive agriculture, based on hand labor, gave back 10 to 100 times more energy in the form of food calories compared to the labor energy input. In Chapter 3, "Agriculture, Technology, and Natural Resources," David Pimentel discusses the serious energy and natural resource limitations of present agricultural technology, together with options for reducing dependence on fossil fuels and other inputs.

Why is this important? No one is arguing that we must go back to subsistence agriculture. The issue here is sustainability. Modern agriculture is in trouble. We are today essentially at the historic peak of world oil production. In the near future, oil will again be a scarce and more expensive resource. Food produced by modern agriculture will be considerably more expensive to produce. Many analysts see a major restructuring of agriculture in the MDCs as necessary. But in the past few decades, a highly technical form of agriculture called the *Green Revolution* (hybrid grains, fertilizers, pesticides, irrigation, and machinery) has been exported to many developing countries in an attempt to keep food production ahead of rapidly rising populations. Indeed, yields have increased dramatically. The Green Revolution, however, has not been so successful in terms of environmental and societal side effects, and there are also problems with its high cost and sustainability. For example, many poorer farmers cannot afford to invest in Green Revolution techniques, leading to even greater inequities between rich and poor. The need to feed rapidly rising populations forces countries to pursue Green Revolution technologies, even though they cannot be continued indefinitely. The increases yielded by Green Revolution techniques seem to be reaching saturation limits. Moreover, the health and ecological costs of widespread use of pesticides and fertilizers are only now becoming apparent to those in LDCs. There are other serious limits. The great majority of the world's potentially arable land is already being cultivated; attempts to increase farming area by removing tropical forests and by expanding into marginal areas have major drawbacks. The age-old dream of increasing ocean productivity is very far from realization; we may have already reached or exceeded limits in terms of sustainable fishery production.

Do we have other options that could resolve these dilemmas of our modern agricultural system? Sustainable agriculture encompasses many current diverse farming systems: a common feature is their ability to sustain productivity over time without large resource inputs, and without destroying the productivity base and the environment. In recent decades, increasing support for sustainable agriculture has come from many places. For example, government studies have consistently shown that sustainable techniques can help farms become more economical: they can provide equal or greater yields per acre with less resource inputs. Today, consumer demand for organic produce is relatively strong and growing, causing many farmers to shift to organic farming techniques. Increasingly, analysts argue that modern agriculture can, and must, be modified if we are to have productive agriculture in the twenty-first century. In Chapter 4, "Sustainable Food Production," Terry Gips outlines a vision of an agricultural future that is not only physically sustainable, but socially sustainable as well.

Appropriate Technology

The questions surrounding development and technology transfer are sensitive issues in the global community. Aid provided by industrialized countries has not been without strings and has often been self-serving. Much of the technology transfer is controlled by the local elites, is inappropriate for the culture, and does not benefit the vast number of people who need help. In the past few decades, a new theme, appropriate technology (AT), has emerged. In AT, the social, economic, political, and environmental implications of a technology are studied in order to determine what the actual benefits of a technology will be. The technologies that are beneficial are not necessarily the same ones that industrialized countries use. In Chapter 5, "Appropriate Technology for Sustainable Development," Eric Hyman and John Ferchak review many of the technologies that have been found to be successful in improving the lives of poorer people in LDCs, or that have the potential for significant benefits. Most of these technologies also alleviate the pressure on life-support systems, thereby enhancing both development and environmental conditions. Because of the importance of small-scale agriculture and decentralized use of natural resources to these countries, it is not surprising that AT focuses on these areas. Perhaps a few examples will demonstrate the often different needs of LDCs.

In many parts of the world, people depend on biomass energy such as wood, crop residues, and cattle dung to provide energy for cooking meals. Traditionally, the dung was used to fertilize fields. As fuelwood becomes more scarce, farmers are forced to use cattle dung. Using cattle dung for cooking, instead of letting it return to the fields, breaks the soil fertility cycle and leads to a vicious downward spiral: the subsistence farmer gets less and less yield, the soil deteriorates and is subject to more erosion, and the farmer ultimately cannot support his or her family on the land. This situation occurred in rural villages in India where there was increasing use of cattle dung for cooking because of shortages of fuelwood. Yields dropped as soil fertility declined. With the energy crisis of the 1970s, artificial fertilizer made from fossil fuels was too expensive. An appropriate solution turned out to be Gobar, or biogas, plants in which the dung is mixed in a slurry and anaerobically decomposed to produce methane gas, which may be used for cooking and other purposes. The remaining slurry is still a good fertilizer. The biogas plants actually provide both energy and fertilizer rather than using the dung for either purpose alone. Within a few short years, tens of thousands of biogas plants were built and have worked successfully. The technology was economically viable, simple to use, environmentally sound, and applicable to the rural setting.

Some attempted applications of appropriate technology have had difficulties. As noted above, fuelwood scarcity is a major problem in many developing countries where a major part of the energy use of families involves cooking food. The poor cannot afford the more expensive options, such as kerosene or charcoal, so they collect whatever wood is available. With increasing population growth and deforestation occurring, both time and distance traveled to do this has increased dramatically, with the chore most often falling on the women and children. The wood burned is often used inefficiently, and it was thought that an improved wood stove program could help these people. Many such programs were started in the past 20 years with minimal success. The reasons were numerous: the new wood stoves often were not very efficient or were too expensive. If the stoves were given to a local populace, they were not valued much, and stoves that were constructed locally often did not meet the critical dimensions needed for good performance. Within the past few years, wood stove programs have been improved and the success rate has been much higher. The focus is now on the urban poor since they are in a monetary economy and must buy fuel, and wood scarcity is greater there. Emphasis has been placed on portable metal stoves instead of the more expensive clay ones. Construction is often done by local artisans to provide quality yet affordable products. Generally, the learning curve for applying AT successfully has been considerably improved in the last decade. It is ready to undergo significant expansion if funding can be found.

Energy

The role of energy in understanding the population, environment, and development dilemmas is critical. The study of energy use can provide great insight about cultures and historical development. It therefore provides a thread for understanding present and future possibilities for developing and industrialized countries.

Historically, the use of energy and technology marked the transition to the industrial age. Inanimate energy sources replaced humans or animals for driving machinery to do work. As long as there was a seemingly inexhaustible supply of cheap energy—the fossil fuels—anything seemed possible. Rapid growth during the past few centuries in the industrialized nations was primarily due to unlimited use of coal, gas, and oil. If each person is equivalent energetically to a 100 watt light bulb, the average American has the equivalent of 100 servants at his or her disposal. Human energy expenditure is now miniscule in the industrialized nations, as machinery and appliances, run mostly by fossil fuels, substitute for human labor and provide the ability to turn raw materials into manufactured products. Although the energy crisis of the 1970s brought home the connections between energy, economics, and environment (the three e's), to the American people it did not fundamentally change American energy policy or its massive dependence on nonrenewable energy sources.

This massive use of nonrenewable energy, however, has consequences: the burning of fossil fuels produces air pollution. It is intimately connected to the acid deposition problem and the global greenhouse effect. Nuclear power was supposed to be the next important energy source, but it has not lived up to its promises. It is too expensive and capital intensive, and the radioactive waste disposal problem is yet to be solved. It is also considered by many to provide unacceptable risk in case of accident. The industrialized nations have not incorporated social, political, and environmental costs into pricing of nonrenewable energy supplies, and have huge and often hidden subsidies supporting their continued use.

A possible solution to the energy question could be the use of renewable energy: solar energy and solar-related technologies, such as hydropower, biomass, and wind energy. While such energy sources have been used extensively in the past and are currently massively used by the poor of many countries, they are not unlimited in amount, and in some cases further research and development efforts are needed for them to compete economically with fossil fuels. Yet renewable energy coupled with increasing efficiency of use represents a viable alternative to the present nonrenewable energy system. In Chapter 6, "World Energy: Sustainable Strategies," Christopher Flavin argues that the joint goals of adequate energy supplies for all nations and a sustainable environment are achievable. He presents an energy path that includes significant improvements in energy efficiency, coupled with a rapid phase-in of renewable energy supplies in the near future.

Since 1973, increasing efficiency has succeeded in limiting energy growth and environmental damage, while still allowing expanded economic development. Increasing efficiency further can stabilize energy growth while allowing for a more equitable distribution of energy between industrial and developing countries. A major benefit will be avoiding the worst consequences of a potential global warming by limiting the carbon dioxide emitted to the atmosphere. In the next few decades, efficiency should be the cornerstone of U.S. and world energy policy.

The use of energy in industrialized countries should be contrasted with the three-quarters of the world's population who are energy poor. Poverty and lack of energy resources are intimately connected, as the use of energy provides heat and hot water, cooking, pure water, and manufactured goods. Development and modernization involve the transfer of technologies and their accompanying energy consumption to these energy-poor countries. That is one of the difficulties. Modern technology is fossil-fuel dependent, yet the source is limited and unevenly distributed throughout the world. Moreover, in the context of developing countries, smaller, more decentralized energy technologies may be more desirable than large, centrally planned ones.

Development and Ecosystem Destruction

Hydro-Development

The issues raised by hydro-development, particularly in tropical regions, intersect with several other important areas. This is partly because hydro-development usually has multiple use objectives—including irrigation for agriculture, electricity production, flood control, tourism, and economic development—and partly because the unintended impacts of large-scale hydro-development are so many and so large.

As a renewable and sustainable form of energy already providing one-fourth of the world's electricity, there is little doubt of hydropower's continued growth; only a small percentage of its potential has been tapped in Asia, Africa, and Latin America. This growth will occur for many reasons, including the prestige accruing to the government, the potential for increasing productivity and jobs, and the ease of financing such projects with international loans. Unfortunately, along with the benefits of such development come a range of undesirable effects, particularly if the full ramifications of the project are not systematically analyzed during the planning stages and mitigation efforts are not implemented.

There is increasing recognition that hydro-development under many circumstances may indeed be a poor choice for providing energy. The amount of land flooded per unit of power generated, the effect on displaced indigenous peoples, changes in ecosystems (both upstream and downstream), health effects on local populations, and silting are only a few of the factors that must be considered. The recent controversy over hydro-development in James Bay in Canada is sensitizing Americans to the fact that impacts may be considerable even in nontropical regions.

In Chapter 7, "Hydro-Development," Eric Karlin, Howard Horowitz, and William Makofske present the two faces of hydropower. On the one hand, it has the potential to provide sustainable energy, particularly to many LDCs who desperately need the energy. On the other hand, the track record of hydro-development, ecologically and socially, leaves much to be desired. By mitigating impacts of large-scale development, and by using small-scale approaches, a much better balance may be achieved.

Biodiversity

Many human activities, such as deforestation, modern large-scale agriculture, and hydro-development, destroy the habitats of many plants and animals. The end result of these actions could be the loss of over half of all existing species in the next few decades, presenting what is termed a *biodiversity crisis*. The present mass extinction of species is unique. It is the direct and indirect result of actions taken by a single species—humans—that is causing the destruction of ecosystems and the concomitant loss of species. This loss is extraordinarily accelerated compared to the natural rate of extinction. In Chapter 8, "The Loss of Biodiversity," Howard Horowitz and Eric Karlin analyze the reasons for the current biodiversity crisis, and more importantly, explore ways that humans can begin to stem these losses before they become a flood.

Fueled by rapidly increasing population and ever more powerful technology, the removal and simplification of ecosystems is rapidly reducing the remaining environmental space for other species. Thus, the primary reason for loss of biodiversity is the loss of habitat due to the growth of human activities that dominate and change the landscape. Even within agriculture itself, there is a major decline in the diversity of domesticated plants and animals utilized.

Today, there are various approaches to both species and ecosystem preservation. Species have been protected by the Endangered Species Act in the United States; by laws, conventions, and treaties for specific species; wildlife refuges; and zoos, aquariums, and botanical gardens. Each species approach, unfortunately, is limited in scope and effectiveness. The alternative approach, preserving representative ecosystems, particularly rarer ecosystems, protects the greatest environmental space for the greatest number of species. A final approach, given the large loss of habitat that has already occurred worldwide, is ecosystem restoration. This is usually difficult and is being investigated in selected areas around the world today.

Although the loss of diversity is serious today, it will become of even greater concern in the near future. This is due to the intensification of human activities as population increases and to increasing global pollution levels. Increased levels of acid deposition, water pollution, UV radiation from ozone loss, and potentially, global warming, will exact a greater toll on life. Humans must begin to take action now to reverse these trends and to prevent this calamity from taking place. A framework for global action was established at the Earth Summit in Rio de Janeiro in 1992. The Biodiversity Convention, when ratified, will require nations to inventory plant and animal species and to slow their extinction.

Tropical Forests

The tropical forests have great ecological value because of their biological diversity. It is estimated that over 50% of the species on Earth live in these regions, and yet these forests are rapidly being destroyed along with their inhabitants. This loss will directly and indirectly affect humans. As yet undiscovered species will have great utility as food crops, in natural pesticides, as disease-resistant plants, and in medicines. Tropical forests have value in preventing erosion, reducing flooding, and preventing silting of tidal wetlands. They hold vast amounts of carbon, which would otherwise contribute to the greenhouse effect. They also serve to moderate and moisturize the climate. There is increasing concern today that deforestation in Africa may be leading to a drier climate and even may be causing increasing drought. Yet, population pressure creates demand for more food products, and forests have been increasingly converted into crop and grazing lands.

Tropical forests may seem to be prime places for agriculture, with their lush growth and high rainfall. Yet attempts to expand agriculture in these areas have had little success. Why? A major reason has to do with soils and climate. A large percentage of the nutrients are in the vegetation and not in the soil, as is the case in temperate soils. Many tropical soils are actually quite infertile. Burning and clearing the forests provide ash, which is high in nutrients. The high rainfall, however, leaches these nutrients from the soil system quickly. After a short time, the soil becomes infertile and industrial fertilizers are required. When exposed to the sun, some tropical soils bake into hard, rocklike laterites. Thus the land quickly becomes unsuitable for agriculture, and the people move on to cut down and destroy more of the forest.

Much effort is presently being expended to stem the loss of tropical forests. A wide range of approaches are being explored, and these range from setting aside biosphere preserves and extractive reserves to the use of more environmentally sound harvesting practices and improved reforestation techniques. Although promising, many implementation details need to be worked out, and successful approaches need to be rapidly propagated to prevent these ecosystems from suffering irreparable losses.

The importance of tropical forests and related issues of protecting climate and biodiversity are becoming recognized as major world concerns today. For example, a general statement of forestry management principles emphasizing sustainable forestry was adopted at the Earth Summit in Rio de Janeiro, Brazil, in 1992. In Chapter 9, "Tropical Forests: Problems and Prospects," Howard Horowitz, Eric Karlin, and Barbara Bramble examine current problems and possibilities for preventing additional losses of tropical forests.

Desertification

Expanding agriculture or intensifying grazing in hilly areas and semi-arid areas poses long-term risks to the environment and ultimately the people. Deforestation of hilly areas for agriculture provides, at best, only temporary yields. It also increases soil erosion and flooding. Expanding cultivation to drier areas usually requires irrigation water, which is not always available. When it is not, wind and water erosion degrade the exposed soil. When irrigation water is available, salinization can cause the eventual loss of the land as a productive system. Desertification does not necessarily mean the creation of absolute deserts, but is a general term for any decline in the biological productivity below the land's potential. It is the result of the sustained overuse of land, and is most frequent in, but not limited to, semi-arid areas. As much as one-third of the land area in the world may be affected by some degree of desertification, including portions of every continent. In Chapter 10, "Desertification," Howard Horowitz, Eric Karlin, and William Makofske analyze the causes, results, and prevention of desertification. It is evident that the implementation of well-known technologies and procedures can halt and potentially reverse the effects of desertification if the political will and resources are made available.

Solutions to desertification include relatively straightforward actions: better forest and rangeland management, use of climatically and regionally appropriate plants and animals, and erosion control. But they do require resources, which many of the most severely affected countries are lacking. The United Nations, a leader in the fight against desertification, estimates an expenditure of perhaps 4.5 billion dollars per year would check and reverse the problem. The benefits would be enormous; the loss due to desertification is conservatively estimated at over 20 billion dollars per year. Desertification illustrates that the health of ecological systems and economic health are intimately connected, a fact that we are only slowly beginning to grasp.

Pollution and Ecosystems: Problems and Alternatives

Ocean Pollution

In the past few years, particularly along the eastern seaboard of New York and New Jersey, the subject of ocean pollution has evoked strong public reaction. Algae blooms, loss of fisheries, garbage washed up on the beach, polluted seafood, and other dislocations have affected the beach-going habits of millions of people and have caused significant economic decline in the economies of coastal communities. Similar problems are occurring throughout the world.

Ultimately, it is what we do on land that adversely affects the ocean. Coastal development leading to runoff of biocides (i.e., pesticides, herbicides), and the emission of industrial and sewage wastes, cause toxic contamination and nutrient loading. Dredging and fresh water diversion, again the result of development, are prime culprits in damaging the estuarine and ocean environment. The root causes of estuary and ocean decline are much the same throughout the world. Along with pollution and deteriorating estuaries, the quest for food, particularly high quality protein, has caused a major impact on fisheries. Present pressure on many fisheries may exceed sustainable yields and preclude even the present harvest of much needed food for the growing world population. In Chapter 11, "Coastal Ocean Resources and Pollution: A Tragedy of the Commons," John Pearce and Angela Cristini examine the role of development in creating nearshore pollution problems and its impact on coastal ecosystems and resources. Solutions involve efforts to manage environmental deterioration and resource overuse, primarily resulting from uncontrolled development and unsustainable pressure on fisheries.

Multijurisdictional management, particularly in Europe and the United States, may provide models of how to develop international plans to ensure the water quality of the ocean environment. Even so, continued population growth, along with accompanying human activities, puts greater pressure on the surrounding bays and estuaries. Much greater efforts at pollution control will be needed just to stay even.

Air Pollution

The atmosphere is another fluid that is shared by the world's inhabitants. Likewise, it has been subject to "the tragedy of the commons." The consumption of fossil fuels by power plants, industry, and transportation provides much of the air quality problems around the world today. Although the emphasis initially was on local air pollution problems, the focus in the past decade has turned to more regional (acid deposition) and even global (climate change and stratospheric ozone depletion: see following sections) problems. It is important, however, to recognize that the various problems are linked in complex ways, and they all are in desperate need of solution.

The very attempt to solve local pollution by increasing the height of smoke stacks actually created regional air pollution in the form of acid deposition. Both sulfur dioxide (SO_2) and nitrogen oxides (NOx's), their secondary aerosol products, have increased the acid loading of the environment, changing the pH of water and soil in sensitive ecosystems. This also causes toxic metals from the environment to become soluble and leach into water. Acidification ultimately changes the species composition of ecosystems.

The major impact of acid deposition has been on lakes and forests in northern Europe, Canada, and the northeastern United States. Although there is little doubt that the acid deposition is to blame, the effect is complicated and in some cases it is difficult to apportion damage, since other forms of pollution (ozone, sulfur dioxide) are often involved. In addition, the acidified environment makes trees more susceptible to other natural stresses. The effects of acid deposition will be intensified as energy use is increased in developing countries where acid-sensitive ecosystems are found. China's rapid coal-based industrial buildup, for example, is causing dramatic pollution of the region's atmosphere.

Of course, acid deposition is only one of several interrelated air pollution issues that includes: regional tropospheric ozone (smog), industrial smog, and visibility problems (haze). In Chapter 12, "World Air Pollution: Sources, Impacts, and Solutions," Eric Karlin and William Makofske examine the major air pollutants, their interaction in creating regional and global air pollution, and policy efforts to alleviate air pollution problems from a national and international perspective.

Stratospheric Ozone Depletion

In 1974, two scientists warned that chlorofluorocarbons (CFCs), synthetic chemicals widely used in aerosols and as refrigerants, might be depleting the stratospheric ozone layer. As the ozone layer thins, more ultraviolet light (UV-B) penetrates through the atmosphere, reaching the Earth's surface with potentially severe consequences. The health effects on humans include significantly increased levels of skin cancer, increased incidence of cataracts, and potential damage to the immune system. For ecosystems, increased UV-B is likely to negatively affect susceptible animal species; at particular risk are single-celled organisms; larval stages of multicellular organisms, including crabs, shrimp, and fish larvae and juveniles; and potentially many amphibians. In addition, many plant species, including algae, food crops, and forest trees, are known to be damaged by increasing UV-B levels.

In spite of extensive scientific study for over a decade (1974–1984), the warnings of ozone depletion remained experimentally unverified until 1985. In that year, data published by a British research team in Halley Bay, Antarctica, indicated a significant decrease in ozone over several years, culminating in a massive 40% decrease in ozone at their site in 1984. These initial findings were quickly verified by Nimbus 7 satellite measurements and the *ozone hole,* a significant depletion of ozone covering millions of square miles and extending halfway to the Equator, was discovered.

Attempts to develop international treaties to protect the ozone layer, begun by the United Nations Environmental Programme in 1977, ultimately led to a successful conclusion. In 1987, the Montreal Protocol, freezing CFC production to 1986 levels by 1990, and then cutting it back by 50% by 1999, was signed by many nations. The ink was barely dry, however, when a NASA report showed losses of from 1.7 to 3% in the Northern Hemisphere, greater than the 2% loss predicted to occur by 2075. Nations reconvened in London in the summer of 1990 and the Montreal Protocol was amended to completely phaseout production of CFCs by the year 2000. Shortly afterward, the phaseout was moved up to 1995, partly in response to the potential for a comparable Arctic ozone hole.

In Chapter 13, "Stratospheric Ozone Depletion," William Makofske and Eric Karlin examine the present scientific and ecological understanding surrounding ozone depletion, and the accompanying development of a global agreement. While the ozone depletion problem is now being addressed by a rapid phaseout of CFCs and halogen production worldwide, it is still uncertain whether the world has acted quickly enough to avoid severe ramifications. The ozone layer in the Northern Hemisphere appears to be depleting twice as fast as was predicted just a few years ago. There is still fear that a significant ozone hole will form in the Northern Hemisphere, and because of the long residence time of CFCs in the atmosphere, depletion will continue for decades to come. Furthermore, our understanding of the processes involved is still not good enough to predict with certainty what the final outcome will be. As many scientists have suggested, we are conducting a massive global experiment on the Earth's atmosphere with an uncertain outcome. We may be in for more surprises.

Climate Change and the Greenhouse Effect

With industrialization and the rapid growth in the use of fossil fuels has come another pollutant, carbon dioxide (CO_2). Although less spectacular in its immediate impact on the environment than acid deposition or ozone depletion, it has potential long-range effects that are perhaps even more severe and more difficult to accurately forecast. The greenhouse effect, a natural and desirable action of certain gases in the atmosphere that absorb and reradiate escaping long-wavelength radiation from the Earth, helps keep the Earth warm so that life as we know it will flourish. Increasing greenhouse gas concentrations in the atmosphere, however, will cause the average global temperature of the Earth to rise. While the greenhouse effect is partly caused by naturally occurring sources, the combustion of fossil fuels together with deforestation and other human activities has added significant quantities of these gases into the environment over the past century (CO_2, ozone, methane, nitrous oxide, and CFCs all contribute). In recent years, scientists have developed better models of the Earth's climate and have been able to simulate the effect of added CO_2 and other gases to the atmosphere. The results from most of the models indicate a rather large warming trend of 1.3 to 4.5°C (2.3–8.1°F) if atmospheric carbon dioxide levels are doubled.

The greenhouse effect poses special questions and dilemmas. The process is gradual, occurring over a time scale of decades. Its onset may be almost imperceptible at the beginning, since there are climate fluctuations that occur normally. Given that predicting longer term climate change is a very tricky business, how well do we believe the models that scientists are using? What is acceptable evidence that the scientists are right? What would it cost to do economic restructuring now to avoid even greater dislocation in the future? What are the obligations of present society to its children and future generations? In spite of uncertainties, global warming was of sufficient concern to make it one of the major issues at the Earth Summit in Rio de Janeiro in 1992. A Global Warming Convention was signed that encourages industrial nations to reduce greenhouse gases to 1990 levels by the year 2000 and encourages developing countries to develop policies to regulate their own emissions of greenhouse gases. Unfortunately, the treaty presently does not require such reductions and lacks a compliance timetable.

In Chapter 14, "The Challenge of Global Climate Change," William Makofske examines scientific and societal issues surrounding climate change from a broad perspective. He summarizes the present state of scientific uncertainty and the implications this has on climate impacts and policy development. In Chapter 15, "Predicting Climate Change: The State of the Art," Kerry Cook reviews the present ability of climate models to project the result of increasing greenhouse gases. Taken together, these chapters provide an overview of the scientific and policy issues that need to be addressed as the world grapples with this complex and potentially devastating issue.

Reexamining Our Relationships to Technology and the Earth

Technological Disaster

If AT seems promising, there are clear instances of inappropriate technology transfers, some of which have led to recognizable disasters such as those at Bhopal, India and Seveso, Italy. The roots of such accidents are complex: failures of management, human error, design and technological failures, poor land use decisions, and even corruption. The point is that such large-scale complex technologies are unpredictable, with multiple avenues leading to unanticipated failure. Victims of such disasters are often exploited and abandoned in the highly charged political aftermath.

Yet such accidents raise many larger issues beyond technological disaster. These include the role and responsibility of multinational corporations, their financial liability, and the disparity between safety and environmental standards in MDCs and LDCs. In recent decades, we have seen relocation of toxic industries to cheaper and less regulated environments, the export of banned or untested chemical products, and the dumping of wastes in accommodating countries. There is often an inability to regulate in developing countries due to government reluctance to control industry, a persistent weak regulatory capability, and a desire for financial gain.

Technological disasters are not only found in developing countries. The nuclear meltdown at Chernobyl will have impacts lasting for many generations due to radioactivity dispersed in the environment. The effect of Chernobyl on many millions of people cannot be overestimated: it has been suggested that Chernobyl acted as a major catalyst for political and economic reform within the former Soviet Union, and in Eastern Europe as well. Nuclear power, already suffering from widespread opposition and economic realities, was dealt a crushing blow. Several countries decided to eliminate their nuclear dependence in the wake of Chernobyl. In Chapter 16, "Disaster Revisited: Bhopal and Chernobyl—What Are the Lessons?," Michael Edelstein compares two of the most serious technological accidents of the past few decades. He challenges us to consider not only what we have learned from these disasters, but perhaps more importantly, what we have yet to learn.

Gaia: The Earth's Regulatory System

Simply stated, Gaian theory views life itself as a critical element in the regulation of the environmental conditions necessary for life. This is because life has a major impact on the composition of the atmosphere and therefore on the temperature regulation of the Earth. In essence, it presents a changed view of the role of life in the evolution of the planet and in geochemical processes. Evidence for this view, according to advocates, is the regulation of oxygen levels and the remarkable constancy of temperature while the sun's intensity gradually increased by over 30% over a few billion year period. Although the theory doesn't address the role of the human species in the Earth's evolution (we've been here too short a time), it does provide a new conceptual framework for addressing research questions on the role of life in what were previously thought of as mere geochemical phenomena.

In Chapter 17, "Gaia, A New Look at the Earth's Surface," Lynn Margulis and Eric Karlin provide an overview of evidence supporting the Gaian theory and its implications for understanding Earth system interactions. In the context of the global environmental changes that humans are now making, the notion of Gaia emphasizes the myriad number of feedback loops that connect all living organisms on the planet. The acknowledgment that life plays a significant role in regulating the global surface environment and the incorporation of that knowledge into our everyday lives represents an essential step towards ensuring the survival of our species.

Conclusions

Although many of the global problems may seem at first glance to be separate, they are indeed very much related. They all stem from outdated worldviews that supported unlimited growth of human population and activities without regard to their consequences for life on Earth. In addition, there is a blind faith in technological progress, and an assumption that we can correct or undo significant damage to the Earth's ecosystems. These views are challenged by the devastating and perhaps irreversible changes that are occurring to the planet today. At the same time, there are new paradigms that offer potential solutions to these problems.

The key concept for balancing human needs with natural Earth systems is sustainability. In essence, humans must learn to live on our earned income rather than using up inherited capital (that is, the life-support systems of the Earth). We can do this by achieving a sustainable population by limiting population growth, by developing a sustainable agriculture, by moving to renewable energy sources and increasing energy efficiency, by recycling resources, by protecting vulnerable and valuable ecosystems, and by reducing pollution.

Although science and technology are a critical part of the solution, they will probably have a lot more limited role than most people realize. Ultimately, it will be the social and cultural systems that generate these environmental problems that will have to change. Paradigms of societies, attitudes toward nature, and current views of progress and economic growth must be examined. A global environmental perspective needs to develop in decision-making on the individual, national, and international levels. The United Nations–sponsored Earth Summit held at Rio de Janeiro in 1992 was an initial attempt to deal with the problems of maintaining a sustainable world in the face of a dramatically expanding human population coupled with economic growth and development. Agenda 21, a blueprint for environmentally sensitive development, dealt with a full range of issues, such as hazardous waste, ocean pollution, human health and poverty, and advancement for women. It is hoped that the chapters in this book provide further insight into what needs to be done to ensure the development of sustainable human societies in the twenty-first century.

Chapter 2

Population Growth and the Global Environment: An Ecological Perspective

Eric F. Karlin

Sometime in the vicinity of 1960, the human population reached 3 billion. I was in fourth grade then and I can distinctly remember my teacher leading a lively class discussion about this fact. Three billion was an enormous number and my classmates and I were amazed that there could be so many people. We concluded that it must have taken just about forever for the human population to reach that size. I now know that it took about 250,000 years, for that is when *Homo sapiens* first appears in the fossil record (Stringer 1990), and that over 90% of the population growth had occurred within the last 2000 years. In 1993, just 33 years later, the human population had reached 5.5 billion (Table 2.1), with the 6 billion mark projected by 1998. Adding the second 3 billion people to the planet will have taken only 38 years. Not hundreds or thousands of years, but just a few decades. It is hard to believe that the number of people on Earth will have almost doubled in such a brief period of time. It is mind numbing to realize that the human population will continue to have explosive growth well into the twenty-first century, but that is the prospect that lies before us.

Rapid population growth is a global problem of enormous proportions. Although only one of many environmental issues, the immense size of the current human population and its rapid rate of growth is without a doubt one of the most critical environmental issues confronting humanity. The number of humans is projected to continue to increase until sometime near the end of the next century, when it is projected that the human population will stabilize at around 12 billion (Table 2.2). This is more than twice as many people as there are now. These projections are based on a number of assumptions: that rates of population growth will decline to replacement level in all nations in the not too distant future, that the Earth has sufficient resources to sustain greater numbers of people, and that the technological and social innovations needed to support these populations will be developed and successfully implemented within a short period of time. Although few observers question that more people will be added to the present population, the key questions are how many more, at what quality of life, at what rate of growth, and for how long?

Table 2.1

Total Global Population and Annual Growth (both in millions) in 1993 by Region

Region	Total Population	Percent Growth Rate	Annual Growth
World	5,506	1.635	90.0
LDCs	4,276	1.982	84.8
MDCs	1,230	0.428	5.3
Asia (w/o China)	2,079	2.051	42.7
Africa	677	2.916	19.7
China	1,179	1.164	13.7
Latin America	460	1.928	8.9
North America	287	0.755	2.2
Former USSR	285	0.564	1.6
Europe	513	0.181	0.9
Oceania	28	1.156	0.3

Because of rounding, summing the data for regions in each column will not equal the data listed on a global basis (the first line of the table).

Source: 1993 World Population Data Sheet (data disk), Washington DC: Population Reference Bureau, 1993.

Carrying Capacity

The maximum number of individuals of a given species that can be sustainably supported by a given ecosystem is referred to as its *carrying capacity*. Carrying capacity not only varies among different ecosystems (100 hectares of Alaskan tundra can support many more arctic foxes than 100 hectares of the Florida Everglades could), it also changes as the environment of an ecosystem changes (an intact wetland supports many more ducks than does a drained wetland). With its great diversity of ecosystems, determining the Earth's carrying capacity for humans is not an easy task. A further complication is that humans have the capacity to modify the global carrying capacity through technology. Without technology (particularly the innovations of fire and clothing) the human species would probably number about 1 million and be limited to tropical Africa and tropical Asia. Pre-agricultural technology (hunter-gatherer societies) allowed humans to migrate throughout the world and led to a global carrying capacity of about 4 to 5 million about 10,000 years ago (Deevey 1960; McEvedey and Jones 1979; Ehrlich and Ehrlich 1990). As the current human population is three orders of magnitude greater than what pre-agricultural technology could support, it is clear that advances in technology have played a significant role in harnessing and channeling an ever increasing percentage of the Earth's resources to meet human needs. At the present time, humans and their livestock directly consume (as food, fodder, and timber) 4% of the net primary production (NPP) on land (3% of the global NPP). If all of the plant mate-

Table 2.2
World Population Over Time

Date	Global Population (in millions)	Years to Add 1 Billion
250,000*	1	—
10,000*	5	—
1800	1,000	>250,000 .
1930	2,000	130
1960	3,000	30
1975	4,000	15
1987	5,000	12
Projected		
1998	6,000	11
2009	7,000	11
2020	8,000	11
2033	9,000	13
2046	10,000	13
2066	11,000	20
About 2100	12,000	34

*Years before present.

Source: Population Reference Bureau, *1991 World Population Data Sheet,* Washington DC: Population Reference Bureau, 1991; Deevey, E., "The Human Population," *Scientific American* 203 (3), 1960, pp. 194–204.

rials that were used indirectly by humans (clearing land for development and farming, NPP on rangelands and so forth) were to be included, it would be found that humans utilize approximately 31% of the NPP on land (19% of the global NPP) (Vitousek et al. 1986). Some believe that humans could successfully exploit an even greater percentage of the Earth's resources and that, if the available resources of the Earth were to be exhausted, technology could provide extrinsic resources (water and minerals from other planets and asteroids, growing food in space) to support immense human populations.

But this viewpoint assumes that the impact of our activities on the Earth are minimal or easily mitigated by technological fixes. One needs to remember, however, that the Earth is a finite planet. The supplies of fresh water, food, arable land, wood, minerals, and fossil fuels are not limitless and many resources are not renewable. Many are already in short supply and some (such as oil) are expected to be economically depleted within the next few decades. For instance, 26 nations do not have sufficient water resources to support their present populations (Holden 1992a; Postel 1993). In many countries, the quality of the water that is available is also an issue. In the LDCs, water quality is often impaired by a lack of sanitation systems. Consequently, a host of water-borne diseases (such as schistosomiasis, giardi, and diarrhea) are a major problem in many LDCs. Water quality is also an issue in the MDCs, where pollution from the activities of industrialized society causes many problems. Pesticides, road salts, motor oil, excess fertilizer, PCBs, gasoline leaking from underground storage tanks, and excessive quantities of oxygen-depleting wastes are just a few of the many substances that enter and degrade the water systems of MDCs (see Chapter 11, "Coastal Ocean Resources and Pollution: A Tragedy of the Commons"). Ecosystems are also being used in a nonsustainable way. It is projected that almost all of the tropical wet forests will be destroyed within the next few decades and that hundreds of thousands of species associated with them will have become extinct (see Chapter 9, "Tropical Forests: Problems and Prospects"). As the percentage of the Earth's resources that are utilized by humans increases, there is a corresponding decrease in resource availability for other species and natural ecosystems. This will lead to increased rates of extinction and extensive ecosystem degradation, and ultimately result in a lowering of the Earth's carrying capacity for humans (see Chapter 8, "The Loss of Biodiversity"). These are irreplaceable losses for which there is no technological fix. Clearly, there are limits to what technology can achieve, even with the use of extrinsic resources.

As we transform the world into a progressively more human dominated planet, we are finding that our actions are having an increasingly negative impact on the ecosystems that sustain us. In addition to resource use, extensive pollution and accelerated soil erosion lead to further ecosystem degradation and destruction, and lower the global carrying capacity, not only for humans but also for many other species. Even as we strive to develop the new technologies and social systems needed to support a larger human population, the requirements of the current population and our misuse of technology is damaging, and even annihilating, the global environment that makes our civilizations possible.

This should not be surprising, because one of the basic concepts of ecology is that each organism has the capacity to degrade its environment. An animal takes in resources (oxygen, food) and releases wastes (carbon dioxide, feces, urine) back to the environment. In a balanced ecosystem, these wastes then serve as resources for other species (carbon dioxide used by plants, flies lay their eggs in feces) and are eventually transformed (recycled) back into resources for the first species to use again. This is one of the fascinating aspects of ecosystems: they allow for the long-term sustainability of life. No species (except perhaps a very few primitive prokaryotes) can survive in the absence of a supporting ecosystem. *Apatosaurus* (formerly *Brontosaurus*) was one of the largest of the dinosaurs, and it must have deposited massive piles of dung on a regular basis (cow and horse droppings would be insignificant in comparison). Such enormous dung piles represented a major waste disposal problem for the ecosystems in which they were deposited. If they weren't processed and decayed quickly, they would have eventually covered the landscape. The degraded ecosystem that would have resulted would ultimately be unable to support the dinosaurs that produced the dung. As large herbivorous dinosaurs like *Apatosaurus* were present on Earth for many millions of years, a highly efficient group of dinosaur dung processing species must have existed. No species can survive, whether it be a dinosaur, mammal, or microbe, without the presence of a waste recycling system.

As a population grows, its impacts on the ecosystems that sustain it become greater. As long as the population remains below the carrying capacity, these impacts do not degrade the ecosystem. Any increase in population size above the carrying capacity, however, results in ecosystem degradation because resources are being harvested at a nonsustainable rate and/or more waste is being produced than the ecosystem can process. Consequently, exceeding carrying capacity does not simply result in the deaths of the excess individuals, it also results in ecosystem degradation. This in turn may lead to a lowering of the carrying capacity. Although ecosystems can support populations in excess of the carrying capacity, they cannot do so in a sustainable fashion. Eventually, environmental limitations will lead to a population decline.

Suppose that 200 deer were introduced into a forest that had a carrying capacity of 100 deer. It may very well be that the forest could support all 200 of the deer during the first summer and fall (when there would be an abundance of food), but this would be only at the expense of having an excessive harvest of plants by the deer, with even the marginally edible plant species that deer normally avoid eating being consumed. With the coming of winter, food would be much less available and many deer would die of starvation. It is the lack of food in the winter that primarily limits the forest's carrying capacity to 100 deer. This process would be repeated for a few years until the deer population was in equilibrium with the carrying capacity, but the carrying capacity would now be less than 100 deer. The excessive browsing and extensive damage to the plants caused by the deer, especially when they were starving in the winter, would have greatly depleted the supply of food plants in the forest; the forest's carrying capacity would be significantly diminished. This is the price that has to be paid when populations greatly exceed carrying capacity.

Ecosystem degradation is minimal and short term when populations do not significantly exceed carrying capacity. When the carrying capacity is greatly exceeded, however, or is exceeded slightly for a long time, the impact on the ecosystem may be quite significant and long term. For instance, intensive logging, burning, and grazing since prehistoric times have degraded extensive areas of England and Scotland that were formerly forested into shrub and graminoid-dominated ecosystems called *heaths* and *moors* (Figure 2.1). The degradation has been so severe in some cases that reforestation is not possible (Walter 1985). Another example can be found in Africa, where elephant populations in many parks have greatly exceeded carrying capacity. Because the parks are one of the few places that elephants can find some measure of protection from hunting and poaching, many have migrated to the parks and spend a large percentage of their time there. Consequently the forests are being ripped apart by the overabundant animals (Bonner 1993). Certainly humans are not exempt; as our population increases, so will our impacts on the Earth. Although the Earth currently supports over 5.7 billion people, we have to seriously consider whether or not our population has already exceeded the global carrying capacity.

Cultural and Social Carrying Capacity

Since global carrying capacity refers to the maximum number of people that the world could support, a population at carrying capacity would mean that each per-

(a)   (b)

Figure 2.1. Ecosystem degradation in Dartmoor National Park, England. (a) An example of the deforested landscape that dominates Dartmoor National Park. (b) One of the few intact forested areas in the park.

son would have just the bare minimum for survival: bread, water, one set of clothes, a space to sleep, and so forth. If a higher standard of living is desired (eating meat, having a car, having ten changes of clothes, having a pet), then the resources required to maintain this life-style would have to be subtracted from the resource base supporting the carrying capacity. The diminished carrying capacity that results from the use of resources to support a life-style other than basic survival is called *cultural carrying capacity* (Hardin 1986). The higher the standard of living that is desired, the lower the cultural carrying capacity will be.

Then there is the question of how effectively social systems can meet the needs of the people (social carrying capacity). Because social systems are not perfect, the social carrying capacity is also less than the carrying capacity. Although global food production is currently sufficient to support a population of over 5.7 billion, 40 million people die each year because of starvation and hunger-related diseases and hundreds of millions suffer from malnutrition and undernutrition (Kaufman and Franz 1993). Insufficient food supplies, a lack of potable water, and the absence of effective sewage treatment facilities all lead to extensive health problems in many LDCs. Debilitating diseases such as diarrhea, a water-borne disease, are responsible for the high death rate of children under 5 years of age in developing nations. It is estimated that up to 2 million children under the age of two die each year from diarrhea alone (Krejs 1992). Access to adequate health care is a major issue for a large percentage of the human population, even in some MDCs. In many LDCs, people may have to walk great distances to locate a clinic with even minimal medical facilities. The larger medical centers that have the capacity to deal with a wide range of health problems are usually found in cities and are largely unavailable to the average person, both in terms of distance and financially. Adequate health care is a problem in many MDCs as well. The United States, with reputedly the best medical system in the world, does not have universal health care coverage. Advanced medicine is available, but it is often prohibitively expensive. Unable to afford a private doctor, many of the poor in the United States use hospital emergency rooms for their routine medical needs. Although the Clinton administration initiated a major campaign to remedy these problems, no significant changes had been made by the

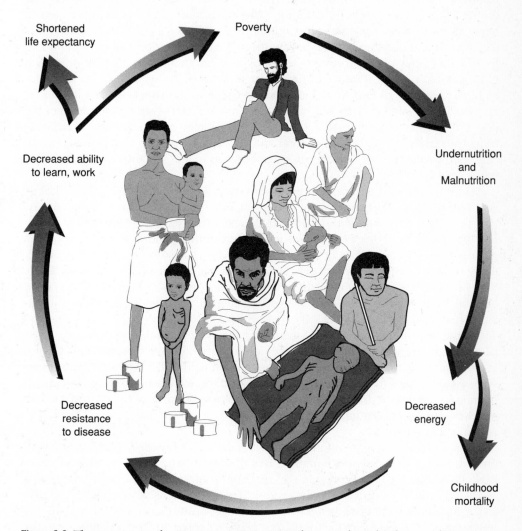

Figure 2.2. The poverty cycle. Poverty creates an insidious cycle. It leads to undernutrition and malnutrition, which in turn lead to decreased resistance to disease and decreased energy. Illness and lack of energy lead to decreased ability to learn and work, which perpetuates poverty. Poverty also results in shortened life expectancy. Unless something is done to break the cycle of poverty, poor children are doomed to a life of hardship and deprivation simply by virtue of having been born poor. Adapted from Kaufman, D. G. and C. M. Franz, *Biosphere 2000: Protecting Our Global Environment,* New York: HarperCollins College Publishers, 1993.

end of 1994. As these examples show, our social systems are not even capable of adequately supporting the world's current population.

Cultural and religious practices and beliefs also have their impacts, particularly on women and children. In many developing nations pregnant women are expected, by tradition or superstition, to avoid many nutritious foods, such as eggs, certain vegetables and fruits, fish, and even milk. This affects the health of both the mother and the developing fetus. Another common practice in many LDCs is to allow men to have access to the largest and most nutritious portions of food. Consequently, it is the women and children who suffer the most from malnutrition and undernutrition. In addition, boys often receive more food, medical attention, and education than girls (Jacobson 1992). Because of the elevated status of the male in China, the one-child per family goal that was established in China in the late 1970s resulted in increased female infanticide there (Ehrlich and Ehrlich 1990). A strong indication of the plight of women in developing countries is the declining ratio of females to males in India (Jacobson 1993). Clearly, cultural and social systems have channeled a significant percentage of the impact of overpopulation and resource depletion onto women and children.

How nations and people will respond to dwindling resources is also a critical issue. For instance, the groundwater system underlying the West Bank drains into Israel, providing 40% of the groundwater that Israel uses each year. One reason that Israel wants control of the West Bank is to ensure that the groundwater system there is not depleted (Homer-Dixon et al. 1993; Gleick 1994). Insufficient resources are already leading to violent conflict, especially in many less developed countries (LDCs). Violence and war will only escalate as resources become more scarce, and it may be that the framework of many societies will disintegrate long before the human population reaches 12 billion.

Some believe that standards of living increase with expanding populations. They posit that people are a resource; that we are producers as well as consumers (Simon 1989). Many advances in agriculture, industry, and medicine have been made in response to the pressure of population growth, and these have led to the development of modern amenities. In turn, improvements in education, health, and income can foster desires for smaller families. Limited population growth may have a positive impact in some situations.

Rapid population growth (annual growth rates $\geq 2\%$), however, is associated with environmental and social deterioration. It is the major issue for many nations, especially those that lack the economic and natural resources, political organization, and social infrastructure necessary to deal with it. The populations of many LDCs are projected to double within one generation—an almost insurmountable challenge to any government's ability to provide for the needs of its people. In these countries, rapidly increasing numbers are not going to stimulate more production; instead, lowered standards of living and more suffering will result. In many cases, the combined impacts of insufficient food, disease, and lack of education results in a continuing cycle of poverty (Figure 2.2). Because a significant percentage of their citizens are debilitated by disease and/or cannot function on even a grade-school level, development in LDCs is a very difficult and problematic process. Development has not kept pace with the rapidly expanding populations of the LDCs since the 1940s, even in those

Table 2.3

Mean Annual World Grain Production (in millions of metric tons) and Yield Per Capita (in kilograms) Based on 5 Year Periods

Date	World Harvest			Per Capita		
	Total (Mean)	Change (Mean)	Percent Change	Total (Mean)	Change (Mean)	Percent Change
1950–54	681	—	—	257	—	—
1955–59	804	123	18.1	277	20	7.8
1960–64	878	74	9.2	273	−4	−1.4
1965–69	1,019	141	16.1	291	18	6.6
1970–74	1,188	169	16.6	306	15	5.2
1975–79	1,369	181	15.2	322	16	5.2
1980–84	1,526	157	11.5	329	7	2.2
1985–89	1,642	116	7.6	325	−4	−1.2
1990–92*	1,740	98	6.0	323	−2	−0.6

*Based on a 3 year period, with the 1992 data being preliminary.

Source: Brown, L. R., "Grain Production Rises," In L. R. Brown et al. (ed.), Vital Signs 1993, New York: W. W. Norton, 1993, pp. 26–27.

having free market systems. If it did, Bangladesh, India, Kenya, Nigeria, Brazil, and Mexico would all be among the world's richest countries by now. Instead, this list includes some of the world's poorest countries, and even those making encouraging progress have been slowed by the weight of rapid population growth. Just providing jobs for the expanding population is daunting! In the next two decades, hundreds of millions of additional jobs will have to be created in the developing countries for the some 630 million children who will reach adulthood and seek employment. That represents the creation of more jobs than currently exists in all more developed countries (MDCs). In addition, this large pool of young adults will also produce the next generation of children. Rapid population growth will continue to be a problem for many decades; it will not disappear overnight.

Although there have been significant gains in global food production, economic growth, and medicine over the past 30 years, direct comparisons between these types of growth can be misleading. As most of the growth of goods and services occurred in the MDCs and most of the population growth occurred in the LDCs, the rich got richer and the poor became poorer during this time. More equitable sharing of resources and reductions in population growth are both required in order to significantly diminish the gulf between the rich and poor. Complicating this is the fact that some technological advances are no longer meeting the needs of the rapidly expanding human population. For instance, global grain production on a per capita basis (Table 2.3) has been declining since 1984 (Brown 1991, 1994).

As the central component of the concept of carrying capacity is sustainability, it is quite obvious that the human population has exceeded the current global carrying capacity (Daily and Ehrlich 1992; Holden 1992b; Postel 1994). Human activities are causing extensive damage to the Earth's ecosystems, and this is resulting in a major restructuring of the global environment. Our interactions with the environment are not on a sustainable basis, even in countries where rapid population growth is not an issue. The extensive erosion of soils, the degradation and destruction of ecosystems, mass extinction of species, pollution of all kinds, and warfare all point to a human environment that is steadily deteriorating. Another sign that carrying capacity has been surpassed is the fact that many countries (both LDCs and MDCs) are unable to provide the food, shelter, education, stability, health services, energy resources, and jobs needed by their current populations (Daily and Ehrlich 1992). Although the Earth might be able to sustainably support larger human populations in theory, further population growth without concomitant changes in social systems and technology can only result in a progressively degraded planet and a lower quality of life (Bongaarts 1994a); ultimately this would lead to a massive and disruptive population decline. Stopping population growth is only part of the answer, however. All nations also have to learn how to interact with the ecosystems that support them in a sustainable way. After all, the demands of the current population are destroying the world as we know it. This mandate is even more critical in light of the fact that most LDCs are seeking to achieve a standard of living similar to that currently enjoyed by the United States. The Earth could not even come close to supporting 5.7 billion people at such a high standard of living.

LDCs vs. MDCs

Some blame the LDCs for the most extreme environmental abuse. To be sure, poor policies, lack of sensitivity, and shortsightedness have marked the response of many LDC governments to warnings about damage to ecosystems. Having roughly 78% of the human population and the highest rates of population growth (averaging 2.0%), LDCs certainly face many complex environmental issues. Their populations are expected to increase by almost 3 billion by 2025 (going from 4.3 to 7.1 billion), while the populations of MDCs are expected to increase by only 0.2 billion (going from 1.2 to 1.4 billion) during the same time period (Population Reference Bureau 1993). There is no doubt that meeting the needs of roughly twice as many people within a 30 year period represents a formidable, if not impossible, problem for LDCs, especially when they are unable to meet the needs of a substantial percentage of their current populations. Indeed, extensive ecosystem destruction and degradation are already diminishing the carrying capacities of the very ecosystems needed by LDCs to sustain their growing populations.

There is another dimension, however, to the population issue: the amount of resources used by people. A small population with a high standard of living is capable of consuming more resources than a large population having a much lower standard of living. The MDCs, which have only 22% of the world's population, consume 80% of the resources utilized globally each year (Miller 1992; Kaufman and Franz 1993); an amount equivalent to that needed to support 19.2 billion people at the standard of living typical of LDCs! A more specific example involves a comparison between the energy use of India and the United States. Per capita energy consumption in the United States is 66 times greater than it is in India (Kaufman and Franz 1993). Assuming no change in its average standard level of living, India would need to have a population of 56.7 billion in order to consume the same amount of energy as the United States.

This is a complex issue, for not only are LDCs experiencing very high rates of population growth, they are also seeking to improve the standard of living for their populations. Many LDCs would like to achieve the standard of living that is enjoyed by the MDCs, particularly the United States, where the per capita use of resources is one of the highest in the world (Kaufman and Franz 1993). But as the above numbers clearly indicate, there is no way that the Earth could support 5.7 billion people if they all consumed resources at the same rate and in the same way that the MDCs currently do. In order to achieve a more equitable international sharing of resources, all nations must learn to consume resources in a more environmentally sound manner. It is also essential that population growth in all nations be significantly reduced.

In addition to the amount of resources used by a population, we also have to consider the entire spectrum of environmental impacts incurred by the harvesting (i.e., mining, logging, damming of rivers), processing (i.e., oil refining, smelting of ores, production of paper), distribution (i.e., electric power lines, oil tankers, freight trains), and use (i.e., combustion of fossil fuels, construction of buildings) of resources. An increase in resource consumption will be accompanied by a corresponding increase in the associated environmental problems. Consequently, it is the MDCs, that use 80% of the resources consumed each year, that are responsible for a large percentage of the global environmental issues associated with resource consumption. The LDCs, however, also have considerable environmental damage resulting from resource consumption. In particular, extensive tracts of natural ecosystems are destroyed and agricultural lands are overburdened in order to help meet the food and fuelwood needs of their immense populations, leading to severe problems with deforestation and desertification. In addition, many LDCs harvest large quantities of natural resources (such as oil, lumber, and bauxite) for export to MDCs, and these activities often create large-scale environmental degradation.

War and the threat of war place an inordinate demand on natural and economic resources and also represents immense environmental disruption. Nations have amassed and used a vast array of weapons. The Iraqi invasion of Kuwait, the subsequent Operation Desert Storm, and the bloody tragedy that is currently occurring in Bosnia are stern reminders that large-scale military actions are still a global reality. Global military expenditures exceeded $1 trillion dollars a year from 1985 to 1988, the highest amounts ever expended. Even though the breakup of the former Soviet Union has allowed for a significant reduction in the arms race, massive amounts of money (about $935 billion in 1990) are still being allocated for military expenditures and enormous stockpiles of weapons still exist (Renner 1992). The social and environmental impacts resulting from any major war range from calamitous, if only conventional weapons are utilized, to incalculable in the event of a full scale nuclear war. Even if not currently at war, weapons design and production accounts for a significant percentage of most country's budgets and also involves the use of large amounts of scarce and non-renewable resources. Imagine what could be done for humanity and the global environment if the creative efforts and resources that are devoted to weapons were focused on peaceful development. Unfortunately, violent conflict is bound to increase as resource availability declines and the environment deteriorates (Homer-Dixon et al. 1993).

What is Needed

Small is Beautiful

Two requirements are essential in order for the population problem to be brought under control. The first is that people must believe that it is in their best interest to have smaller families. This can only happen when a number of issues are satisfied. One critical issue is that a basic human need has to be fulfilled: to have a sense of worth. One way to partly fulfill this need is to have children. Parents play a major role in their children's lives, and a part of each parent lives on in the child. In many societies, children provide work, which aids family survival and also support their parents in their old age. If there is a high probability that one's children will survive into adulthood and have full and productive lives, then there is little reason to have a large family. In the absence of this, one has no choice but to have as many children as possible and hope that at least one or two will survive. But this provides only one dimension of worth; other dimensions are also important for the development of a well-balanced person. A sense of worth can also be gained by having a chance to make contributions to society. Knowing that your presence will be noted and remembered by others is important to many people. If there is little chance that one can meaningfully participate in society, then this avenue to gain self-worth is not a possibility. Many of those living in poverty face this dilemma, particularly women (Jacobson 1993). Another aspect of self-worth is having the freedom to make one's own decisions and to be acknowledged as an equal to others. This is an extensive problem, especially for women and in race relations. It is essential that women not only have the right to control their fertility, but that they also feel empowered to do so. *All* should have an equal opportunity to be educated, have a career, and be economically secure. Until there is a global environment where all races and cultures feel accepted and unexploited, there will always be some resistance to population programs because of the belief that they may be based on racist attempts to control, or even destroy, certain segments of the human species.

Poverty is a major issue that must be addressed (see Figure 2.2). It is estimated that more than 1 billion people (roughly 20% of the total population) live in poverty, with the vast majority of poor being in the LDCs (The World Bank 1990). The combined impacts of larger populations and degraded environments have led to an erosion in the standards of living in many LDCs. Instead of life getting better, the standards of living in many LDCs are actually lower now than they were in the

1960s and 19⁻⁻ ⁻ (The World Bank 1990). Although the percentage of people that are in pov⋯⋯ ⋯⋯jected to decline, the absolute number of people in poverty will probably ⋯⋯ ⋯⋯er in 2000 than in 1985 (The World Bank 1992). Development is seen by n⋯⋯y as a key component in raising the standard of living in the LDCs and also in curbing population growth; extensive efforts have been made toward this end. Unfortunately, much of the development that has taken place has not been environmentally sound and may be self-defeating in the long run.

A stable and equitable socioeconomic system, one in which each person has the opportunity to live a comfortable life, is also an important factor. Given the excessive size of the human population, the development of such an socioeconomic system would have to be based on a stable, or declining, population; it could not be based on a rapidly expanding population. Nations would have to adopt antinatalist population policies and develop strong economic incentives to have smaller families. Cultural aspects are also important. Unless one's culture promotes smaller families, it will be difficult to choose to have a smaller family. Small families will have to become the norm for all cultures. Religions that promote fertility and ban birth control are no longer ethically affordable. It is becoming increasingly more difficult to defend high birth rates when they primarily result in lives of misery and early death.

Family Planning

The second requirement needed to achieve population control is the ability to effectively plan for family size. A belief that smaller families are desirable is not enough. It is also essential that safe, effective, and easy-to-use contraceptives be readily available to all. Family planning programs provide information about contraception and contraceptive devices to the general population of many nations, especially to women. Strong and effective national family planning programs are considered to be essential for controlling fertility, having equal importance with economic development (Robey et al. 1993). Two organizations that have played a significant role in bringing family planning services to LDCs are the United Nations Fund for Population Activities (UNFPA) and the International Planned Parenthood Federation (IPPF). The mass media, especially radio and television, is also playing a major role in disseminating information about family planning and the benefit of smaller families to many millions of people in the LDCs. This rapid and direct transmittal of information allows people to quickly become aware of different life-styles, to learn about contraception, and to better understand national and global issues. The exposure to information provided by the mass media, in conjunction with the availability of effective contraceptives, has played a significant role in the declining birth rates experienced by the LDCs over the past two decades, a decline that occurred in spite of the absence of improved living conditions (Robey et al. 1993).

The introduction and acceptance of modern contraception has played a major part in the improvement of the lives of poor people in LDCs. Family planning allows people to make choices about one of the most significant aspects of their lives—reproduction. When women have been educated about birth control, have access to contraceptives, and understand the advantages of limiting the number of children, they are granted a degree of control over their future and the future of their immediate family (Nadakavukaren 1990; Brown 1991; Jacobson 1992). By controlling family size, parents are better able to provide the balanced and nurturing environment needed for the optimal growth and development of children. This is becoming more widely recognized in LDCs, and many couples are expressing a desire for smaller families. Unfortunately, a large percentage do not have access to adequate family planning services (Fornos 1991; Robey et al. 1993; Bongaarts 1994b).

The development of more effective and easier methods of contraception would have substantial benefits for the citizens of all nations. In recent years, two new and promising methods of birth control have been developed: Nor-Plant and RU-486 (Nadakavukaren 1990). Nor-Plant was developed to meet the need for a safe, long-acting, and dependable product, especially appropriate for the women of LDCs who have little access to medical services. Nor-Plant consists of narrow, 3 cm long, permeable rods that contain a timed-release dose of progestin. Several rods are surgically implanted (in a very simple operation) in the skin of a woman's arm under a local anesthetic. The supply of progestin lasts an average of 5 years, providing effective birth control with no known major side effects. It is a great advantage to any woman who would have to travel long distances to renew prescriptions for contraceptives and who would have to remember to take them on a timely basis. Produced by a French pharmaceutical company (Roussel-Uclaf), RU-486 is used to interrupt pregnancy. RU-486 simplifies the abortion process because it can be taken as a pill in the privacy of a doctor's office and does not require a surgical process. Although not yet legal in the United States, RU-486 has recently been approved for study by the Clinton Administration. It is already available in France, Sweden, and the United Kingdom.

Family planning programs have had considerable success in several countries, including China, Indonesia, Sri Lanka, and Thailand. Because of antinatalist population policies and extensive family planning programs, crude birth rates (live births per 1,000 people/year) in Indonesia dropped from 45 in the late 1960s to 26 in 1993, and in China they declined from 44 in 1963 to 18 in 1993 (Johnson 1987; Population Reference Bureau 1993). On the other hand, family planning has met with moderate to minimal success in many other nations (India, Egypt, Pakistan, and many Latin American and African nations). Insufficient funding for the family planning programs in these nations has played a large part in their lack of effectiveness.

Perhaps the most effective, and the most controversial, family planning program has been the one mounted by China, which was the first nation to seek an end to population growth and a reduction in population size. Having lost about 30 million people to famine between 1958 and 1962, the Chinese government was forced to realize that its population had surpassed the nation's carrying capacity (Kaufman and Kranz 1993). Being a dictatorship, China was able to develop and implement a vigorous population program, including easy access to free contraceptives, urging one-child families, and requiring one parent in a two-child family to be sterilized. Many have criticized the Chinese program for being coercive and have cited incidents of forced abortion and involuntary sterilization. China has denied that its program is coercive, but in the face of strong international pressure and problems with enforcing the program, especially in rural areas, China has relaxed its regulations since 1985. The rate of population growth, which had reached a low of 1% in the early 1980s, now stands at 1.2% (Population Reference Bureau 1993). Although this is one of the lowest growth rates found in LDCs, it still means that the population is growing. China's population currently stands at 1.18 billion, and even with its modest rate of growth, its population will surpass 1.5 billion by 2025 (Population Reference Bureau 1993). This means adding 300 million additional people (in just three decades) in a country where the current population has already greatly exceeded the carrying capacity. This additional population is greater than the current population of the United States.

As the situation in China all too clearly establishes, even significant reductions in population growth rates will not immediately stop population growth. The larger a population is when family planning programs are implemented, the harder it will be to control population growth and the longer it will be before the population stabilizes. Because of this, many nations not only have to make every effort to stabilize population growth, they will also have to reduce their populations in order to achieve a sustainable relationship with the ecosystems that support them. All nations must respond to the population crisis immediately; it will be too late if they wait for major breakdowns to occur. We must face the fact that the Earth is already being seriously degraded by the human population, and that it will soon be experiencing even more extensive disruption from a larger number of humans. How many more people and how much damage depends upon the actions that the human species takes now and in the next few decades.

The Role of the United Nations

When the United Nations Population Commission first met in 1947, demographic statistics were so incomplete that it was difficult to assess global population problems. The UN played a significant role in the development of a database on global population, especially by promoting the gathering of population statistics by member nations. As this information was amassed it became clear that overpopulation was a critical issue. Unfortunately, little was done by the UN to deal directly with population growth until 1966, when the General Assembly passed Resolution 2211 (XXI) on Population Growth and Economic Development. This resolution contained language to the effect that the UN was to provide assistance in developing and promoting population related services. Based upon voluntary contributions, a UN Trust Fund was established in 1967, which was to be used to finance this mandate. This fund was to become the United Nations Fund for Population Activities (UNFPA). The first major programs undertaken by UNFPA started in 1969 and by 1974 UNFPA had supported population projects in approximately 90 nations (Johnson 1987).

Although most nations acknowledge that rapid population growth is a major global problem, no universal agreement has been established on how to deal with it. At the World Population Conference (held in Bucharest in August 1974), two different points of view surfaced. The MDCs argued that rapid population growth was an impediment to economic and social development, and wanted targets to be set for population growth. The LDCs, on the other hand, argued that rapid population growth was a symptom of underdevelopment and that the wealthy nations should provide increased support for development programs in LDCs in order to curb their rapidly expanding populations. A second conference (the International Conference on Population) was held in Mexico City in August 1984. By this time, the rate of the global population growth had dropped to about 1.7%, down from the 2.0–2.2% growth rates experienced during the 1960s and early 1970s (Starke 1993), and family planning programs had experienced some success. The participating nations agreed that governments should address problems with population growth and that each person should have access to family planning resources. The conference also addressed the issues of urbanization, illegal immigration, and environmental degradation associated with overpopulation (Johnson 1987).

Although population issues were not a primary focus of the United Nations Conference on Environment and Development (the Earth Summit) held in Rio de Janeiro, Brazil, in June 1992, the relationship between development and population growth was central to Agenda 21. The goal of Agenda 21 was to have MDCs provide funding for environmentally sensitive development in the LDCs. Again a dichotomy developed between the LDCs (which gave higher priority to economic survival than to environmental issues) and the MDCs (which gave top priority to environmental integrity). The LDCs were willing to implement environmentally sensitive development, but only if the MDCs provided funding to help them do so. The MDCs, on the other hand, advised the LDCs to minimize environmental degradation by lowering their rates of population growth and development (Raven et al. 1993).

The third U.N. population conference (International Conference on Population and Development) was held in Cairo, Egypt on 5–13 September 1994. One of its major accomplishments was the development of a global mission statement on population containment, the final draft of which was unanimously accepted by the participating nations. In contrast to the previous conferences, where issues dividing MDCs and LDCs dominated the discussions, the sharpest debates associated with this conference were centered on abortion and women's rights.

Unfortunately, population issues were downplayed by the Reagan and Bush Administrations. The United States representatives to the International Conference on Population (1984) stated that population growth was a neutral factor in development and that the United States would no longer fund any family planning programs that promoted abortion. This reversal of policy stunned those in attendance, since the United States had been one of the leading advocates for family planning and population control up until this point. The United States greatly reduced its financial support for the International Planned Parenthood Federation (IPPF) and in 1985 eliminated all of its funding to the United Nations Fund for Population Activities (UNFPA), claiming that both organizations were associated with pro-abortion family planning programs in some LDCs (Johnson 1987; Fornos 1991). This policy was maintained until the end of the Bush Administration. As the United States had been the primary financial supporter of these two organizations, this lack of funding seriously undermined their effectiveness. Fortunately, other nations did not waiver in their commitment to address the population crises; Canada, Germany, Japan, the United Kingdom, and the Scandinavian countries increased their donations to UNFPA (Johnson 1987; Fornos 1991, 1993). The Clinton Administration has taken steps to restore the United States' role in global population issues. These have included the reestablishment of United States funding for UNFPA and the significant role that the United States had in the 1994 International Conference on Population and Development. Such actions represent an acknowledgment of the population problem and hopefully signal a long-term commitment by the United States to actively participate in solving it.

A major effort needs to be made toward the development of effective policy approaches to population problems. As each socioeconomic and cultural setting creates a different demographic challenge, there are no simple and universal solutions. Extensive collaboration among people with concerns about the global environment is required in order to address population and development issues. In particular, the development community needs to utilize an integrated outlook that includes resource management, environmental concerns, and population planning.

A Uniquely Human Problem

Humans face an issue that no other species does. We could choose to be like other species and to simply expand in numbers and resource use until global level environmental restraints limit our population. At that point, natural limitations will decide for us, and the results will most probably be quite catastrophic (Ehrlich and Ehrlich 1990). This, however, is what all other species do. They have no choice in their population size or fluctuation; their populations simply track environmental limitations. If environmental limitations are minimal, populations expand. As environmental resistance to growth increases, populations stabilize or decline. But humans theoretically have another option, something which is a uniquely human process. Because of technology and culture, we have the option of choosing what the carrying capacity of the world should be (within the limits set by technical, social, and environmental parameters) and the ability to work toward achieving it. No other species can establish its carrying capacity by conscious decision.

Our culture and technology have given us the ability to transcend local environmental limitations. Although we are utterly dependent on nature for our survival, we have also grown locally independent of nature in terms of the determination of local carrying capacity. For instance, there is absolutely no way that the millions of people who live in New York, London, or Beijing could be supported by the resources provided by the small land area in which they live. The United States consumes far more oil than it can produce, and a significant portion of its oil must be imported from the Middle East. But we are still bound and limited to the constraints of the global environment in which we live.

In order to establish a sustainable population, it is necessary to not only recognize environmental limitations at both the local and global levels, but also to learn to live within them. This can only be done if humans choose to limit both their population size and use of resources. As noted above, this decision would not be immediately and directly imposed on us by the environment, it would come from human interpretation of what the environmental limitations are. It is ironic that the myth that we are apart and independent of nature is in one sense true: the acknowledgment of our connections with the ecosystems that are essential for our survival and the decision to live with them in a sustainable way will flow from the human mind, which is both a part of, and apart from, nature. In a sense it will be an extra-environmental process.

We excel in evolving cultural and technical mechanisms to circumvent local limits to growth. This has allowed us to transcend uncountable local carrying capacities, degrading and destroying many ecosystems in the process. But this ability may also be our undoing, because it has put us out of balance with population-limiting parameters at the global level. The choice is ours and it is very clear: either we come to grips with population growth and resource consumption or natural limitations will eventually take their toll. Do we strive for a world where a diversity of life exists and where all humans have a comfortable and satisfactory life? Or do we settle for letting "nature take its course" and end up (in less than 100 years) with a significantly degraded planet, a world in which human existence is barely tenable. Believe it or not, it is a choice that will be made by you and by me. I wonder how the vote will turn out.

References

Bongaarts, J., "Can the Growing Human Population Feed Itself?" *Scientific American* 270(3), 1994a, pp. 36–42.

Bongaarts, J., "Population Policy Options in the Developing World," *Science* 263, 1994b, pp. 771–776.

Bonner, R., "Crying Wolf Over Elephants," *New York Times Magazine,* 7 February 1993.

Brown, L. R., "Feeding Six Billion," *In* L. R. Brown (ed.), *The World Watch Reader on Global Environmental Issues,* New York: W.W. Norton, 1991, pp. 147–164.

Brown, L. R., "Facing Food Insecurity," *In* L. R. Brown et al. (eds.), *State of the World 1994,* New York: W. W. Norton, 1994, pp. 177–197.

Brown, R. E., "Medicine and World Health: An Overview of the Situation," *In* W. J. Makofske, H. Horowitz, E. F. Karlin, and P. McConnell (eds.), *Technology, Development and the Global Environment,* Mahwah, NJ: Institute for Environmental Studies, Ramapo College, 1991, pp. 26–35.

Daily, G. C. and P. R. Ehrlich, "Population, Sustainability and Earth's Carrying Capacity," *Bioscience* 42, 1992, pp. 761–771.

Deevey, E., "The Human Population," *Scientific American* 203 (3), 1960, pp. 194–204.

Ehrlich, P. R. and A. H. Ehrlich, *The Population Explosion,* New York: Simon and Schuster, 1990.

Fornos, W., "Population Politics," *Technology Review* (Feb/Mar), 1991, pp. 43–51.

Fornos, W., "Factor in Population Growth," *The (Middletown, NY) Times Herald Record,* 8 July 1993.

Gleick, H. P., "Water, War and Peace in the Middle East," *The Environment* 36(3), 1994, pp. 6–15, 35–42.

Hardin, G., "Cultural Carrying Capacity: A Biological Approach to Human Problems," *Bioscience* 36(9), 1986, pp. 599–606.

Holden, C.,"Technology Not Enough For a Thirsty Planet," *Science* 258, 1992a, p. 1435.

Holden, C.,"Scientists Campaign to Save Earth," *Science* 258, 1992b, p. 1433.

Homer-Dixon, T. F., J. H. Boutwell, and G. W. Rathjens, "Environmental Change and Violent Conflict," *Scientific American* 268(2), 1993, pp. 38–45.

Jacobson, J. L., "Improving Women's Reproductive Health," *In* L. R. Brown et al. (eds.), *State of the World 1992,* New York: W.W. Norton, 1992, pp. 83–99.

Jacobson, J. L., "Closing the Gender Gap in Development," *In* L. R. Brown et al. (eds.), *State of the World 1993,* New York: W.W. Norton, 1993, pp. 61–79.

Johnson, S. P., *World Population and the United Nations: Challenge and Response,* New York: Cambridge University Press, 1987.

Kaufman, D. G. and C. M. Franz, *Biosphere 2000: Protecting Our Global Environment,* New York: HarperCollins, 1993.

Krejs, G. J., "Diarrhea," *In* J. B. Wyngaarden, L. H. Smith, and J. C. Bennett (eds.), *Cecil Textbook of Medicine, 19th ed., Vol. I,* Philadelphia: W. B. Saunders, 1992, pp. 680–687.

McEvedy, C. and R. Jones, *Atlas of World Population,* New York: Facts on File, 1979.

Miller, G. T., *Living in the Environment, 7th ed.,* Belmont, CA: Wadsworth, 1992.

Nadakavukaren, A., *Man and Environment: A Health Perspective,* Prospect Heights, IL: Waveland Press, 1990.

Population Reference Bureau, *1993 World Population Data Sheet,* Washington, DC, 1993.

Postel, S., "Facing Water Scarcity," *In* L. R. Brown et al. (eds.), *State of the World 1993,* New York: W.W. Norton, 1993, pp. 22–41.

Postel, S., "Carrying Capacity: Earth's Bottom Line," *In* L. R. Brown et al. (eds.), *State of the World 1994,* New York: W.W. Norton, 1994, pp. 1–21.

Raven, P. H., L. R. Berg, and G. B. Johnson, *Environment,* Philadelphia: Saunders College Publishing, 1993.

Renner, M., "Military Expenditures Falling," *In* L. R. Brown, C. Flavin, and H. Kane (eds.), *Vital Signs 1992,* New York: W.W. Norton, 1992, pp. 84–85.

Robey, B., S. O. Rutstein, and L. Morris, "The Fertility Decline in Developing Countries," *Scientific American* 269 (6), 1993, pp. 60–67.

Simon, J. L., *Population Matters: People, Resources, Environment and Immigration,* New Brunswick, NJ: Transaction, 1989.

Starke, L., "Population Growth Sets Another Record," *In* L. R. Brown, H. Kane, and E. Ayres (eds.), *Vital Signs 1993,* New York: W.W. Norton, 1993, pp. 94–95.

Stringer, C. B., "The Emergence of Modern Humans," *Scientific American* 263 (6), 1990, pp. 98–104.

The World Bank, *World Development Report 1990: Poverty,* New York: Oxford University Press, 1990.

The World Bank, *World Development Report 1992: Development and the Environment,* New York: Oxford University Press, 1992.

Vitousek, P. M., P. R. Ehrlich, A. H. Ehrlich, and P. A. Matson, "Human Appropriation of the Products of Photosynthesis," *Bioscience* 36, 1986, pp. 368–373.

Walter, W., *Vegetation of the Earth and Ecological Systems of the Geo-biosphere, 3rd ed.,* New York: Springer-Verlag, 1985.

Chapter 3

Agriculture, Technology, and Natural Resources[1]

David Pimentel

$\mathbf{M}$ost people of the world would acknowledge that we owe much of our improved standard of living to the many technological advances that have been developed over the centuries. Some of these have increased food and fiber production and stabilized their supplies. Others have helped control serious human diseases, thereby significantly improving public health. Improved public health combined with increased food supplies are major contributors to the rapid population growth (see Chapter 2, "Population Growth and the Global Environment: An Ecological Perspective").

Growth in the world population has contributed to the increasing food shortages. Although more people are being fed in the world today, more people are malnourished than ever before in the history of humankind (Swaminathan 1983). At present nutritionists report that 1 billion people are malnourished and the problem is growing in severity (Latham 1983). During the early 1900s most nations were self-sufficient in food production. Today the majority of the world's 183 nations are major food importers (FAO 1982), underscoring a growing disparity in food resources (Swaminathan 1983). This problem has persisted and in some cases worsened despite an increased pace of development (Latham 1983).

Worldwide efforts are being made to maintain human food supplies adequate enough to meet the needs of the ever-expanding population. In addition to using more fossil energy for fertilizers and pesticides, more arable land is being brought under cultivation and more water is being used to irrigate crops. The use of extensive mechanization, fossil fuels, artificial fertilizers, pesticides, hybrid crops, and irrigation is often called the *Green Revolution*.

The use of new technologies, especially those powered by fossil energy, has enabled humans to secure better control of their environment. To augment agricultural

[1] Portions of this article were previously published in McLaren, D. J. and B. J. Skinner (eds.), *Resources and World Development,* New York: Wiley and Sons, 1987, pp. 679–695.

I thank M. Pimentel for her suggestions in helping to strengthen this manuscript. The preparation of this paper was supported in part by National Science Foundation Grant BSR-8500312.

acreage, forests have been cut, and land cleared and planted to crops or converted to pasture for livestock. At present most of the highly productive land already has been cleared for crop and livestock production. In most regions only marginal land remains to be used for future production, but this will require high energy inputs.

Land Resources

About 97% of all food that humans consume comes from the land, while only about 3% comes from aquatic habitats (CEQ 1980). Of the 35% or 4.6 billion hectares of the Earth's land area that is used for agricultural production, about 1.5 billion hectares are in crops and about 3.1 billion hectares are used by grazing livestock (FAO 1982).

These lands, though productive, are being degraded by mismanagement. Poor agricultural technologies and practices are responsible for much of the destruction of valuable land resources that already has occurred and continues to take place at an alarming rate (Pimentel et al. 1976). For instance, soil degradation by erosion, salinization from irrigation, and other factors are causing the irretrievable loss of an estimated 6 million hectares (mha) of land worldwide each year (UNEP 1980; Dudal 1981; Kovda 1983). In addition, the yearly crop productivity on about 20 mha is reduced to almost zero or has negative net economic return because of intense soil degradation (UNEP 1980). According to the Food and Agricultural Organization of the United Nations (1982), the net expansion of total world cropland is about 3.9 mha per year. Thus, to compensate for the losses (6 mha) and yearly expansion (3.9 mha) about 10 mha of mostly forest land is being converted to cropland each year.

The degradation of agricultural land by erosion and other factors reduces the productivity of the land as measured in kilograms of food. At present, however, degradation is being offset by technologies that rely on increased use of fertilizers, pesticides, and other inputs. These strategies for maintaining and/or increasing production are relying on a presently cheap and abundant supply of fossil fuel, a nonrenewable resource. For example, in U.S. corn production, an additional 920 kcal/ha of fossil energy are required to offset the reduction of 1 kg of corn yield (Pimentel et al., unpublished manuscript).

Water Resources

Next to sunlight and the land itself, water is the most vital resource needed to sustain plant growth. Sufficient rain falls on most arable agricultural land, but periodic droughts continue to limit yields in diverse areas throughout the world. Crops require and transpire massive amounts of water. For example, a corn crop that produces 7,000 kg of grain per hectare will take up and transpire about 4.2 million liters of water per hectare during one growing season (Leyton 1983). To supply this much water about 10 million liters of rain must fall per hectare, and this must be evenly distributed during the year, including the growing season. This amount is equivalent to about 100 cm of rain per year.

A reduction in rainfall of only 5 cm at a critical time in the growing season will reduce corn production by about 15%. Decreasing rainfall by 30 cm reduces yields to only about one-fifth of that produced by 100 cm of rainfall (Finkel 1983). Although other crops such as sorghum and wheat could be substituted for corn because they require less water, these crops produce about one-third less grain than corn per hectare (USDA 1984).

When rainfall cannot reliably supply moisture needed for crop production, irrigation must be employed. Irrigated crop production requires large quantities of water. For example, using irrigation to produce 1 kg of food and fiber products requires 1,400 liters for corn, 4,700 for rice, and 17,000 for cotton (Ritschard and Tsa 1978). In the United States, agricultural irrigation presently consumes 81% of the total 360 bld (billion liters per day) that is consumed (Viessman and Hammer 1985).

In addition to water use, irrigated crop production is costly in terms of energy. In Nebraska, for example, rain-fed corn production requires about 630 l/ha of oil equivalents, whereas irrigated corn requires more than three times as much energy (Pimentel and Burgess 1980). In fact, nearly one-fifth of all energy consumed in U.S. agricultural production is used to pump irrigation water (USDA 1974).

Energy Resources

Of all resources, the one most responsible for rapid population growth and environmental destruction is fossil energy, used to manipulate and manage ecosystems. Energy-resource use and new energy-dependent technologies have been responsible for raising crop and livestock production yields per hectare.

The solar energy reaching a hectare of land during the year in temperate North America averages about 14×10^9 kcal (Reifsnyder and Lull 1965). During a 4-month summer growing season in the temperate region, nearly 7×10^9 kcal reach an agricultural hectare. Under favorable conditions of moisture and soil nutrients, corn is considered one of the most productive food and feed crops per unit area of land and is used in this discussion to illustrate energy expenditure in crop production (Pimentel 1980). For example, high-yielding corn grown on the good soils of Iowa can produce about 7,000 kg/ha (112 bu/acre) of corn grain plus another 7,000 kg/ha of biomass as stover (remainder of the corn plant after the ears are harvested). Converted to heat energy this totals 63×10^6 kcal (heat energy = 4,500 kcal/kg) and represents about 0.5% of the solar energy reaching the hectare during the year or 1% during the growing season.

The efficiency conversion for other crops is typically much less than that for corn. For example, potatoes, with a total biomass produced of 12,000 kg/ha and an energy value of 54×10^6 kcal, have a 0.4% conversion efficiency. Or a wheat crop, yielding 2,700 kg/ha of grain, produces a total of 6,750 kg biomass/ha, which has a heat-energy value of 30×10^6 kcal, making the conversion efficiency only 0.2%. Agricultural ecosystems as a whole, including pastures and rangeland, have an average conversion rate of about 0.1% per year, which is similar to that of natural vegetation associated with many terrestrial systems (Pimentel et al. 1978). In these analyses only the energy in the biomass itself was compared to the solar energy reaching the hectare. Agricultural crop production, however, requires additional energy inputs for tillage, seeds, weeding, and harvesting.

In the following analysis several different farming systems and technologies are examined along with their relative efficiencies in producing food-corn energy. Corn produced by hand in Mexico employing swidden or cut/burn agricultural technology requires only a person with an axe, a hoe, and some corn seed (Table 3.1) (Pimentel and Pimentel 1979). A total of 1,144 hours of labor is required to produce 1,944 kg/ha. This 1,144 hours represents about 57% of the total labor output for an adult per year. Assuming the individual is from the rural area of a developing country and consumes about 3,000 kcal of food per day, the labor input for a hectare of corn has been calculated to be 589,160 kcal. When the energy for making the axe and hoe, and producing the seed, is added to the human power input, the total energy input needed to produce corn by hand is about 0.6 million kcal/ha. With a corn yield per hectare of 1,944 kg, or 8.7 million kcal, the output/input ratio is 13.6 (see Table 3.1).

The energy inputs in tractor-powered agriculture are considerably greater than those of hand-powered systems (see Tables 3.1 and 3.2) and those of animal-powered systems. Typically U.S. corn production relies heavily on machinery for power. The total human power input is dramatically reduced to only 10 hours compared with 1,144 hours for the hand-powered system such as that in Mexico (see Tables 3.1 and 3.2). The input for labor in the U.S. system, which is only 5,250 kcal for the entire growing season, is minimal compared with the other inputs (see Table 3.2).

But balanced against this low–human power input is the significant increase in fossil-energy input required to run the machines that replace human labor. In 1983

Table 3.1
Energy Inputs and Outputs in Corn (Maize) Production in Mexico Using Only Human Power

Item	Quantity/ha	kcal/ha
Input		
Labor	1,144 h[a]	589,160[c]
Axe and hoe	16,570 kcal[b]	16,570[d]
Seeds	10.4 kg[b]	36,608[d]
Total		642,338
Output		
Total yield	1,944 kg[a]	8,748,000
kcal output/kcal input		13.62

[a] Lewis, O., 1951.
[b] Estimated.
[c] See text for assumptions for calculating kcal input.
[d] Pimentel and Pimentel (1979).

the energy inputs (mostly fossil fuel) required to produce a hectare of corn averaged about 11.0 million kcal/ha or the equivalent of about 1,136 liters of oil (see
Table 3.2). Based on a corn yield of about 7,000 kg/ha, or the equivalent of 31.5
million kcal energy, the output/input ratio is about 2.9:1. Note that the fossil-
energy input in this system represents 18% of the solar energy captured by the
above-ground corn biomass (63×10^6 kcal).

Corn yield is now at about 7,000 kg/ha in the United States (see Table 3.2).
How high can these yields go? In general it appears that corn yields have tended to
rise more slowly and even level off since 1970, while fluctuations in yields have
been increasing in amplitude since 1945. This is expected when the production
system is stressed by the large inputs of fertilizers and pesticides. The result of such
stress is that any limiting factor, such as moisture, affecting the system is more apt
to cause a major decline in yield per hectare.

Table 3.2
Energy Inputs and Outputs per Hectare for Corn (Maize) Production in the United States

Item	Quantity/ha	kcal/ha
Inputs		
Labor	10 h	5,250
Machinery	55 kg	1,018,000
Gasoline	40 l	400,000
Diesel	75 l	855,000
Irrigation	–	2,250,000
Electricity	35 kwh	100,000
Nitrogen	152 kg	3,192,000
Phosphorus	75 kg	473,000
Potassium	96 kg	240,000
Lime	426 kg	134,000
Seeds	21 kg	520,000
Insecticides	3 kg	300,000
Herbicides	8 kg	800,000
Drying	3,300 kg	660,000
Transportation	300 kg	89,000
Total		11,036,650
Outputs		
Total yield	7,000 kg	31,500,000
kcal output/kcal input		2.85

Source: Pimentel, D. and W. Dazhong. "Technological Changes in Energy Use in U.S. Agricultural Production," *In* S. R. Gliessman (ed.), *Research Approaches in Agricultural Ecology*. New York: Springer-Verlag, 1987.

This fact emphasizes the importance of using an integrated production system to achieve maximum corn yields. The corn variety must be a suitable genotype that will grow best under high fertilizer inputs, effective pest control, ample moisture, suitable temperature levels, and a long growing season.

With a suitable corn hybrid and ideal growing conditions, corn yields may reach 20,000 kg/ha of grain (about 320 bu/acre). For a total biomass of 40,000 kg/ha, this represents 180 million kcal of energy or conversion of about 1.3% of annual solar energy into corn biomass. This would be about ten times greater than the conversion of solar energy by the hand-powered system described earlier.

The limitation of the approach of increasing one input while holding all others constant has been well illustrated with nitrogen. In Iowa, for example, when no nitrogen was applied, the yield was about 2,200 kg/ha. Corn yields increased to 6,900 kg/ha as nitrogen applications rose to about 230 kg/ha. With further applications of nitrogen, however, corn yields actually declined (Munson and Doll 1959).

The unfavorable yields associated with high nitrogen application might be offset by increasing the applications of phosphorus and potassium while at the same time increasing the amount of water available to the corn crop. Also desirable would be having a corn hybrid that is tolerant of high fertilizer and herbicide levels.

Improved Technologies and Resource Management

The aim of present and future agricultural production programs is to maintain and augment crop yields while conserving soil, water, energy, and biological resources—all of which have limited supply and availability.

Soil, Nutrients, and Alternative Fertilizers

Significant quantities of nutrients that once were in the soil are removed whenever crops are harvested because they have been incorporated into the plant material as it grew. These must be replaced if a high yield is to be achieved in subsequent years. Often a more significant loss occurs from soil erosion by wind and rain (Pimentel and Krummel, unpublished manuscript). Clearly, then, a practical alternative would be to employ soil-conservation practices to reduce nutrient losses.

As a substitute for some or all commercial fertilizer, livestock manure can be used. Manure not only is a source of nutrients that crops need, but in addition it helps to reduce soil erosion and improves soil structure (Neal 1939; Zwerman et al. 1970). Although more than 70% of this collected manure is applied to land, it provides agriculture with only about 8% of the needed nitrogen and 20% of the needed phosphorus and potassium.

The amount of nitrogen from manure applied to U.S. lands could be doubled if proper management practices were employed in handling livestock manure. When manure is left standing, most of the nitrogen is lost as ammonium. To prevent this loss, manure should be collected promptly and stored in large tanks or ponds, and then immediately covered with soil when application is made to the land (Muck, personal communication). Even though distribution of manure requires energy, its use is cheaper than commercial fertilizer (Linton 1968; Pimentel et al. 1983).

Before commercial fertilizers were so universally used in corn production, corn often was planted in rotation with a legume crop (Pimentel 1981). When sweet clover is planted in the fall after the corn is harvested and plowed under 1 year later, nearly 170 kg of nitrogen is added per hectare (Willard 1927; Scott, personal communication). However, because 2 ha of land must be cultivated to raise 1 ha of corn, this practice became a problem when land prices increased. When legume rotations are not feasible, legumes can be planted between corn rows in late July or early August and then this legume planting, considered "green manure," is plowed under in early spring when the field is being prepared for seeding. Winter vetch and other legumes, for example, planted in this manner yield about 150 kg/ha of nitrogen (Mitchell and Teel 1977; Scott, personal communication). Also, the use of a vetch cover crop protects the soil from wind and water erosion during the winter and has the additional advantage of adding organic matter to the soil.

Substitute Cultural Practices

Relying on a no-till or minimum-till system would reduce the tractor fuel inputs normally required to plow and disc soil by conventional tillage and produce a saving of about 45 liters or 513,630 kcal of fossil fuel per hectare for tractor fuel if a no-till system were adopted. About 60 liters of diesel fuel are used to till 1 ha of soil with a 50-hp tractor, compared with about 15 liters for no-till planting. On the other hand, with a no-till system, herbicide must be increased to control weeds, and pesticides have to be applied to control increased insect and slug problems that often occur in no-till systems (USDA 1975). With no-till, pesticide use frequently doubles, thereby making the total energy inputs for no-till greater than those for conventional tillage.

Whether energy inputs are greater or about the same for no-till compared with conventional tillage will depend on the crop and the various practices used, tractor size, type of soil, abundance of pests, and environmental factors. No-till, however, has advantages over conventional tillage because it reduces human power inputs, decreases soil erosion, and helps conserve soil moisture.

Alternative Pest Control

Each year the 500,000 tons of pesticides applied to U.S. cropland (USDA 1981) represent an energy expenditure of about 43×10^{12} kcal. A wide variety of alternative pest controls can be used as substitutes for pesticides (Pimentel et al. 1982). One of these is biological control, which uses parasites and predators to control insects and weeds. If successful, as was the case when the Vedalia beetle was introduced into California to control cottony-cushion scale on citrus, then no further use of insecticides is required for this particular pest.

Many times heavy pesticide use has eliminated or decreased the effectiveness of parasites and predators and other natural controls. An alternative approach, called integrated pest management (IPM), consists of a wide array of different pest control strategies, including making maximum use of natural enemies (Pimentel 1982). In IPM programs, both pest and natural enemy populations are monitored and pesticides are applied only when the pest population will escape natural-enemy control and when economic damage will take place in the crop. The use of natural enemies and other biological controls is ideal from the standpoint of energy. Because parasites and predators obtain their food fuel directly from the pests, they are in fact solar powered.

Another successful way to control pests without pesticides is breeding food and fiber crops that have resistance against plant pathogens, pest insects, and weeds (Pimentel et al. 1973, 1982). Most crops now planted in the United States have had some degree of resistance to plant pathogens bred into them. Some major pests controlled by host-plant resistance include cabbage yellows and the Hessian fly pest of wheat.

In addition to the alternatives mentioned, various agronomic or cultural practices in crop production may be modified to enhance pest control. These include crop rotation, timing of the planting to elude pests, tilling of the soil to destroy weed and insect pests, irrigation-water applications to control growth sequences of crop and pests, plus the judicious use of organic and plastic mulches (Pimentel et al. 1982). All of these techniques have energy and environmental costs, but for certain pests in a particular crop and region they may be highly beneficial in terms of energy and costs.

Water Supplies

Looking to the future and considering the decline of supplies of cheap fossil fuel, it will have to be decided whether dry lands will be irrigated or used for production of dryland crops. Attention must be focused on the more basic problem of groundwater supply, which collects from precipitation and is held in the vast underground aquifers. Concern is growing in many quarters about the overuse or "mining" of these vital reserves (CEQ 1980). Surely conservation of water deserves high-priority attention from all sectors of society.

Many techniques are available for use in conserving water and, not surprisingly, most of these are similar to those for preventing soil erosion. These include using organic mulches as in no-till crop culture, terracing of crops, planting strip crops, and planting crops on the contour. All help retain the water on the land for growing crops.

Livestock Production

Animals used for human food are fed grains, forages, and grasses. Although forage and grasses are unsuitable for human consumption because they cannot be digested, grains surely are excellent human food. In this country about 90% of all grain is fed to livestock that provide meat, milk, and eggs to American consumers (Pimentel et al. 1980).

Energetically, anywhere from 7 to 88 kcal of fossil energy are required to produce 1 kcal of animal protein (Table 3.3). This contrasts with the production of

Table 3.3

Energy Inputs and Returns per Hectare for Various Livestock Production Systems in the United States

Livestock	Animal Product Yield (kg)	Yield in Protein (kg)	Protein as kcal × 10³	Fossil Energy Input for Production in kcal × 10⁶	Ratio of kcal Fossil Energy Input/kcal Protein Output	Labor Input (human hours)
Broilers	2,000	186	744	7.3	9.8	7
Eggs	910	104	416	7.4	17.8	19
Pork	490	35	140	6.0	42.9	11
Sheep[a]	7	0.2	0.8	0.07	87.5	0.2
Dairy	3,270	114	457	5.4	11.8	51
Dairy[a]	3,260	114	457	3.3	7.2	50
Beef	60	6	24	0.6	25.0	2
Beef[a]	54	5	20	0.5	25.0	2
Catfish	2,783	384	1536	52.5	34.2	55

[a]Grass fed

Source: Pimentel, D., P. A. Oltenacu, M. C. Nesheim, J. Krummel, M. S. Allen, and S. Chick, "Grass-Fed Livestock Potential: Energy and Land Constraints," *Science* 207, 1980, pp. 843–848.

protein in soybeans and corn grain that require only 0.7 to 3 kcal of fossil energy to produce 1 kcal of plant protein (Table 3.4). The major reason that animal-protein products are significantly more energy expensive than plant-protein food is that first the forage and grain feeds have to be grown and then consumed by the animal to produce the desired animal food. In addition, forage and grain have to be fed to the breeding herd to maintain them.

In the United States over 420 million ha of land are cultivated to provide the forage and grains fed to livestock. It is interesting to speculate what could happen if the United States were to switch to a grass-fed livestock system and eliminate the use (and associated energy expenditures) of grains and forage. Estimates are that about 2.8 million tons of animal protein could be produced annually or slightly more than half that of the 5.4 million tons currently produced. The 2.8 million tons of animal protein (meat, eggs, and milk) would provide 40 g protein per day per person. On average, each person in the United States consumes about 32 g of plant protein per day (Pimentel et al. 1980). Thus, the 40 g of animal protein per day from a grass-fed system added to the 32 g of plant protein would total 72 g of protein per person per day. This is greater than the recommended daily allowance (RDA) of about 46 to 54 g per person per day (NAS 1980).

Because animal-protein products require significantly larger inputs of energy than plant-protein production (see Tables 3.3 and 3.4), some would argue in favor of using grains directly as food so more people would have food. One estimate is that if the 135 million tons of grain annually cycled through U.S. livestock were fed

Table 3.4

Energy Inputs and Returns for Various Food and Feed Crops Produced per Hectare in the United States

Crop	Crop Yield (kg)	Yield in Protein (kg)	Protein as kcal × 10⁶	Fossil Energy Input for Production (10⁶ kcal)	Ratio of kcal Fossil Energy Input/kcal Protein Output	Labor Input (human hours)
Corn (U.S.)	7,000	630	2.5	6.9	2.8	12
Wheat (North Dakota)	2,022	283	1.1	2.5	2.3	6
Oats (Minnesota)	2,869	423	1.7	2.1	1.2	3
Rice (Arkansas)	4,742	272	1.1	12.5	11.4	30
Sorghum (Kansas)	1,840	202	0.8	1.5	1.9	5
Soybean (Illinois)	2,600	885	3.5	2.3	0.7	8
Beans, (Michigan)[a]	1,176	285	1.1	3.1	2.8	19
Peanuts (Georgia)	3,720	320	1.3	10.9	8.4	19
Apples (N.E. U.S.)	41,546	83	0.3	26.2	87.3	176
Oranges (Florida)	40,370	404	1.6	11.8	7.4	210
Potato (New York)	34,468	539	2.2	15.5	7.0	35
Lettuce (California)	31,595	284	1.1	19.7	17.9	171
Tomato (California)	49,620	469	1.9	16.6	8.7	165
Cabbage (New York)	53,000	1,060	4.2	16.8	4.0	289
Alfalfa (Minnesota)[a]	11,800	1,845	7.4	3.6	0.5	12
Tame hay (New York)[a]	2,539	160	0.6	0.6	1.0	7
Corn silage (N.E. U.S.)[a]	9,400	753	3.0	5.2	1.7	15

[a] Dry weight

Source: Pimentel, D. (ed.), *Handbook of Energy Utilization in Agriculture*, Boca Raton, FL: CRC Press, 1980; Pimentel, D. and M. Pimentel, *Food, Energy and Society*, London: Edward Arnold Pub., Ltd., 1979.

to humans, 400 million people could be sustained for 1 year on a vegetarian diet providing about 80 g of plant protein per person per day (Pimentel et al. 1980).

Again, it should be emphasized that both livestock-based foods and plant foods contribute not only calories but also important vitamins and minerals to the human food supply. While animal foods contain fairly high levels of saturated fats and cholesterol, their protein contribution is substantial and of higher quality than that of plant protein.

Livestock production has a major advantage in the food-production system because livestock can be maintained on marginal land that is too poor to grow food crops. This pasture and rangeland can be utilized effectively to produce valuable livestock protein (Pimentel et al. 1980), and hopefully in the future this type of animal-production system will be improved by using combinations of livestock types on the same pastures and ranges.

Conclusions

Technologies have enabled humans to use natural resources and to improve the standard of living for humanity. More people, for example, are being fed than ever before in history. Unfortunately, at the same time more people than ever before are malnourished.

This suggests two important principles. First, for development it is essential that we have both the technology and resources. Technology without resources is worthless. For example, nations with little or no fresh water and no fossil energy have an almost impossible task for further development. These nations will continue to be dependent on other nations.

Second, resources can be degraded and destroyed—this includes arable land, fresh water, energy, and biota. Topsoil can be eroded, water polluted or groundwater mined, fossil energy mined rapidly, and biological diversity reduced. Thus, even resources that are renewable, such as arable land, water, and biota, can be degraded through mismanagement so that they are no longer available for development. In fact, the degradation of the world's environment appears to be common (CEQ 1980).

Although some technologies can improve the efficient use of resources, limits for most technologies and resources exist. Thus, these technologies and resources can only support a limited number of humans. World resources and technology can support an abundance of humans, for example, 10 to 15 billion humans living at or near poverty, or support approximately 1 billion humans with a relatively high standard of living.

The decision as to whether human society lives in poverty or in relative wealth depends on humans themselves. To live with a high standard of living and have a population of about 1 billion will require that humans control their numbers. If the human population exceeds the carrying capacity, permanent damage to renewable resources will occur (see Chapter 2, "Population Growth and the Global Environment: An Ecological Perspective"). It is clear that if humans do not control their numbers, nature will.

References

CEQ, *The Global 2000 Report to the President. The Technical Report, Vol. II*. Washington, DC: U.S. Government Printing Office, 1980.

Dudal, R., "An Evaluation of Conservation Needs," *In* R. P. C. Morgan (ed.), *Soil Conservation Problems and Prospects*. Chichester, UK: John Wiley and Sons, 1981, pp. 3–12.

FAO, *1981 Production Yearbook*. Rome: Food and Agricultural Organization of the United Nations, 1982.

Finkel, H. J., "Irrigation of Cereal Crops," *In* H. J. Finkel (ed.), *CRC Handbook of Irrigation Technology, Vol. II,* Boca Raton, FL: CRC Press, 1983, pp. 159–189.

Kovda, V. A., "Loss of Productive Land Due to Salinization," *Ambio* 12, 1983, pp. 91–93.

Latham, M. C., "International Nutrition and Problems and Policies," *In World Food Issues,* Ithaca, NY: Center for the Analysis of World Food Issues, International Agriculture, 1983, pp. 55–64.

Lewis, O., *Life in a Mexican Village: Tepoztlan Restudied,* Urbana: University of Illinois Press, 1951.

Leyton, L., "Crop Water Use: Principles and Some Considerations for Agro-Forestry," *In* P. A. Huxley (ed.), *Plant Research and Agroforestry.* Nairobi, Kenya: International Council for Research in Agroforestry, 1983, pp. 379–400.

Linton, R. E., "The Economics of Poultry Manure Disposal," *Cornell Ext. Bull.* 1195:1–23, 1968.

Mitchell, W. H. and M. R. Teel, "Winter Annual Cover Crops for No-Tillage Corn Production," *Agron. J.* 69, 1977, pp. 569–573.

Munson, R. D. and J. P. Doll, "The Economics of Fertilizer Use in Crop Production," *In* A. G. Norman (ed.), *Advances in Agronomy XI.* New York: Academic Press, 1959, pp. 133–169.

NAS (National Academy of Sciences), *Recommended Dietary Allowances, 9th ed.,* National Research Council, Washington, DC: National Academy of Sciences, 1980.

Neal, O. R., "Some Concurrent and Residual Effects of Organic Matter Additions on Surface Runoff," *Soil Sci. Soc. Am. Proc.* 4, 1939, pp. 420–425.

Pimentel, D. (ed.), *Handbook of Energy Utilization in Agriculture,* Boca Raton, FL: CRC Press, 1980.

Pimentel, D., "The Food-Land-Fuel Squeeze," *Chem. Tech.* 11, 1981, pp. 214–215.

Pimentel, D., "Perspectives of Integrated Pest Management," *Crop Prot.* 1, 1982, pp. 5–26.

Pimentel, D. and M. Burgess, "Energy Inputs in Corn Production," *In* D. Pimentel (ed.), *Handbook of Energy Utilization in Agriculture,* Boca Raton, FL: CRC Press, 1980, pp. 67–84.

Pimentel, D. and W. Dazhong, "Technological Changes in Energy Use in U.S. Agricultural Production," *In* S. R. Gliessman (ed.), *Research Approaches in Agricultural Ecology.* New York: Springer-Verlag, 1987.

Pimentel, D., S. Fast, and G. Berardi, "Energy Efficiency of Farming Systems," *Agr. Eco. Env.* 9, 1983, pp. 359–372.

Pimentel, D., C. Glenister, S. Fast, and D. Gallahan, *Environmental Risks Associated with the Use of Biological and Cultural Pest Controls,* Final Report to the National Science Foundation. Report No. PB-83-168-716, Springfield, VA: National Technical Information Service, 1982.

Pimentel, D., L. E. Hurd, A. C. Bellotti, M. J. Forster, I. N. Oka, O. D. Sholes, and R. J. Whitman, "Food Production and the Energy Crisis," *Science* 182, 1973, pp. 443–449.

Pimentel, D., D. Nafus, W. Vergara, D. Papaj, L. Jaconetta, M., Wilfe, L. Olsvig, K. French, M. Loye, and E. Mondoza, "Biological Solar Energy Conversion and U.S. Energy Policy," *Bioscience* 28, 1978, pp. 376–382.

Pimentel, D., P. A. Oltenacu, M. C. Nesheim, J. Krummel, M. S. Allen, and S. Chick, "Grass-Fed Livestock Potential: Energy and Land Constraints," *Science* 207, 1980, pp. 843–848.

Pimentel, D. and M. Pimentel, *Food, Energy and Society,* London: Edward Arnold Pub., Ltd., 1979.

Pimentel, D., E. C. Terhune, R. Dyson-Hudson, S. Rochereau, R. Samis, E. Smith, D. Denman, D. Reifschneider, and M. Shepard, "Land Degradation: Effects on Food Energy Resources," *Science,* 194, 1976, pp. 149–155.

Reifsnyder, W. E. and H. W. Lull, "Radiant Energy in Relation to Forests," *Tech. Bull No. 1344,* U. S. Dept of Agriculture Forest Service, 1965.

Ritschard, R. L. and K. Tsa, *Energy and Water Use in Irrigated Agriculture during Drought Conditions.* USDOE LBL-7866. Berkeley: University of California, Lawrence Berkeley Laboratory, 1978.

Swaminathan, M. S., "Our Greatest Challenge–Feeding a Hungry World," *In* G. Bixler and L. W. Shemilt (eds.), *Perspectives and Recommendation. Chemistry and World Food Supplies: The New Frontiers.* Philippines: International Rice Research Institute, 1983, pp. 25–46.

UNEP, *Annual Review.* Nairobi, Kenya: United Nations Environment Programme, 1980.

USDA, *Energy and U.S. Agriculture: 1974 Data Base Vols. I and II.* Office of Energy Conservation and Environment, State Energy Conservation Programs, Washington, DC: Federal Energy Administration, 1974.

USDA, *Minimum Tillage: A Preliminary Technology Assessment.* Office of Planning and Evaluation, Washington, DC: U.S. Department of Agriculture, 1975.

USDA, *Agricultural Statistics 1981,* Washington, DC: U.S. Government Printing Office, 1981.

USDA, *Agricultural Statistics, 1984,* Washington, DC: U.S. Government Printing Office, 1984.

Viessman, W., Jr. and M. J. Hammer, *Water Supply and Pollution Control., 4th ed.,* New York: Harper and Row, 1985.

Willard, C. J., "An Experimental Study of Sweet Clover," *Ohio AESR B* 405, 1927, pp. 1–84.

Zwerman, P. J., A.B. Jones, G. D. Jones, S. D. Klausner, and D. Ellis, "Rates of Water Infiltration Resulting from Applications of Dairy Manure," *Proceedings of the Second Cornell Agricultural Waste Management Conference,* 1970, pp. 263–270.

Chapter 4

Sustainable Food Production

Terry Gips

The present worldwide agricultural scenario is distressing. Soil erosion, contamination of food and water, pesticide poisonings, pest and bacterial resistance, depletion and salinization of groundwater, fossil fuel dependence, and the loss of germplasm are but a few of the plagues that have already begun to confront us. If left unchecked, these factors, coupled with rapidly growing populations, loss of tropical rainforests, desertification, ozone depletion, acid rain, carbon dioxide buildup, and global warming, will create a bleak future for human civilization as we know it. In fact, it can be argued that such deterioration must be fundamentally and swiftly reversed worldwide in order to avoid locking into long-term destruction.

In agriculture, all the signals are there, whether it is environmental destruction, deterioration of health, loss of family farms, or the faltering farm economy. Worldwide, while agricultural production has increased, per capita food production has been decreasing over the past decade. Agricultural production in industrialized countries remains heavily dependent on rapidly dwindling supplies of oil. Our feelings and observations are detecting a dangerous reality that is not indicated by the economists' projections, and many believe that we are in far worse shape than our political leaders would have us believe. But, are we going to heed these warnings and change direction?

Roughly three decades ago, Rachel Carson's *Silent Spring* provided a beacon of hope, pointing the way to another path, a new way of farming without toxic chemicals. She did not possess a Ph.D. but she had the common sense to open her eyes and observe what was taking place. Though "experts" ridiculed her, she had the resolve to stand up and speak her mind. It is now clear that her vision has stood the test of time. This paper provides another vision—a positive perspective on new possibilities in agriculture.

What is Alternative Agriculture?

The term *alternative agriculture* means different things to different people. To some, it implies "back to the land" rural communes growing organic sprouts; to others, it implies fully controlled indoor hydroponic systems where chemicals are fed to plants in a sterile growing medium. A contrasting image is provided by development agencies seeking to replace ancient seed varieties in the Third World with the Green Revolution's hybrid seeds, synthetic fertilizers, and pesticides.

Others imagine a seemingly limitless, brave new world in which genes can be spliced, allowing humans to engineer new, more productive animal species or plants that can resist herbicides. And for others, alternative food production means growing a fresh, ripe New Jersey tomato on their farm or in their garden without hazardous pesticides and synthetic fertilizers so that it tastes like a real tomato.

It appears that the only thing these "alternatives" have in common is that they differ from conventional practices. But that does not tell us much because what is conventional now will not be in the future. For example, "organic farming," which includes the use of animal manures, crop rotations, and proper cultivation to control pests and maintain fertility, was how humans had farmed for centuries and was thus "conventional." That is, until just after World War II when organic practices were quickly replaced by newly developed, "alternative approaches," including the use of hybrid seeds requiring large amounts of chemical fertilizers and hazardous pesticides, such as parathion, which originally was developed by the Nazis as a lethal nerve gas. Within 10 years, these new "alternatives" became conventional and remain so today. Now, organic practices have been labeled *alternative*.

Clearly, *alternative* is a relative term that changes over time, just like the vagaries of fashion. The term *alternative* tells us very little of substance, particularly when we are discussing agriculture. Instead of reacting to what is, we should look ahead and discuss what should be. What is our goal? Where do we want agriculture to be, not just next year or even next generation, but a century from now?

"Sustainability" as an Organizing Concept

Such questions would suggest that what is needed is an agriculture that is sustainable, an agriculture that can be passed on to future generations. It is a central component of traditional Iroquois thinking to base decisions on what their effects will be on the seventh generation to come. While such thinking about sustainability has always been central to Native American tradition, it took the deterioration of various social, economic, political, and biological systems before the importance of sustainability as a necessary, fundamental concept was recognized. The evolution in thinking was furthered by the first photos taken of Earth from space, which many feel brought about a shift in consciousness.

Since the mid-1970s, the concept of sustainability has received increasing attention worldwide, ranging from discussions of how to create a sustainable society (Clark 1977; Meadows 1977; Brown 1981; Coomer 1981) to a focus by the World Bank, the United Nations, and other agencies on sustainable development (Clausen 1981; Tolba 1986). A common thread is the concern that the systems succeed and continue in both the short and long terms.

Definitions of Sustainable Agriculture

As the use of the term *sustainable agriculture* has become widespread, it has become important to have a clear definition. Various definitions have been provided for what constitutes a sustainable agriculture, ranging from a narrow focus on economics or production to the incorporation of culture and ecology (Altieri et al. 1983; Rodale 1983; Douglass 1984; Vogtmann 1984; Hill 1985; Sustainability of California Agriculture 1986). One of the earliest definitions was provided by Fisher (1978) at the 1977 IFOAM (International Federation of Organic Agriculture Movements) conference. He pointed out eight basic components for sustainability: systemic dynamism, harmony with nature, diversity, renewable resources, personal involvement, nutrition, community, and aesthetics. He also noted that economic implications could be added.

American farmer/poet Wendell Berry has combined many of these ideas in his definition: "A sustainable agriculture does not deplete soils or peoples" (Jackson et al. 1984). The same simple, yet elegant concept might also be expressed: A sustainable agriculture nourishes the people and the entire agroecosystem. Gordon Douglass (1986) has discussed economics, ecology, and community as three components of a sustainable agriculture.

The International Alliance for Sustainable Agriculture and other researchers, farmers, policymakers, and organizations have developed a definition that unifies many diverse elements into a widely adopted, comprehensive, working definition:

> A sustainable agriculture is ecologically sound, economically viable, socially just and humane (Gips 1984).

This definition clearly establishes four essential goals or criteria for sustainability that can be applied to all aspects of any agricultural system, from production and marketing to processing and consumption. Rather than dictating which methods can and cannot be used, it establishes four basic standards by which widely divergent agricultural practices and conditions can be evaluated and modified, if necessary, to create sustainable systems. The result is an agri*culture* designed to last and be passed on to future generations.

It should be pointed out that a sustainable agriculture represents a never-ending, ongoing process whose measured achievement of sustainability at any particular point is only the groundwork for its future. Also, what is sustainable in one area or country may not be sustainable in another, so systems must be locally adapted (Gutierriz and Dahlsten 1986).

Principles of a Sustainable Agriculture

Ecological Soundness

The first criterion for sustainability is that the system be ecologically sound. This applies to the vitality of the entire agroecosystem, from humans and plants to wildlife and soil organisms. The naturalist Aldo Leopold summed up this concept quite simply:

> A thing is right when it tends to preserve the integrity, stability and beauty of the biotic community. It is wrong when it tends otherwise (Leopold 1949).

The concept of the *agroecosystem* is essential to a sustainable agriculture, and central to this is the component *eco,* which is derived from the Greek *oikos,* meaning house or household. In current usage, *eco* implies the wisdom and authority to manage in the best interests of the household. In *agroecosystem* the idea of an orderly household is expanded to include the managed environment, thus composing the cultivated land and its adjacent surroundings, the plants contained or grown thereon, and all the animals associated with it (Wilhelm 1976).

There are two components necessary to achieve a whole, healthy agroecosystem, both of which are based on basic biological processes in nature: self-regulation and resource efficiency (Harwood 1985). Such a system will maintain the health of people, flora, and fauna and be designed in such a way to handle pest problems and nutrient needs internally, instead of through costly, time-consuming external interventions.

It will also be resource efficient in order to conserve precious resources, avoid system toxicity, and decrease input costs. It respects complex chemical, energy, and water cycles, acknowledging that, "Any action that disrupts the cycling processes is destructive to life and threatening to the sustainability of specific individuals and potentially threatening to species" (Moles 1986). Consequently, an ecologically sound system should be bioregionally based and designed with closed resource cycles or loops so that nutrients, and other resources are recycled and not lost, with an emphasis on renewable resources that allow greater self-reliance. For example, a comparison of resources consumed by the present American diet compared with alternative diets indicates a vast potential for improvement (Figure 4.1).

Not only will an ecologically sound agroecosystem be able to adapt, grow, and continue into perpetuity, but its health will provide the basis for the health of all of its parts, including humans. The agroecosystem should produce nutritious foods that nourish humans in all respects: physically, mentally, and spiritually. Thus, such an agriculture is both sustainable and sustaining.

Economic Viability

The second test of a sustainable agriculture is that it be economically "viable," or "able to take root and grow." It must be able to maintain itself and grow over both the short and long terms. Essential to this perspective is that there be a positive net return, or at least a balance, in terms of resources expended and returned. The return must be sufficient to warrant the labor and costs involved. If there is a net drain, the system will not continue. Of course, how the costs and returns are determined is based on the relative values of those assessing them.

In addition to short-term market factors relating to supply and demand, real viability requires an understanding of a number of other considerations, includ-

• Length of time world's petroleum reserves would last if all human beings ate meat-centered diet	13 years
• Length of time world's petroleum reserves would last if all human beings ate vegetarian diet	260 years
• Pounds of beef that can be produced on 1 acre of land	165
• Pounds of potatoes that can be grown on 1 acre of land	20,000
• Percentage of protein wasted by cycling grain through livestock	90
• Number of people who could be adequately fed by the grain saved if Americans reduced their intake of meat by 10%	60 million
• Water needed to produce 1 pound of meat	2,500 gallons
• Water needed to produce 1 pound of wheat	25 gallons
• Amount of U.S. cropland lost each year to soil erosion	4,000,000 acres
• Percentage of U.S. topsoil loss directly associated with livestock raising	85

Source: Diet for a New America, by John Robbins, *Carrying Capacity,* March 1989.

Figure 4.1. Potential adverse effects of meat-eating habits.

ing relative risk and qualitative factors (e.g., security, beauty, and satisfaction), which are often ignored in economist's models because they are difficult to quantify (Bird 1988). Finally, there are additional costs and subsidies, often hidden, that are not accounted for in the determination of economic viability. Outside costs or "externalities," such as loss of wildlife and health care costs from chemical exposure, usually are not considered in decision-making because their determination is difficult.

In only a short time, conventional agriculture has nearly used up nature's historic "capital investment" through its depletion of millions of years of fossil energy, erosion of thousands of years' development of topsoil, and loss of invaluable, centuries-evolved germplasm. Burning the candle at both ends, conventional agriculture has borrowed against the future by passing on tremendous public health and environmental costs in the form of various "toxic time bombs."

Also ignored in current accounting are numerous subsidies that make agriculture appear economically viable, including, for example, taxpayer-subsidized research and extension efforts with agrochemical and biotechnology, an indemnity program paying beekeepers millions of dollars for their massive loss of bees from pesticide spraying, and U.S. taxpayer-funded irrigation schemes that provide water to California farmers at extremely low prices.

If all of the above costs and subsidies were fully accounted for, conventional agriculture would not be considered economically viable. Such a system can only continue by hiding the true costs and passing them on to future generations. Clearly, a holistic system of accounting is needed to ensure economic viability.

Social Justice

The third requirement for a sustainable agriculture is that it be socially just. Quite simply, the system must assure that resources and power are distributed equitably so that the basic needs of all are met and their rights are ensured.

This standard is frequently overlooked, often because an assessment of power, privilege, and exploitation is considered an uncomfortable subject and threatens the status quo. Yet we must confront the issue because we are all interdependent. There are two essential components of social justice: equitable control of resources and full participation. Regarding the first, access to land is necessary in order for a majority of the world's population to escape poverty and grow the food they require. As important as equitable land tenure is the availability of adequate resources to succeed in this effort, including capital, technical assistance, and market opportunities. At the same time, the rights of landless farm workers and the urban poor must be recognized. This requires fair wages, a safe work environment, proper living conditions, and the right to nutritious, healthy food.

Unfortunately, there is widespread injustice. Today, in a majority of Third World countries, 80% of the land is controlled by only 3% of the population (*Facts About Hunger* 1985). Although women provide 66% of the total working hours, they receive only 10% of the world's income and own only 1% of the world's property (Oxfam, USA 1985). A 1988 study found that North American farm workers faced working conditions that were as hazardous as their Third World counterparts, with 43% of those interviewed reporting they had been poisoned by pesticides and only half receiving medical help (Moses 1988).

The second essential component of social justice is the right to full participation. Whether in the field, market, or voting booth, all people must be able to participate in the vital decisions that determine their lives. This right is particularly important in the case of women, indigenous people, and others who historically have been discriminated against and shut out of the decision-making process. Farm workers also must have the right to organize, and the public should be able to participate in the society's dominant institutions. In addition, the right of self-determination must also apply to entire countries.

The United Farm Workers Union has been forced to carry out a boycott of California table grapes because the government has refused to enforce the California Labor Relations Act. In many countries, federal farm policy is designed to eliminate family farms while providing subsidies to large, corporate farms. Farmers have been unable to gain increases in minimum prices for agricultural commodities, and vertically integrated agribusiness increasingly controls marketing and prices of many products. Farmers and consumers have little influence on commodity pricing policies or on agribusiness boards.

Such lack of control is even felt by the governments of developing countries. As the Third World debt crisis has grown and tough sanctions have been imposed by international lending agencies, the importance of national self-determination has become underscored. Industrialized nations like France have developed cash crop industries in the Third World (such as peanuts in Senegal) but blocked the creation of value-added processing (such as peanut oil), keeping that privilege for themselves and selling the finished product back for a profit. Transnational corporations have controlled the policies of Central American "banana republics" and freely exploited Third World genetic resources, patenting hybrids and selling them back at high profit.

Clearly, there must be a shift to a more just international economic order. In some cases, new, bioregionally based organizational structures must be established to allow greater participation. Cooperatives, for example, are based on a set of principles assuring recognition of basic rights, including democratic voting and shared ownership. Decentralized, worker-owned institutions are another viable form. Not only is greater democracy needed, but ultimately, a sustainable agriculture requires a shift to an equitable or democratic economy (Lappe 1987).

Humaneness

The fourth and final requirement for sustainability is that it be humane. Most often, the term is applied to our treatment of animals. While this is certainly an important component, a humane agriculture must embody our highest values in all aspects: from respect for life to the protection of diverse cultures. The doctrine of *ahimsa,* or noninjury to sentient beings, was originally specified in the *Vedas,* India's 5,000-year-old Sanskrit scriptures. While *ahimsa* loosely translates as "nonviolence," in the Vedic tradition it has a much broader meaning: "Having no ill feeling for any living being, in all manners possible and for all times" (Patanjali Yoga Sutras 1987). The Vedic viewpoint is that a person should see the same life force in all living entities and that those who cannot understand the principle of life in lesser beings may then eventually misunderstand what the life force is altogether and lose their sense of humanity. Comparable teachings of respect for life can be found in religious teachings from all over the world, in traditions as diverse as Judaism, Christianity, and native American beliefs.

Although humans clearly have an interdependent relationship with animals—from their physical labor and companionship to their invaluable recycling of organic matter and provision of foods—too often animals are seen only as objects to be exploited. Humane agriculture must be based on a fundamental respect for animals and a recognition of their rights.

In agribusiness, animals have been subjected to inhumane conditions on "factory farms," with laying hens jammed into tiny battery cages, veal calves kept in small, dark stalls and fed anemic diets, and hogs confined in unhealthy buildings and fed antibiotics to ensure extra weight gain (Mason 1985). While countries such as Switzerland and Sweden have banned battery cages and legislated other reforms, inhumane practices still persist worldwide.

It is equally important that the highest values apply to human interactions as well. The fundamental dignity of all human beings must be recognized, and both relationships and institutions must incorporate such essential human values as trust, honesty, self-respect, cooperation, self-reliance, compassion, and love. We must balance the society's predominant, aggressive patriarchal traditions with others, such as nurture, feeling, and reverence, which are drawn from our female sides, or what Native Americans term *The Great Earth Mother*.

For an agri*culture* to be sustainable, the cultural integrity of its society must be preserved and nurtured. Cultural roots are as important to agriculture as plant roots. Without strong communities and vibrant cultures, agriculture will not flourish. The Amish have often been cited as an example of such a sustainable community (Berry 1977).

The increasing substitution of the term agri***business*** for *agri**culture*** reflects a fundamental shift to a monetized economy in which everything, including human beings, is assigned a certain value. These "inputs" are viewed as expendable and replaceable. The human is lost. Farmers become competing production units with the sole goal of efficiency in a system rewarding those who can manage to maximize short term profits.

Rather than encouraging a sense of community, such a system leads to an increased sense of competition, isolation, and alienation. As rural societies break down, their values are lost as the backbone to the larger society. Without such a backbone, agriculture is neither humane nor sustainable.

Aldo Leopold cited the lack of any ethic dealing with the relationship of humans to land and to the animals and plants that grow upon it: "There is as yet no social stigma in the possession of a gullied farm, a wrecked forest, or a polluted stream, provided the dividends suffice to send the youngsters to college." The need for such an ethic is essential to counter economic self-interest, which "tends to ignore, and thus eventually eliminate, many elements in the land community that lack commercial value, but that are (as far as we know) essential to its healthy functioning" (Leopold 1949).

But to embody the highest human values requires even more. Wendell Berry has said: "An agriculture that is whole nourishes the whole person, body and soul. We do not live by bread alone." (Fukuoka 1985). Music, art, poetry, literature, and dance are all integral to such nourishment, just as the aesthetics of the farmscape. Japanese farmer/philosopher Masanobu Fukuoka has shared a related notion: "The ultimate goal of farming is not the growing of crops, but the cultivation and perfection of human beings" (Fukuoka 1985).

Agriculture is not just a job, but a personal discipline and a spiritual path. In fact, agricultural practices are often a part of spiritual practices, as in the case of the Hopi Indians, who incorporate sacred Kachina dancing in their planting rituals.

A New Perspective Is Required

To meet the challenge of creating a truly sustainable agriculture is not easy, not just because of big agrochemical companies, consumer behavior, or government policy, but because of how we have been raised and trained. We lack a basic relationship with nature.

Unfortunately, the more mechanistic world view of Francis Bacon, René Descartes, and Isaac Newton has dominated. This mechanistic model is reductionist, emphasizing dissection instead of interrelatedness. This approach not only misses the dynamics of the ecosystem, but the essence of the organism being studied. Given such thinking, it is easy to see why it is that what often begins as the optimistic introduction of a promising new technology, such as pesticides, frequently fails to accomplish its purpose and results in unforeseen disasters. By lacking a fundamental understanding of how the ecosystem functions, the reductionist model ends up asking the wrong questions and providing wrong answers.

For example, the simplistic thinking of the reductionist approach leads to the "magic bullet" solution of spraying pesticides to kill all "pests" (any insect or plant not desired by humans) because they seemingly interfere with a narrowly defined goal of maximum yield. It assumes that if the pest can be eradicated, the problem will be solved. The pesticides, however, may not only kill an essential food source for beneficial insects, but likely the beneficial insects as well, thus leading to secondary pest outbreaks. In addition, many pesticides are biologically magnified up the food chain and adversely affect other species (Figure 4.2).

Also, while the pesticides certainly eliminate the symptoms, they may mask more serious, underlying problems with soil management and agroecosystem balance. For example, a lack of balanced soil nutrition contributes to unhealthy, weak plants to which insects are attracted. Ironically, it has been shown that some pesticides weaken plant immune systems and make them so attractive to pests that even more pesticides must be used (Chaboussou 1986).

The dominant worldview not only separates nature from humans and contributes to an expensive and hazardous "pesticide treadmill," but it leads to an even more damaging result: nature is seen as a brutal force that must be fought, dominated, and exploited. Wendell Berry has addressed "the agricultural crisis as a crisis of culture," noting that, "We are all to some extent the products of an exploitive society, and it would be foolish and self-defeating to pretend that we do not bear its stamp." (Berry 1977). As Rachel Carson wrote in *Silent Spring:*

> The "control of nature" is a phrase conceived in arrogance, born of the Neanderthal age of biology and philosophy, when it was supposed that nature exists for the convenience of man . . . It is our alarming misfortune that so primitive a science has armed itself with the most modern and terrible weapons, and that in turning them against the insects it has also turned them against the earth (Carson 1962).

It is clear that we must rethink not only our approach to pest control, but to agriculture and life as we know it. Rather than being captivated by maximum yields and short-run return, we need optimally designed, sustainable agroecosystems that will assure a healthy, productive future while respecting all life.

A New, Common Ground—Sustainable Agriculture

Sustainable agriculture offers the possibility of a new, common ground. The concept is not partial to any particular farming system. Rather, it seeks to ensure that all agricultural systems meet the four basic principles for sustainability. If an approach fails to

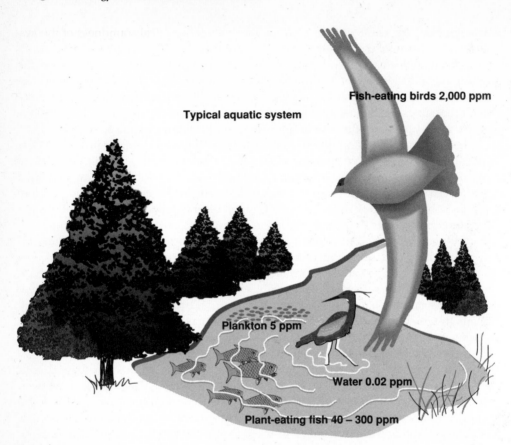

Figure 4.2. Biological magnification of DDD shown for a generalized aquatic system. The DDD concentration data taken from Clear Lake, California. (From Turk, A., J. Turk, J. T. Wittes, and R. W. Wittes, *Environmental Science,* Philadelphia: Saunders College Publishing, 1974.)

meet one or more of the criteria, a set of positive guidelines is available to assist in making the changes necessary for sustainability.

Conceived in this sense, sustainable agriculture represents a positive response to the limits and problems of both traditional and modern agriculture. It is neither a return to the past nor an idolatry of the new. Rather, it is a balance of old and new. It seeks to take the best aspects of both traditional wisdom and the latest scientific advances. This results in integrated, nature-based agroecosystems designed to be self-reliant, resource conserving, and productive in both the short and long terms.

Sustainable agriculture encompasses a surprising diversity of farming systems. These systems possess different names, and their accepted practices vary greatly. This is a function of their unique histories, geographic locations, and cultural associations. Some of the most common names for such systems are organic, biological, alternative, ecological, resource efficient, low input, natural, biodynamic, regenerative, and permaculture.

The first four place a primary emphasis on the ecological soundness of the system, although their proponents often incorporate the other principles of sustainable agriculture. Resource-efficient and low-input systems focus even more narrowly on the actual input use. Finally, the proponents of biodynamic, natural, regenerative, and permaculture approaches address many, if not all, of the sustainable agriculture principles in a holistic manner.

Sustainable agriculture also embodies components from many disciplines and concepts, including agroecology, farm systems research, agroforestry, and integrated pest management. These systems and approaches are briefly described in *Breaking the Pesticide Habit* (Gips 1988a).

It is important to note that there are examples of traditional farming systems that are sustainable, and that such systems are serving as important models and need to be preserved. For example, Dr. Stephen Gliessman (1983) found that traditional farmers in Mexico intercrop certain plants to control weeds. These plants release chemicals that inhibit the growth of weeds through an allelopathic effect, acting as natural herbicides.

Squash (*Curcurbita pepo*) is interplanted in corn-bean polycultures primarily for weed control, with harvested fruit just an added bonus. The leguminous plant, velvet bean (*Stizolobium deeringianum*) has been used in rotation with corn on some plots continuously for 20 years, maintaining yields at close to 2,678 pounds per acre (3 metric tons per hectare) without the need for additional fertilizer applications or weedings other than the initial ground preparation with a machete at the time of corn planting (Gliessman 1983).

While there are numerous such examples, many have become unsustainable for various reasons, from population pressure and lack of productivity to government policies. As the decline of the great Mayan and Mesopotamian civilizations has shown, traditional agriculture is not inherently sustainable and can lead to ecological ruin.

To help describe sustainable agriculture, two completely different operating systems are briefly described, one from a temperate industrialized country and one from a developing country in the tropics.

Organic Weed Control and Healthy Animals in Iowa

On a typical, All-American farm in the heart of Iowa's cornbelt, Dick and Sharon Thompson successfully grow 300 acres of corn, soybeans, oats, and hay, while managing a 90-sow farrow-to-finish hog operation and a 90-head cow-calf herd. While their yields are the same as their neighbors—120 to 145 bushels per acre of corn and 40 to 45 bushels per acre of soybeans—their costs are up to $90 lower per acre with less than half the soil erosion.

What's their secret? Regenerative agriculture. In 1968, the Thompsons changed from 15 years of farming with heavy fertilizer and pesticide applications that had led to leveling yields, mounting agrochemical bills, worsening weed problems, and growing animal illnesses. Their shift from continuous corn production to a rotation of corn-soybeans-corn-oats-hay, along with the adoption of ridge tillage and other sophisticated planting practices, works to control weeds and insects while maintaining yields.

Dick Thompson actually feels weeds can be used as an asset: "Not only will they help hold soil, water and plant nutrients, but we're learning that their roots (particularly foxtail) may chemically inhibit the germination of broadleaf weed-seeds in the lower soil profile." (Thompson and Thompson 1983). He then uses his "Buffalo-Till" planter to remove the top two inches of soil from the ridge in order to both form the seedbed and push the foxtail between the rows, where it can be controlled later by a cultivator. He plants high densities of fast-growing soybean varieties in order to create a dense canopy for shading out weeds.

The Thompsons also changed their animal-rearing practices from total confinement, in which Dick said animals "sleep on slatted, concrete floors over fuming manure pits in buildings with no sunlight and stale, moist air" and "are fed a steady diet of corn and soybeans mixed with a variety of drugs, antibiotics and synthetic vitamins and minerals." (Thompson and Thompson 1983). Instead, they built open-air finishing units (with straw bedding instead of concrete) at a 75% savings compared to the cost of a conventional $80,000 confinement facility. By using lime in the ground corn-cob bedding material and feeding a diversified ratio containing grasses, legumes, and beneficial lactobacillus bacteria, the Thompsons have eliminated the need for iron shots, antibiotics, and disinfectants. A single dust bag—filled only once a year—is hung in the gateway for beef cows and calves to control face flies and grubs. The result is that their animals have become much healthier and their farm has become a model of stewardship that is visited by more than 600 farmers every year.

Agroforestry Saves Soil and Farmers in Africa

Six million people live in the Central African Highlands nation of Rwanda, where lush forests have disappeared as a rapidly increasing population has sought fuelwood and building materials. Heavy tropical rains erode the barren topsoil, leaving behind scarred hillsides, environmental destruction, and ultimately, hunger. In response to this situation, one of the world's foremost "ecodevelopment" projects was initiated with local farmers, the Rwandan government and the German Agency for Technical Cooperation. Project Agro-Pastoral (agroforestry) recognized the importance of indigenous systems and sought to examine all aspects of the agroecosystem to determine what was ecologically sound and should be maintained, and what was not and needed corrective action (Neumann 1983; Janssens et al. 1986; Gips 1988a).

Innovative mini-farms, each the size of a typical family farm in Rwanda (2.5 acres), were designed to take care of an entire family's needs, from food and fodder to building materials and cash crops. Maize, beans, and sweet potatoes were closely interplanted to shade out weeds, reduce erosion, and produce an optimum yield. Another section was devoted to coffee, which served as the family's cash crop for market purchases. Fodder crops also were grown to feed the family's cows, goats, and rabbits, which were penned rather than allowed to graze freely in order to halt hillside erosion, reduce disease, improve nutrition, and recycle animal waste. Perhaps the most distinguishing feature was the use of agroforestry, the planting of fruit and leguminous trees amongst the plants. Varieties were chosen that did not shade out the plants and met a number of needs, including mulch, animal feed, firewood, building materials, soil nutrition, and food. This well-designed, complex agroecosystem not only provided for its own sustainability, but met the family's diverse needs at the same time. Project Agro-Pastoral has shown that creative, sustainable approaches can help address some of the difficult problems facing the Third World today. Unfortunately, recent turmoil and tragedy in Rwanda challenges the long-term success of such projects which depend heavily on societal stability.

Paths to Sustainable Agriculture

Based on these two examples, it is clear that there are many paths to a sustainable agriculture. It is not so important which path is selected, but rather, that the path lead to the desired goal: a healthy agri*culture* that can be passed on to future generations.

In 1986 California became the first state to pass sustainable agriculture legislation (Johnson, 1986a). Since then, many other states, such as Minnesota, Wisconsin, and Iowa, have each created multimillion dollar sustainable agriculture programs (Johnson 1986b, 1987; Gips 1988a). With funding from the Agricultural Productivity Act, the Federal Government has now established a $4 million "low-input, sustainable agriculture" program. The University of Maine and the University of Vermont have both established 4-year, sustainable agriculture programs. More than a hundred of North America's leading scientists have signed the June 5, 1986 Ottawa Declaration, which cited the need "to completely transform prevailing methods of food and fiber production" in favor of "sustainable agricultural methods" (Gips 1988b).

Religious organizations, such as the National Catholic Rural Life Conference, have devoted publications and programs to sustainable agriculture (Catholic Rural Life 1987). Finally, a number of organizations, such as the California Steering Committee on Sustainable Agriculture and the International Alliance for Sustainable Agriculture, were created in the early 1980s to carry out education and policy efforts. The numbers have mounted, with *Healthy Harvest—A Directory of Sustainable Agriculture and Horticulture Organizations* (1989) now listing more than 1,000 such groups. Additional organizations and their farming practices are described in *Planting the Future—A Resource Guide to Sustainable Agriculture in the Third World* (IASA 1989).

As many as 100,000 of America's 2.1 million farmers are practicing sustainable agriculture, and their numbers are growing rapidly as farmers struggle to cut costs and reduce soil erosion, health risks, and environmental contamination (Schneider 1987). These farmers range in size from a 2,100-acre North Dakota wheat grower and 1,270-acre California grape grower to 1-acre truck farmers. There are thousands more in Europe, Japan, Australia, and other industrialized countries, and *Planting the Future* documents the impressive developments in the Third World (IASA 1989).

The movement has received a major boost from mounting public concern about pesticide residues in food, as reflected in a Harris Poll conducted in November 1988 that found 84.2% of Americans said they would buy organic food if available, and 49% would pay more for it (Nazario 1989). As a result, some 20 big U.S.

supermarket chains have started stocking organic produce recently. While organic produce still accounts for less than 1% of the nation's sales, some distributors have more than doubled their sales in the past year, and some are having difficulty obtaining an adequate supply. Estimates on the size of the national organic industry vary from $1 billion annually to as high as $10 billion (Burros 1988).

In response to consumer demand, farmers have begun a shift toward organic farming, which encourages proper soil health and forbids the use of all synthetic pesticides and fertilizers. There are at least 30,000 American farmers using organic practices, with thousands more in other countries (Harwood 1983). Several large Kern County, California growers have converted to organic practices on more than 1,000 acres, and big suppliers such as Sunkist Growers, Inc. and Castle & Cooke's Dole Foods subsidiary are starting to grow organic produce (Nazario 1989).

What Can We Do?

Fortunately, there are some very basic steps we can each take to help create a sustainable agriculture, reduce agricultural impacts on the Earth, and achieve a better balance between humans and nature. The first step involves educating yourself about sustainable agriculture, its goals, methods, and techniques. Educating others is the next step; for example, you can write a letter to the editor of your local paper about sustainable agriculture or perhaps organize educational trips to local organic farms. You can support and participate in groups working for sustainable agriculture, such as organic food coops, farmers markets, and farming groups. You can also buy certified organic food. Ask your local market to carry it if they do not; food dollars are votes for the kind of food system you want. It is also possible to encourage and work with your organizations, school, and church/synagogue to provide organic food at events. Once knowledgeable, you can grow an organic garden in your backyard and avoid household pesticides. Be sure to ask your local agricultural extension service for information and help; they need to know of your interest. Finally, you can write your legislators to support various laws (labeling, certification) that support sustainable agriculture goals in your state. Of course, there are many other actions that can be taken, but perhaps one of the most important is the need to imagine a sustainable world.

President John Kennedy once dared to share a seemingly impossible dream: to land a human being on the moon. Many laughed at the impracticality but Kennedy was able to inspire people to join together to make the dream become a reality. Now it is time for us to accomplish another dream: the creation of a sustainable future for the human species. This dream *can* and *must* be achieved if humans are to continue living on Spaceship Earth.

References

Altieri, M. A., D. K. Letourneau, and J. R. Davis, "Developing Sustainable Agroecosystems," *BioScience* 33, 1983, pp. 45–49.

Berry, W., *The Unsettling of America: Culture and Agriculture,* San Francisco: Sierra Club Books, 1977, p. 22.

Bird, E. R., "Why 'Modern' Agriculture is Environmentally Unsustainable: Implications for the Politics of the Sustainable Agriculture Movement in the U.S.," *In Global Perspectives on Agroecology and Sustainable Agricultural Systems, Proceedings of the Sixth International Scientific Conference of the International Federation of Organic Agriculture Movements,* Santa Cruz, CA: Agroecology Program, University of California, Santa Cruz, 1988.

Brown, L. *Building a Sustainable Society,* New York: W.W. Norton and Company, 1981.

Burros, M., "Organic Food: Now the Mainstream," *New York Times,* March 29, 1988, p. 17.

Carson, R., *Silent Spring,* Boston: Houghton Mifflin, 1962.

Catholic Rural Life, "Sustainable Agriculture: An Ethical Approach," February 1987.

Chaboussou, F., "How Pesticides Increase Pests," *The Ecologist* 16, 1986, pp. 29–35.

Clark, D. (ed.), *The Sustainable Society,* New York: Praeger Publishing, 1977.

Clausen, A. W., President, The World Bank, "Sustainable Development: The Global Imperative," The Fairfield Osborn Memorial Lecture, Washington, DC, November 12, 1981.

Coomer, J. C. (ed.), *Quest for a Sustainable Society,* New York: Pergamon Press in cooperation with The Woodlands Conference, 1981.

Douglass, G. K. (ed.), *Agricultural Sustainability in a Changing World Order,* Boulder, CO: Westview Press, 1984.

Douglass, G. K., "Sustainability of What? For Whom?," *Sustainability of California Agriculture,* Davis, CA: University of California Cooperative Extension, January, 1986, pp. 29–47.

Facts About Hunger, Elkhart, Indiana: Church World Services, 1985.

Fisher, C., "Introduction to the Conference Theme: Towards a Sustainable Agriculture," *In* J. M. Besson and H. Vogtmann (eds.), *Toward a Sustainable Agriculture,* 1977 IFOAM International Conference, Sissach, Switzerland: Verlag, 1978, p. 241.

Fukuoka, M., *The One Straw Revolution,* Toronto: Bantam Books, 1985, p. xiii.

Gips, T., "What is Sustainable Agriculture?" *Manna,* July/August 1984, p. 2. This definition was developed in collaboration with N. Perlas, T. Fricke, and N. Herzberg, along with W. Winter.

Gips, T., *Breaking the Pesticide Habit: Alternatives to 12 Hazardous Pesticides for Sustainable Agriculture, 2nd ed.,* Minneapolis, MN: International Alliance for Sustainable Agriculture, 1988a, pp. 86–90.

Gips, T., "$3.45 Million for Minnesota Sustainable Agriculture," *Manna* 5, December 1988b, p. 1.

Gliessman, S. R., "Allelopathic Interactions in Crop-Weed Mixtures: Applications for Weed Management," *Journal of Chemical Ecology* 9, 1983, pp. 993–995.

Gutierrez, A. P. and D. L. Dahlsten, "Emerging Problems in Pest Management," *In Sustainability of California Agriculture,* Davis, CA: University of California Cooperative Extension, January, 1986, p. 124.

Harwood, R. R., "International Overview of Regenerative Agriculture," *In Proceedings of Workshop on Resource-Efficient Farming Methods for Tanzania,* Emmaus, PA: Rodale Press, 1983.

Harwood, R. R., "What is Different About Sustainable Agriculture?" Paper presented to the International Institute for Environment and Development Meeting, Washington, DC, July 9, 1985.

Healthy Harvest—A Directory of Sustainable Agriculture and Horticulture Organizations, Washington, DC: Potomac Valley Press and the International Alliance for Sustainable Agriculture, 1989.

Hill, S., "Redesigning the Food System for Sustainability," *Alternatives* 12, Spring-Summer 1985, pp. 32–36.

International Alliance for Sustainable Agriculture, *Planting the Future: A Resource Guide to Sustainable Agriculture in the Third World,* Minneapolis, MN: International Alliance for Sustainable Agriculture, 1989.

Jackson, W., W. Berry, and B. Colman (eds.), *Meeting the Expectations of the Land: Essays in Sustainable Agriculture and Stewardship,* San Francisco: North Point Press, 1984.

Janssens, M. J. J., I. F. Neumann, and I. Froidevaux, "Low-Input Ideotypes," *In* H. Vogtmann, B. Englehardt, and F. Enke (eds.), *The Importance of Biological Agriculture in a World of Diminishing Resources,* Proceedings of the 5th IFOAM International Science Conference at University Kassel (22–30 August 1984), Witzenhausen, Germany: Verlagsgruppe Witzenhausen, 1986, pp. 282–296.

Johnson, P., "California Supports Sustainable Agriculture," *Manna* 3, October-December 1986a, p. 1.

Johnson, P., "Minnesota Governor Supports Sustainable Agriculture Plan," *Manna* 3, October-December 1986b, p. 1.

Johnson, P., "Minnesota Creates Sustainable Agriculture Chair," *Manna,* April-June, 1987, p. 1.

Lappe, F. M., "Marx Meets Muir—Toward a Synthesis of the Progressive Political and Ecological Visions," *In Global Perspectives on Agroecology and Sustainable Agricultural Systems, Proceedings of the Sixth International Scientific Conference of the International Federation of Organic Agriculture Movements,* Santa Cruz, CA: Agroecology Program, University of California, Santa Cruz, 1988. Also see Lappe, Hamline University Commencement Address, St. Paul, MN, May 23, 1987.

Leopold, A., *A Sand County Almanac with Essays on Conservation from Round River,* Oxford, Oxford University Press, 1949.

Mason, J., "Is Factory Farming Really Cheaper?" *New Scientist,* March 28, 1985. Also see Mason J. and P. Singer, *Animal Factories,* New York: Crown Publishers, 1980.

Meadows, D. (ed.), *Alternatives to Growth I: A Search for Sustainable Futures,* Cambridge, MA: Ballinger, 1977. Papers adapted from the 1975 Woodlands Conference.

Moles, J. A., "A Prologue to the Future: Searching for Wisdom Which Will Lead to a Sustainable California and World Agriculture," *In Sustainability of California Agriculture,* Davis, CA: University of California Cooperative Extension, January, 1986, p. 252.

Moses, M., *Pesticide Poisonings and North American Farm Workers,* San Francisco: Pesticide Action Network, 1988.

Nazario, S. L., "Big Firms Get High on Organic Farming," *Wall Street Journal,* March 21, 1989, p. B1.

Neumann, I. F., "Use of Trees in Smallholder Agriculture in Tropical Highlands," *In* W. Lockeretz (ed.), *Environmentally Sound Agriculture,* New York: Praeger, 1983, pp. 351–374.

O'Keefe, M., "Regenerative Agriculture in Iowa: An Interview with Dick Thompson," *Manna* 3, January-March 1986, p. 4 and personal interviews by T. Gips, September 1985.

Oxfam, USA, "Women Creating a New World," *Oxfam Facts for Action,* Publication No. 3, Boston, MA: Oxfam America Publications, 1985.

Patanjali Yoga Sutras, 2.30, *In* S. Rosen, *Food for the Spirit: Vegetarianism and the World Religions,* New York: Bala Books, 1987, p. 72.

Rodale, R., "Breaking New Ground: The Search for a Sustainable Agriculture," *The Futurist* 17, February 1983, pp. 15–20.

Schneider, K., "Farming Without Chemicals: Age-Old Technologies Becoming State of Art," *New York Times,* August 23, 1987, p. 10.

Sustainability of California Agriculture, Davis, CA: University of California Cooperative Extension, January, 1986.

Thompson, D. and S. Thompson, "Naturally Healthy Livestock," *The New Farm,* November/December, 1983, *In The Thompson Farm, Nature's Ag School,* Emmaus, PA: Regenerative Agriculture Association, 1985, p. 20.

Tolba, M. K., Under Secretary General of the United Nations and Executive Director, U.N. Environment Program, "Partnership Among People to Replenish the Earth," Speech to the Annual Society of American Foresters Meeting, Birmingham, Alabama, 1986, Nairobi, Kenya: UNEP, undated.

Vogtmann, H., "Trends of Development in Ecofarming," *In* B. Glaeser (ed.), *Ecodevelopment: Concepts, Projects, Strategies,* Oxford: Pergamon Press, 1984, p. 197.

Wilhelm, S., "The Agroecosystem: A Simplified Plant Community," *In* J. L. Apple and R. F. Smith (eds.), *Integrated Pest Management, Vol. 6,* New York: Plenum, 1976, pp. 59–70.

Chapter 5

Appropriate Technology for Sustainable Development

Eric L. Hyman and John D. Ferchak

After World War II, the modernization and industrialization of less developed countries (LDCs) became a major focus; the goal was to rapidly make these countries into more developed countries (MDCs) through the transfer of large-scale industrial technologies. It soon became evident that this approach was failing; the long-awaited "demographic transition" to lower population growth rates and higher living standards never occurred. Schumacher (1973) pointed out that the scale of technology was a critical factor for technology transfer success and that "intermediate technology," with a better fit to the existing social, cultural, and economic characteristics of LDCs, was a more "appropriate" path for development. The search for more appropriate technologies for dissemination was begun, although the conventional large-scale approach was hardly abandoned.

Environmental problems are now occurring on an unprecedented scale (Speth 1990). Yet, conventional foreign aid strategies often contribute to environmental destruction, indebtedness, and unfavorable terms of trade for LDCs (Hyman 1990a; Goodland et al. 1991; Korten 1991–1992). Historically, most economists have downplayed the role of natural resources and have ignored environmental pollution in evaluating economic growth (Hyman 1984; Colby 1990). Today, however, there is widespread recognition that economic growth must be redefined to include externalities previously dismissed. In addition, the cumulative activities of an escalating multitude of small-scale producers in LDCs, which might be viable at low levels of population pressure or in areas with less fragile ecosystems, are no longer sustainable.

The Brundtland Commission defined *sustainable development* as economic growth that "meets the needs of the present without compromising the ability of future generations to meet their own needs" (World Commission on Environment and Development 1987). Fri (1991) noted, "If we accept the need for economic development and the inevitability of some population growth, then we must look to technology as the chief engine of sustainability. . . . Above all, we must invest heavily in creating the new knowledge that we need to increase the productivity of our use of natural and environmental resources."

Technology encompasses knowledge of equipment, tools, products, processes, materials, and skills, as well as the organization of production and marketing (Jeans et al. 1991). Many technologies developed in industrialized countries are inappropriate for an LDC context. LDCs often have a higher population growth rate, less mobile capital and labor, and more people at the subsistence level. More than one-quarter of the population in LDCs live in absolute poverty and cannot afford to take risks (Table 5.1). Consequently, there is usually more uncertainty about probable effects of new technologies in LDCs. In addition, those in MDCs often have little knowledge about the value of many species and practices indigenous to LDCs (Tisdell 1988).

Appropriate technologies for LDCs typically are small in scale and use domestically produced equipment and locally available raw materials. Compared to modern, large-scale technologies, appropriate technologies are less capital intensive, having lower investment costs per job created; less dependent on scarce foreign exchange; easier to operate, maintain, and repair with commonly obtainable skills; and have fewer unintended, negative social or environmental impacts. They often rely on human or animal power, or other renewable energy forms. Yet, appropriate technologies are labor saving in comparison to traditional production methods (Jeans et al. 1991).

The destruction of fragile environments can only be controlled when people recognize that they can adopt less-damaging alternative ways to fully provide for their needs. Access to more environmentally sound technologies depends on information on the benefits and costs, local availability of inputs and equipment, and financing. For example, the use of small loans for grass-roots financing of enterprise development has been successful in relieving poverty in Bangladesh, but technology has remained a critical constraint. Figure 5.1 illustrates how this process works in the community. Better technologies can reduce environmental damage on site and off site and increase incomes of the local people, while preserving natural resources for future use.

Table 5.1
Poverty in the Less Developed Countries, 1990[a]

Region	Number of Poor (millions)	Percentage of Population Below the Poverty Line
South Asia	562	49.0
East Asia	169	11.3
Sub-Saharan Africa	216	47.8
North Africa & Middle East	73	33.1
Eastern Europe[b]	5	7.1
Latin America & Caribbean	108	25.5
All less developed countries	1,122	29.7

[a]The poverty line used is $420 annual income per capita in 1990 dollars.
[b]Does not include the former USSR.

Source: World Bank, *World Development Report 1992,* New York: Oxford University Press, 1992.

Borrowers
Loans are usually given to individuals or groups of individuals who are at the lowest level of the economic spectrum. For an individual loan, a typical amount would be $50 to $300 for six months to a year. Borrowers are charged principal and interest.

Project
The borrowers invest their money in rapid-turnaround projects—small-scale agriculture, livestock, textiles.

Money recycled
If borrowers are working with a microenterprise bank, they can often borrow again once the original loan is repaid. If the loan originates with a donor agency, the repayment money is often funneled into a local facility. That money then becomes available to others in the community.

Profit for family
The larger part of a profit goes to improving the borrowers' daily living conditions. Many programs are designed with a built-in savings mechanism for borrowers.

Profit for repayment
A small percentage of borrowers' profit goes to repay the original loan.

Figure 5.1. Grass-roots loans—Investments in people who can help people help themselves out of poverty and protect the environment. (From Kaufman, D. G. and C. M. Franz, *Biosphere 2000: Protecting Our Global Environment,* New York: HarperCollins College Publishers, 1993.)

Sustainable technologies can (1) enhance the carrying capacity of the natural resource base by making a productive activity more environmentally benign; (2) substitute alternative means of earning livelihoods in place of those that damage critical environments; (3) adapt production to degraded or depleted natural resources; (4) promote the restoration of damaged ecosystems; (5) prevent, control, or mitigate adverse environmental impacts; and (6) indirectly improve natural resource use by reducing poverty, increasing education levels, and stabilizing population growth (Romanoff 1990).

Sustainable technologies have to be economically viable and culturally accept-able. They should avoid any significant adverse environment/natural resource im-pacts, especially risks to public health or safety, irreversible changes in land use, loss of biodiversity, or violations of international environmental agreements. Low-cost techniques are available for valuing environmental and social impacts against eco-nomic benefits and costs (Hyman et al. 1988). The technologies should not have neg-ative social impacts, especially the displacement of vulnerable ethnic minorities.

Where the environmental or natural resource benefits are long term in nature, more immediate economic benefits are needed to provide an incentive for people to carry out the activities. The economic benefits may come from an increase in the quantity or quality of production, lower production costs, a switch to local process-ing, or expanded employment due to adoption of more labor-intensive technolo-gies. Although some technologies will have to be developed, many existing sus-tainable technologies have not yet reached small-scale producers in LDCs.

This chapter focuses on several major areas where existing or new technologies can increase the sustainability of development in LDCs—sustainable agriculture and forestry practices, renewable energy conservation and use, common property re-source management, and pollution prevention and control (as outlined in Figure 5.2).

Sustainable Agricultural and Forestry Practices

Population growth, resettlement policies, land redistribution, and migration have been factors in opening up areas for agriculture that are often marginal or other-wise unsuitable. For example, farmers who are not familiar with the proper sus-tainable techniques might be forced to move to upland areas, or onto fragile soils that previously supported a tropical rain forest. Agricultural technologies devel-oped for fertile and moist soil in level lands are often inappropriate in wetter or drier areas, on steep hillsides, or with infertile soils. This problem is exacerbated by a shift from a diverse mix of subsistence crops to a few major cash crops. Declining yields, brought about by poor agricultural practices that degrade soil and water re-sources, may force farmers to attempt to farm other marginal or fragile lands (Bre-mer et al. 1984). An estimated 370 million people in LDCs farm in areas that are particularly vulnerable ecologically (OTA 1991).

Conservation of Land and Water Resources

Farmers can minimize adverse environmental impacts on their land and off site by adopting technologies for better land and water management. On steep hillsides, farmers can do terracing or contour strip planting, construct soil stabilization barriers, establish erosion control barriers, mulch, maintain a continuous vegetative cover, leave fallow areas, follow low-tillage techniques, and rotate crops. Erosion can be re-duced by planting tree crops or multipurpose grasses. Farmers can add compost and organic fertilizers and pasture animals in fallow fields and harvested plots.

Raised beds or water-tolerant crops can be used in wet areas. In areas of low rainfall, drought-resistant or short-cycle crops can be grown, and the efficiency of

AGRICULTURE	FORESTRY
Erosion control	Reforestation
Contour strip planning	Agroforestry techniques
Microclimate modification	
Small-scale irrigation	**ENERGY**
Controlled release fertilizer	
Crop rotation	Improved wood and charcoal efficiency
Green manure crops	More efficient kilns
Nitrogen-fixing biofertilizers	Biomass for electricity production
Integrated pest management	Human and animal powered machinery
Reuse of biomass waste and residuals	Solar drying
	Small-scale wind, hydro, and photovoltaics
POLLUTION CONTROL	
	COMMON RESOURCE MANAGEMENT
Small-scale pollution control technology	
Small-scale water supply	Extractive reserves
Small-scale sanitation technology	Non-wood forest product development
Livestock nutrition	

Figure 5.2. Some appropriate technologies, techniques, and methods that can be used successfully in LDCs.

water use in agriculture can be increased through drip irrigation. Microclimates can be altered by planting trees, changing the spacing of crops, or using windbreaks or live fences. Pasture rehabilitation and management can reduce the damage from overgrazing and increase yields of livestock products. Hydroponic forage production or fodder from trees could substitute for overgrazed pasture areas with poor soils and low rainfall.

Small-scale irrigation based on individual farmer systems (e.g., treadle pumps) can decrease the need for large-scale dam and canal systems, with their destructive environmental impacts (see Chapter 7, "Hydro-Development"). Over 200,000 manual, low-lift, treadle pumps are in use in Bangladesh. Yet, this technology did not spread spontaneously to African countries. First, the treadle pump had to be adapted to facilitate production in small African workshops and then demonstrated to farmers in Mali and Senegal (Hyman 1993a).

Water conservation technologies can reduce the risk of inadequate rainfall and help rehabilitate marginal or degraded lands for agricultural use. The benefits to farmers from soil conservation, however, may be less apparent and more long term, and these measures can be time consuming and may require short-term sacrifices in production.

Biotechnologies and Integrated Pest Management

Agricultural biotechnologies can improve crop yields and resistance to pests and diseases. There is significant potential for utilizing relatively simple, small-scale agricultural biotechnologies in developing countries for the benefit of resource-poor farmers (Ferchak and Croucher 1992; Ferchak and Ribeiro 1992). For instance, tissue culture makes it relatively easy to propagate many improved plant varieties relatively quickly. It could, however, reduce crop genetic diversity if used to overpromote a limited number of varieties. Vietnam has pioneered simple techniques for micropropagation of virus-free potato plantlets at a startup cost of less than $100.

Biofertilizers

Biofertilizers can reduce the need for expensive inorganic fertilizers and increase yields. Nitrogen-fixing *Rhizobium* bacteria are commonly used to inoculate seeds of legumes in the United States and their advantages have been demonstrated in Thailand and the Philippines. Experiments have also shown the possibility of inoculating nonlegumes, such as wheat and rice, with *Azospirillum* bacteria.

Cyanobacteria (formerly called blue-green algae), free living or in association with the water fern *Azolla,* capture atmospheric nitrogen and may provide other plant growth factors for rice. Because of the wide variation in the environmental requirements and the nitrogen-fixing ability of different species and varieties of cyanobacteria, yields can be increased by inoculating paddies with selected strains of these bacteria. In the United States, a commercial cyanobacteria product is being marketed as a wettable powder for aerial inoculation of rice fields.

Soil fungi that form symbiotic mycorrhizal associations with plant roots can be introduced as an inoculant to make phosphorus more available to plants (Bethlenfalvay and Linderman 1992). There are two major types of mycorrhizal associations: ectomycorrhizae (primarily associated with trees such as oak, willow, walnut, and pine) and endomycorrhize (associated with many crop plants, orchids, and some trees such as apple, maple, and redwood). In ectomycorrhizal associations the fungus covers the surface of the root and grows into the the root cortex, but does not penetrate the root cells. In endomycorrhizal associations the fungus actually penetrates the root cells and there is also no extensive growth of fungus on the root surface. Both ectomycorrhizae and endomycorrhizae can now be readily produced. Methods have been developed at the Indonesian Institute of Agriculture in Bogor for production of endomycorrhizae at the village level.

Many biofertilizers have a limited shelf life, especially in the tropics. Small-scale, decentralized production of biofertilizers allows the inoculant strains to be selected for local crop varieties and soil conditions, and can reduce losses due to transport and storage time. New promotional efforts and farmer training, however, are necessary to encourage adoption of biofertilizers. There is also a need for product certification to protect against inactive inoculants and false labeling of uninoculated carriers. Government subsidies for chemical fertilizers are an important obstacle to increased utilization of biofertilizers in some developing countries.

Integrated Pest Management

Integrated pest management (IPM) can reduce use of costly pesticides that pose risks to the health of farmers and the environment. Pests have developed resistance to many pesticides and the development of new chemicals is expensive. In addition, pests have sometimes increased in number as a result of unintended killing of their predators. With rice, for example, studies now show that insecticides can be greatly reduced from previously used levels (Litsinger 1990). In fact, when the associated health costs due to pesticide poisoning are factored in, it can be more economical for farmers to use natural controls instead of hazardous chemical pesticides (Rola and Pingali 1993). More location-specific research is needed to support recommendations for controlling pests on particular crops, and good extension services are required to train farmers in their use.

Some pesticides derived from plant products, such as pyrethrin and rotenone, are in common use, and others, such as neem seed oil, are beginning to be commercialized. A fungal nematocide for banana cultivation is under development in the Philippines. Another fungus, *Metarhizium flavoviride,* has recently been developed at the International Institute of Tropical Agriculture as an environmentally safe method of controlling locusts and grasshoppers. A large commercial market has developed for a biopesticide derived from strains of the bacterium, *Bacillus thuringiensis,* which is specifically effective against the larvae of various insects, such as cabbage worms and gypsy moth larvae.

Use of Agricultural and Agroindustrial Wastes

Forestry waste, grain wastes, green crop residuals, crop processing wastes, and animal dung are not optimally utilized in LDCs, even though they are potential resources. These residues can be used directly or modified for fertilizer, animal feed, building materials, or fuel. Factors affecting their usefulness include the costs of collection and storage, ease of drying, and physical characteristics. Agroprocessing residues are often the most economical to tap because they have already been gathered at a central location and their productive use can avoid or reduce disposal costs.

The development of new marketable byproducts can increase the profitability of agricultural product processing. For example, rice bran can be used for animal feed or oil extraction, and rice husks can be used for particle board, packaging, poultry bedding, and brick making. Rice straw can be used for making packaging, thatched roofs, cattle feed, a substrate for mushroom growing, fiberboard, or paper (Office of Energy 1989). The spent seed from annatto production and the press-cake left over after extracting oil from sunflower seeds make good animal feed (Hyman et al. 1989; Hyman 1993b).

Natural Forest Management and Tree Planting

Natural forest management is hampered by a lack of skilled foresters and technical knowledge about indigenous tree species. Other barriers to natural forest management include the lack of funding and supportive policies, and the intense pressures on the harvesting of forest and other resources in order to meet the needs of growing populations and to provide foreign exchange earnings. As Vincent (1992) noted, government policies in developing countries have generally reduced the economic benefits and sustainability of tropical timber trade.

At present in the tropics, reforestation rates are only about 1 to 10% of the overall rate of deforestation. Moreover, the replanting of cleared areas with monocultures, often of exotic species, is no substitute for the diverse environment provided by a natural forest. Reforestation of cleared government- or community-owned lands can sometimes have negative social impacts on low-income people who use the land for agriculture, grazing, or collecting nonwood products. Due to a mismatch of technology and the environment, the success rate in reforestation programs has often been poor, especially in arid or semi-arid zones. As a result, economic returns have often been low and there have been serious sociocultural barriers to the use of communal land and labor. Better technologies for mixed species plantations are needed.

Farm forestry on privately owned or rented land may be profitable for polewood, timber, fruit, or nut production. Growing fuelwood, however, is usually less profitable on good land than annual crops, especially since there is a long waiting period before farmers can earn cash from growing wood. Farm forestry for fuelwood production may be economically viable on lands that are marginal for annual crops. Growing trees that can provide fruits, nuts, and leaves for food or animal fodder, in addition to fuelwood, greatly enhances the profitability of farm forestry (Hyman 1985).

Agroforestry combines woody perennials with herbaceous crops and/or animals on the same land, either by sharing space or alternating over time. More research is needed on technologies and crop and tree combinations suitable for agroforestry under various ecological conditions (Foley and Barnard 1984). Critical factors affecting the success of farm forestry and agroforestry projects include existence of a cash market for the produce; access to extension services; availability of seedlings of the species people are interested in planting; financing, land, and labor; and government policies.

Conservation and Renewable Energy Use

Energy use in developing countries has risen fourfold over the past three decades (OTA 1991). Increased production and utilization of renewable energy, as well as conservation, can reduce the depletion of nonrenewable energy resources. Renewable energy technologies with the greatest potential at present include biomass, wind mechanical energy, wind-powered electricity, micro-hydro for mechanical energy, mini-hydro for electricity generation, and photovoltaics. Solar collectors for heating water have significant potential for fuel savings, but are still often constrained by the lack of low-cost designs. Solar drying can be used to reduce drying time and maintain a more sanitary product compared to open air drying or to lower the fuel costs of hot air dryers.

Biomass, primarily as wood, will continue to be the predominant fuel for household and cottage industries in LDCs because it is the lowest cost alternative. Only a small fraction of the heat value of the biomass that is tapped for energy, however, actually gets used. There is a tremendous need to upgrade traditional combustion technologies for households, small industries, and institutional and commercial users. Biomass energy can be conserved by upgrading the fuel efficiency of traditional woodstoves and charcoal stoves, and by switching to more efficient kilns for production of charcoal, lime, pottery, and bricks. An open fire or a pot supported by three stones has a fuel efficiency of 10 to 15%. Simple mud, clay, or metal stoves use 15 to 20% of the heat content of wood effectively, while more efficient stoves can have 20 to 30% efficiency.

Improved woodstove projects, however, have proven difficult to implement in rural areas where fuelwood is collected for free. Many woodstove projects have failed due to lack of consideration of other design aspects that are important to users besides energy efficiency. Many of the early projects adopted a strategy of encouraging user-built stoves, which proved cumbersome and often resulted in poor quality stoves that were not efficient or durable. A simpler approach for the rural poor that has proven successful in Burkina Faso is to promote inexpensive, manufactured pot shields for an open fire. Small industries and institutions that burn relatively large amounts of purchased wood are more likely to buy efficient stoves and kilns than households.

Prospects for improved charcoal stoves are better than for woodstoves. Charcoal users have an economic incentive to conserve this marketed fuel, which can consume a significant share of their household budget. Charcoal stoves with an efficiency of 30 to 35% can be produced and distributed at moderate cost by informal sector artisans. In addition, the logistics are easier for a charcoal stove project since this fuel is predominantly used in urban areas (Hyman 1987; Allen 1991).

Most of the charcoal produced in LDCs is made in primitive earth kilns or mounds that typically yield 12 to 17% of the weight of the wood. Simple charcoal kilns made from used metal drums can yield 22 to 27%, but have a higher capital cost and more labor is required, since the wood has to be cut in smaller pieces. More efficient kilns made of steel, brick, or cement are mainly used for industrial charcoal that has high quality specifications and are too costly for producing household fuel. Due to the energy loss in carbonization, existing users of fuelwood should not be encouraged to burn charcoal instead.

Biomass can substitute for fossil fuels in direct combustion or electric power generation. Decentralized electricity generation and biomass farming can create jobs and spur industrialization in rural areas. If managed on a sustained yield basis, trees or deep-rooted nonwoody perennials can reduce erosion and help maintain watersheds. Sustainable production of wood on land that would otherwise not remain forested would increase carbon fixation to reduce global warming. Biomass fuels are low in sulfur and, because they can be burned at a lower temperature, biomass combustion generates lower amounts of nitrogen oxides than fossil fuels. Consequently, the combustion of biomass does not greatly contribute to the formation of acid deposition. Generating electricity with biomass-fueled power plants, instead of using fossil–fuel–based systems, would reduce the demand for fossil fuels. A reduction in the amount of coal and oil that needed to be extracted, transported, and processed would, in turn, lead to a reduction in the environmental impacts associated with these activities.

Human and animal waste, often supplemented with straw or other cellulosic waste, or waste from food and beverage processing, can be used to generate methane biogas. Biogas energy is a major resource today in rural areas, particularly in countries such as China and India. The biodigestion slurries that remain after methane is produced may be used for fertilizer, which improves agricultural productivity. Biogas technology continues to be improved (Zhao and Zhiheng 1992; Myles et al. 1993). In addition to small village applications, the prospects for biogas may be substantial for agroindustrial applications where pollution control is a concern.

Sugarcane, cassava, and other such inputs can be used for production of ethanol for fuel. Brazil's promotion of alcohol fuel since the mid-1970s has helped to refine the technology and to promote the development of alcohol-fueled engines, although the promise of such fuels remains elusive due to the continuing low prices for petroleum (de Oliveira 1988). Some vegetable oils can be used as a substitute for diesel fuel in engines for transportation, grain mills, irrigation, pumps, generators, and tractors (Shay 1993).

Human labor and draft animal power can substitute for nonrenewable energy. Nonmotorized vehicles, such as bicycles and cycle trailers, remain an important part of urban transit systems in Asia, and their use should be promoted (Intermediate Technology News 1988; Replogle 1992). Decentralized processing of bulky raw agricultural, forestry, or mineral resources near their sources can reduce energy requirements in transport. In rural areas, electricity supplies are often unavailable or irregular, and diesel fuel is expensive and difficult to obtain. Simple hand-operated technologies are available for small-scale grain milling and vegetable oil extraction (Hyman 1990b,c; 1991; 1993b).

Common Property Resource Management

Many LDCs have vast areas of land in the public domain, including national parks, wildlife refuges, and forest preserves. These lands are subject to encroachment by commercial loggers, miners, slash and burn farmers, and poachers. About 500 million people in LDCs are forest dwellers dependent on these lands for their livelihoods (A.I.D. 1991). Trying to police vast remote areas has failed because of the lack of enforcement capability, high administrative costs, and futility of struggling against people fighting for their survival. For effective protection of these areas, it is necessary to develop alternative means for people to earn a living in buffer zones outside the designated protected areas.

The key to common property resource management is an institutional arrangement with rules for resource management, a known and enforced group size, incentives for co-owners to follow the accepted arrangements, and sanctions to ensure compliance (Bromley and Cernea 1989). The breakdown of traditional systems of authority and the migration of non-indigenous people can make it difficult to successfully manage a common property resource. If carefully managed, ecotourism associated with protected areas represents one way that local communities, as well as national interests, can benefit from preservation efforts (Lindberg 1993; Raslan 1993).

Extractive Reserves

Extractive reserves are "specially designated forest areas inhabited by extractive populations with long-term, usufruct rights to the forest resources they collectively manage" (Geisler and Silberling 1991). They are usually areas of social or ecological interest. Allowing controlled exploitation of these resources can make local people "special inspectors, who through their economic actions, guarantee the conservation of the forest without public spending" (de Beer and McDermott 1989).

Extractive reserves usually involve (1) an available natural resource with commercial value in a protected area of nonprivate land, (2) a user group with restricted use rights to the resource, and (3) a local management system that regulates extraction rates. One of the most noteworthy examples of extractive reserves was developed around wild rubber trees in Brazil. A social movement, the Alliance of the Peoples of the Forest, was organized to represent the interests of the rubber tappers. Government policies and nongovernmental organizations assisted this movement (Schwartzman 1989).

While the extractive reserve concept is new and offers large potential benefits, it may carry some risks if the common property resources are not well regulated. As incomes increase, farmers may be more able to convert forest land for farming or obtain additional livestock to graze in open-access areas. The result could be removal of most of the natural forest vegetation other than the particular species that provides the valuable commercial product.

Also, if extraction of the resource only employs a relatively small number of people in the area, other local people who do not benefit from the extractive reserve may continue to encroach on the land. Moreover, it can be difficult to organize and work with indigenous groups who have low levels of literacy and education, long-established traditional systems, and may speak tribal languages that are not widely understood in the country. (See Chapter 9, "Tropical Forests: Problems and Prospects," for further discussion on extractive reserves.)

Nontimber Forest Products

Nontimber forest products (NTFPs) were previously called "minor forest products" and often neglected in development assistance activities, but their total value in some locations can be high. Uses of NTFPs include food and beverages, forage and fodder, fuel, medicines and cosmetics, chemicals, clothing and ornamentation, and decorative or utilitarian household products (Falconer 1990).

The sale of unprocessed NTFPs to higher valued markets can provide an incentive for maintaining natural forests. Recent studies indicate that the economic value of some tropical rain forest products, especially medicinal plants and pharmaceutical derivatives, can exceed that of alternative uses of the land for agriculture and plantation forestry (Mendelsohn 1992; Roberts 1992). Processing activities that add value to these resources locally could expand the number of people who have an interest in the continued availability of the resource. More direct and equitable linkages between producers, processors, and traders could increase the effectiveness of marketing and the share of the profits retained locally. Pro-environment labeling can be a good selling point for tropical forest products and might make it possible to obtain a premium price for products.

There is often a lack of information on the location, density, yields, quality, preparation, and use of NTFPs. Comparatively little research has been done on selection and cultivation of these resources. Consequently, NTFPs that grow in the wild are susceptible to inadvertent or deliberate overharvesting. To ensure favorable environmental impacts, it is important to maintain sustainable rates of extraction and production, protect other important species and habitats in the area, and conserve soil and water resources. There is a risk that natural forests could be cleared in order to plant the commercially valuable species, with a consequent loss of biological diversity (see Chapter 8, "The Loss of Biodiversity"). The cultivation or regeneration of many NTFPs is hindered by a lack of technical knowledge.

Other potential marketing problems for NTFPs need to be investigated. For example, large seasonal variations often occur in their supply due to climate or changes in the availability of labor for collection or processing. The supply variations may result in a low capacity use rate at processing plants in the off season. Some NTFPs are too scarce or dispersed for economical collection or processing. Efficient technologies may still have to be developed for processing, storage, and packaging of the products. The marketing of NTFPs can be difficult if the resources are located in remote areas poorly served by transport and distribution systems. Local markets are not very remunerative where incomes are low. Urban markets are more distant and might be unfamiliar with these products. Export markets can be hard to tap due to poor logistics, lack of information, and high quality requirements.

Pollution Prevention and Control

Reduction of Pollution from Small-Scale Industries

Pollution is largely uncontrolled in LDCs, even when standards exist on paper. Pollution tends to concentrate along with industry in areas with a disproportionate share of gross domestic product, energy consumption, and population density. Locations particularly susceptible to environmental stresses include river basins and coastal plains, deltas with soils that are unstable for construction, and valleys that trap air pollution. Water pollution can be particularly severe in arid and semi-arid areas, where there is relatively little dilution of contaminants in surface waters.

Small-scale industries in LDCs can generate substantial quantities of pollution in the aggregate. Common sources of water pollution include slaughterhouses, leather tanneries, pulp and paper mills, textile industries, and food and minerals processing. On-site processing by small-scale placer miners causes mercury and cyanide contamination in downstream rivers. Small-scale industries, however, generally are not the most significant sources of air pollution. As is the case for MDCs, electric power plants, transport vehicles, and large-scale industries are the major sources of air pollution in LDCs.

Most technologies for pollution control have been developed to suit large-scale industries in industrialized countries. Most have capital and operating costs and skilled labor requirements that are too high for small-scale enterprises in LDCs. For instance, air pollution control technologies such as scrubbers and electrostatic precipitators are often expensive and complex. The private sector in industrialized countries and governments in LDCs are not supporting the research and development needed to develop cost-effective technologies for small-scale firms. In some cases, simple technologies for pollution control exist, but potential users are unaware of them.

Alternatives to in-plant, unit-level treatment of liquid and solid wastes include centralized effluent treatment plants, process modifications, and changes in ultimate disposal of wastes. Some changes in technology that reduce pollution also increase productivity or reduce costs through reuse or recycling of chemicals and materials, use of a smaller quantity of raw materials to produce the same amount of product, substitution of less hazardous substances, or recovery of marketable byproducts or energy.

Where firms in a subsector congregate in a geographic area, some pollution-abatement measures could be undertaken jointly by all of the firms, making effective treatment or disposal more feasible than they would be for the firms individually. For example, small-scale textile producers in Rajasthan, India, have been paying a voluntary levy per bale of cloth purchased for water pollution abatement. The funds collected from the cess have been kept in a bank for lack of knowledge of feasible pollution control technologies. This industry is presently under a moratorium restricting further construction and access to formal sector credit until its water pollution problems are solved (Ray 1991).

Domestic Water Supply and Sanitation

Providing widespread water supply and sanitation in LDCs through conventional, large-scale alternatives would cost $600 billion (Water and Sanitation Division, 1988). That is roughly $300 per capita, equivalent to or exceeding the per capita GDP in many LDCs. Only $10 billion per year is now being spent on public water supply and sanitation in LDCs. The large-scale public infrastructure approach to domestic water supply and sanitation has high operating, maintenance, and replacement costs and is often unaffordable or incompatible with the preferences of potential users. Decentralized, small-scale technologies under private industry or household management may have better prospects.

Considerable work has been done on the development of low-cost, small-scale technologies in this area (Francis and Mansell 1988; Water and Sanitation for Health Project 1990). Appropriate designs for hand pumps, such as the Indian Mark II pump, are available, but greater attention has to be paid to manufacturing issues, training of users, and ensuring a system for maintenance and repairs. Low-cost water storage systems using ferrocement tanks have been successful in Thailand for collecting and storing rainwater in the wet season for dry season use. Simple latrines have been designed for low-cost sanitation. Less costly water filtration techniques need to be developed. Prospects for commercialization of technologies for safe drinking water supply may be better than those for sanitation, which requires more behavioral changes.

Reduction of Methane Emissions from Ruminant Livestock

About 16% of the global warming from greenhouse gases has been attributed to emissions of methane. Half of the total methane emissions are from agriculture and one-quarter are from deforestation (WRI/UNEP/UNDP 1990). Ruminants may produce from 15 to 25% of the global methane. More than half of the world's cattle are in LDCs (Leng 1990). Small-scale livestock producers in these countries have emphasized increasing the number of animals they hold, rather than the yields per animal because they lack access to improved technologies.

Excessive emissions of methane from ruminants are one of the sources that is easiest to control in terms of technology. Ammonia levels in the rumen are critical for efficient microbial growth and fermentation of the feed. When cattle are fed diets high in agricultural byproducts that have little digestible protein and energy, their methane production may amount to 15% of the food's caloric value. Improved ruminant nutrition and management techniques can increase the efficiency of feed conversion, reducing methane emissions while boosting yields of dairy or meat products for millions of small-scale livestock producers.

One simple approach for improving the nutrition of livestock that depend on free grazing or stall feeding of agricultural residuals is to use multinutrient molasses urea blocks as a feed supplement. These blocks can decrease the amount of methane generated per unit of digested feed by 30 to 50% of the digestible energy. Experiments with buffalo in India found that the animals ate a larger quantity of straw and increased their daily milk production from 1.5 to 2.4 liters when given the feed supplement. This is a well-established technology in the United States, but it is not in common use by small-scale producers in LDCs. Molasses urea blocks have been manufactured in some countries such as India, but obstacles to their widespread commercialization need to be overcome (Bowman et al. 1991).

Conclusions

Poverty is a major cause of environmental degradation in LDCs, and it is the poor who are most directly affected by environmental problems. Environmental protection efforts are only likely to be effective if basic human needs are met and local participation is encouraged. On the margin of subsistence, people have little choice about tradeoffs that will affect them in the long term or affect others in the short term. Increasing the productivity of small-scale producers can raise their incomes and/or save labor time, making it more feasible for environmental and natural resource factors to be considered in production and consumption decisions.

Appropriate technology is one of the keys to sustainable development. In many cases, more downscaling, adaptation, and demonstration of existing technologies are necessary. It would be desirable to reduce the duplication of efforts that could occur if each industry or firm in a multitude of countries has to "reinvent the wheel." Successful examples of indigenous knowledge about resource use may provide lessons that are transferable to other areas. Because women are often the primary users of natural resources in many LDCs, particular attention has to be paid to gender issues in technology applications.

An emphasis on multiple interventions to overcome bottlenecks in a whole commodity sector system can improve the effectiveness of technology transfer (Budinich 1993). Dissemination of a technology takes more than just providing information to potential suppliers and users. Effective extension services are needed to convince farmers, small firms, and households to adopt new techniques.

Better resource management is profitable to society in the long run. Many environmentally preferable technologies, however, have higher capital or operating, maintenance, and replacement costs than other alternatives. Special financing arrangements are needed for pollution abatement because commercial banks are reluctant to provide loans for purposes that do not directly generate profits. Long-term financing is necessary for environmental investments that have a long payback period. Unfortunately, small-scale producers lack access to financing for these costs at the present time.

Many other areas for application of sustainable, small-scale technologies exist or have potential. For example, low-input aquaculture can increase the yield of fish to be used for human consumption without further depletion of overexploited fishery stocks. Ecotourism can provide a financial incentive for preserving nature reserves and parks. Game ranching and farming could give local people a stake in maintaining wildlands and reduce the gains from poaching. Although much of the glass and metal in municipal solid waste in LDCs is separated for re-use by households before collection and by scavengers at dumps, technologies for low-cost, small-scale recycling of plastics need to be developed.

In all cases, the application of new technologies must take into account both the needs, resources, and cultural context of the population that will be using them as well as the environmental setting in which they will be established.

References

A.I.D., *Environmental Strategy Framework,* Washington, DC: U.S. A.I.D., 1991.

Allen, H., *The Ceramic Jiko: A Manual for Stovemakers,* London: I.T. Publications, 1991.

Bethlenfalvay, G. and R. Linderman (eds.), *Mycorrhizae in Sustainable Agriculture,* Madison, WI: American Society of Agronomy Special Publication No. 54, 1992.

Bowman, R., J. Croucher, and M. Picard, *Assessment of the Pre-Feasibility of Strategic Supplementation for Reducing Methane Emissions in Gujarat India Ruminants (Cattle, Deer, Goats, Sheep, Buffalo),* Washington, DC: Appropriate Technology International, 1991.

Bremer, J., T. Babb, J. Dickinson, P. Gore, E. Hyman, and M. Andre. *Fragile Lands: A Theme Paper on Problems, Issues, and Approaches for Development of Humid Tropical Lowlands and Steep Slopes in the Latin America Region,* Washington, DC: Development Alternatives Inc., 1984.

Bromley, D. and M. Cernea, *The Management of Common Property Natural Resources: Some Conceptual and Operational Fallacies,* Washington, DC: World Bank, 1989.

Budinich, V., *ATI's Strategic Subsector Approach: Classes of Small Producers and the Value Added Chain,* Washington, DC: Appropriate Technology International, 1993.

Colby, M., *Environmental Management in Development: The Evolution of Paradigms,* Washington, DC: World Bank, 1990.

de Beer, J. and M. McDermott, *The Economic Value of Nontimber Forest Products in Southeast Asia, with Emphasis on Indonesia, Malaysia, and Thailand,* Amsterdam: Netherlands Committee for IUCN, 1989.

de Oliveira, A., "Choosing Energies in Brazil—Sugar or Oil?," *Appropriate Technology* 15, 1988, pp. 21–23.

Falconer, J., *The Major Significance of "Minor" Forest Products: The Local Use and Value of Forests in the West African Humid Forest Zone,* Rome: FAO, 1990.

Ferchak, J. and J. Croucher, "Appropriate Technology International (ATI) Program for Small-Scale Biotechnology in Asia," *In Biotechnology and Development: Expanding the Capacity to Produce Food,* New York: UN Pub., ATAS Issue 9, Winter 1992.

Ferchak, J. and S. Ribeiro (eds.), *Lab to Land: Agricultural Biotechnologies for Sustainable Development in Asia,* Washington, DC: Appropriate Technology International, 1992.

Foley, G. and G. Barnard, *Farm and Community Forestry,* London: IIED, 1984.

Francis, A. and D. Mansell, *Appropriate Engineering Technology for Developing Countries,* Victoria, Australia: Research Publications, 1988.

Fri, R., "Sustainable Development: Principles Into Practice," *Resources* No. 102 (Winter), 1991, pp. 1–3.

Geisler, C. and L. Silberling, *Extractive Reserves: Amazonia and Appalachia Compared,* Ithaca, NY: Department of Rural Sociology, Cornell University, 1991.

Goodland R., H. Daly, and M. El Serafy, *Environmentally Sustainable Economic Development: Building on Brundtland,* New York: Oxford University Press, 1991.

Hyman, E., "Natural Resource Economics: Relevance in Planning and Management," *Resource Policy* 10, 1984, pp. 163–176.

Hyman, E., "The Monitoring and Evaluation of Forestry Projects for Local Community Development," *Agricultural Administration* 19, 1985, pp. 139–160.

Hyman, E., "The Strategy of Production and Distribution of Improved Charcoal Stoves in Kenya," *World Development* 15, 1987, pp. 375–386.

Hyman, E., "An Assessment of World Bank and A.I.D. Activities and Procedures Affecting Urban Environmental Quality," *Project Appraisal* 5, 1990a, pp. 198–212.

Hyman, E., "The Choice of Technology and Scale in Coconut Processing in the Philippines," *Oléagineux* 45, 1990b, pp. 279–294.

Hyman, E., "An Economic Analysis of Small-Scale Technologies for Palm Oil Extraction in Central and West Africa," *World Development* 18, 1990c, pp. 455–476.

Hyman, E., "A Comparison of Labor-Saving Technologies for Processing Shea Nut Butter in Mali," *World Development* 19, 1991, pp. 1247–1268.

Hyman, E., *The ATI/USAID Technology Transfer Project in Senegal,* Washington, DC: Appropriate Technology International, 1993a.

Hyman, E., "Production of Edible Oils for the Masses and by the Masses: The Impact of the Ram Press in Tanzania," *World Development* 21, 1993b, pp. 429–443.

Hyman, E., R. Chavez, and J. Skibiak, "Reorienting Export Production to Benefit Rural Producers: Annatto Processing in Peru," *Journal of Rural Studies* 6, 1989, pp. 85–101.

Hyman, E., B. Stiftel, D. Moreau, and R. Nichols, *Combining Facts and Values in Environmental Impact Assessment: Theories and Techniques,* Boulder, CO: Westview Press, 1988.

Intermediate Technology News, "Cycle Trailer Manufacture Accelerates in India," *Appropriate Technology* 15, No. 1, June 1988.

Jeans, A., E. Hyman, and M. O'Donnell, "Technology—The Key to Increasing the Productivity of Micro-Enterprises," *Small Enterprise Development* 2, 1991, pp. 14–23.

Korten, D., "Sustainable Development: A Review Essay," *World Policy Journal,* 9, (Winter 1991–92), pp. 157–190.

Leng, R., *Improving Ruminant Production and Reducing Methane Emissions From Ruminants by Strategic Supplementation,* Washington, DC: U.S. Environmental Protection Agency, 1990.

Lindberg, K. (ed.), *Ecotourism: A Guide for Planners and Managers,* North Bennington, VT: The Ecotourism Society, 1993.

Litsinger, J., "Integrated Pest Management in Rice: Impact on Pesticide," *Workshop on the Environmental and Health Impacts of Pesticide Use in Rice Culture,* 28–30 March 1990, IRRI, Los Banos, Philippines.

Mendelsohn, R., "Assessing the Economic Value of Traditional Medicines from Tropical Rainforests," *Conservation Biology* 6, 1992, pp. 128–130.

Myles, R., T. Kumar, and R. Das, "Biogas in Cambodia—Skills Transferred From South to South," *Appropriate Technology* 19, March 1993, pp. 34–35.

Office of Energy, *The A.I.D. Approach: Using Agricultural and Forestry Wastes for the Production of Energy in Support of Rural Development,* Washington, DC: U.S. A.I.D., 1989.

OTA, *Energy in Developing Countries,* Washington, DC: U.S. Congress Office of Technology Assessment, 1991.

Raslan, K., "Jungle Treasures: Sabah Seeks to Exploit and Conserve its Natural Resources," *Far Eastern Economic Review* 1, 1993, pp. 32–33.

Ray, D., *Initial Environmental Report for Pollution Control Technology Assessment for Small-Scale Textile Industries at Jodhpur, Pali, Jaipur (Rajasthan State, India),* Atlanta, GA: Metcalf and Eddy, Prepared for Appropriate Technology International, 1991.

Replogle, M., *Non-Motorized Vehicles in Asian Cities,* Washington, DC: World Bank Technical Paper No. 162, 1992.

Roberts, L., "Chemical Prospecting: Hope for Vanishing Ecosystems?" *Science* 256, 1992, pp. 1142–1143.

Rola, A. and P. Pingali, "Pesticides, Rice Productivity and Health Impacts in the Philippines," *In* Faeth, P. (ed.), *Agricultural Policy and Sustainability,* Washington, DC: World Resources Institute, 1993, pp. 47–57.

Romanoff, S., *ATI's Environmental Activities and Possible Initiatives,* Washington, DC: Appropriate Technology International, 1990.

Schumacher, E., *Small is Beautiful: Economics As If People Mattered,* New York: Harper and Row, 1973.

Schwartzman, S., "Extractive Reserves: The Rubber Tapper's Strategy for Sustainable Use of the Amazon Rainforest," *In* J. Browder (ed.), *Fragile Lands of Latin America: Strategies for Sustainable Development,* Boulder, CO: Westview Press, 1989.

Shay, E., "Diesel Fuel From Vegetable Oils: Status and Opportunities," *Biomass and Bioenergy* 4, 1993, pp. 117–242.

Speth, J., "Needed: An Environmental Revolution in Technology," *In Proceedings of the Symposium Toward 2000: Environment, Technology, and the New Century,* Washington, DC: World Resources Institute, 1990.

Tisdell, C., "Sustainable Development: Differing Perspectives of Ecologists and Economists, Relevance to LDCs," *World Development* 16, 1988, pp. 373–384.

Vincent, J., "The Tropical Timber Trade and Sustainable Development," *Science* 256, 1992, pp. 1651–1655.

Water and Sanitation Division, *Water and Sanitation: Toward Equitable and Sustainable Development: A Strategy for the Remainder of the Decade and Beyond,* Washington, DC: World Bank, 1988.

Water and Sanitation for Health Project, *Lessons Learned From the WASH Project: Ten Years of Water and Sanitation Experience in Developing Countries,* Washington, DC: U.S. A.I.D., 1990.

World Commission on Environment and Development, *Our Common Future,* Oxford: Oxford University Press, 1987.

WRI/UNEP/UNDP, *World Resources, 1990–1991,* New York: Oxford University Press, 1990.

Zhao, Y. and X. Zhiheng, "Digesting Past Lessons: China's Experience With Biogas is Starting to Pay Off," *Ceres* 133, 1992, pp. 29–32.

Chapter 6

World Energy:
Sustainable Strategies[1]

Christopher Flavin

The importance of the world's energy path can scarcely be overstated: it is central to many of our most critical problems. Our economies are powered by fuels that are not only nonrenewable, but that damage lakes, forests, and human health. In addition, our energy systems are irrevocably altering the climate by adding billions of tons of carbon to the atmosphere each year, more than a ton for each person on the planet (Rotty 1987). Simply put, an environmentally sound energy strategy is a prerequisite to a sustainable society. In few other areas is change so essential if we wish to create a healthy and prosperous world.

Increasingly, it seems neither oil, nor coal, nor nuclear power can be counted on to meet future energy needs. A decade ago, the mandate was relatively simple: reduce oil dependence. That goal has been partly achieved, though not by the grandiose schemes to replace oil that were trumpeted by politicians at that time. Rather, greater energy efficiency has been the key to success. Meanwhile, programs to produce millions of barrels of synthetic fuels from coal have never made it off the drawing board. Nuclear power now supplies just 5% of world energy, and plans for expansion are being scaled back in the face of cost overruns and safety concerns raised by the 1986 accident at Chernobyl (British Petroleum Company 1987; see Chapter 16, "Disaster Revisited: Bhopal and Chernobyl—What Are the Lessons?").

The question that comes up among policymakers again and again is: if not coal, and if not nuclear, then what? It is the central energy question today, and it does have an answer. The key to resolving the coal-nuclear conundrum is simple but potentially revolutionary: greatly improved energy efficiency in the short run, complemented by renewable energy in the long run.

Energy efficiency has a demonstrated economic capability to substitute for 25% of projected energy supply by the year 2000—at less than the cost of new supplies. Indeed, improved energy efficiency should be a centerpiece of all government en-

[1] Portions of this article also appear in *State of the World, 1988,* Chapters 2 and 3.

ergy policies during the next two decades. It can be used to begin immediately limiting the economic and environmental damage caused by today's energy systems—and to buy time for the development of other sources.

Renewable energy technologies are for the most part not as economical as improved efficiency, but their costs are falling and reliability is improving. During the nineties and beyond, renewables could make a substantial contribution, fitting well with the much more efficient energy systems now in place. Wind, geothermal, solar thermal, photovoltaics, and various biomass technologies are all positioned to move strongly into the market in the nineties. Wind power is now commercially competitive, with generating costs of 6 to 8 cents per kilowatt-hour and could provide 10% of the world's electricity by 2030. Geothermal power now supplies the world with over 5,000 megawatts of installed capacity at costs of 4 to 8 cents per kilowatt-hour. Solar thermal systems, based on mirrored troughs that focus sunlight onto oil-filled tubes, can convert sunlight to electricity at a cost that is projected to soon reach 8 cents per kilowatt-hour. Solar photovoltaic systems are expected to advance rapidly and to compete with conventional sources for peak power production in the late nineties. Biomass sources already supply 12% of the world's energy, and with careful use, can play an even more important role (Flavin 1990). With a revived commitment to research and development, by the end of the century we could have available a host of new renewable-energy–based technologies, most of them small and decentralized.

Developing a viable energy system while limiting fossil fuel use is feasible, but there is no guarantee that it will take place. Energy policy is often a morass of contradictory incentives that are not easily reformed. Many governments, for example, subsidize coal mining while paying large sums to clean up the air pollution from coal burning. Yet change is possible. The deregulation of oil and natural gas prices in the United States led to substantial efficiency gains. Government-sponsored home weatherization programs in Canada and Sweden helped reduce energy losses. Programs to spur the development of technologies, such as solar photovoltaics and efficient light bulbs, have met with success.

Many of these lessons were learned the hard way, but as a result, it is now possible to put together the outline of an energy policy that makes sense. Governments must create the conditions needed for continued innovation and the commercialization of new energy technologies. They must ensure that energy markets work far more effectively and that the broad environmental implications of our energy choices be explicitly taken into account. But millions of private companies and individual consumers have the responsibility of actually carrying out the needed changes. Given the dangers of continuing on the current path, the cost of inaction is likely to be far greater. This article examines how we can begin to implement a sustainable energy path, one that is both environmentally sound and economically feasible.

Protecting Air Quality and Climate

World energy use is the primary contributor to air pollution problems. Any energy path will need to substantially reduce present impacts on the atmosphere, both regionally and globally. While regional air pollution includes such issues as acid rain and tropospheric ozone (see Chapter 12, "World Air Pollution: Sources, Impacts, and Solutions"), one of the most pressing issues for the future is the rapidly increasing greenhouse gas concentrations, especially the carbon dioxide being formed as some 6 to 8 billion tons of carbon are added to the atmosphere annually by deforestation and fossil fuel combustion. This rise in carbon dioxide from fossil fuel combustion is illustrated in Figure 6.1 (see Chapter 14, "The Challenge of Global Climate Change," and Chapter 15, "Predicting Climate Change: The State of the Art").

The Greenhouse Effect

The annual report of the International Energy Agency (IEA) published in 1987 is the first by that organization to include a reference to the greenhouse effect. But the sometimes alarmist IEA is sanguine about the global climate, concluding that the "situation does not justify hasty measures aimed at rapid reductions in fossil fuel use but . . . a well coordinated international effort aimed at gathering further information is certainly indicated" (IEA 1987a).

The IEA's indifference to the possibility of severe climate change reflects the attitude taken by top energy officials in most countries—that this is still a scientific issue of little practical relevance. Governments throughout the world have not only been slow to respond to new environmental threats, they are actively pursuing energy policies that aggravate them. Ignoring the greenhouse problem reflects a deeper failing of energy policy: fixation on the short-run concerns of

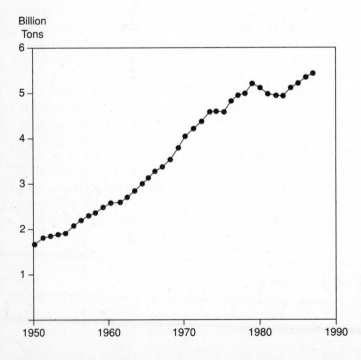

Figure 6.1. World carbon emissions from fossil fuels, 1950–1987. (From Brown L. R. et al., State of the World 1988, New York: W.W. Norton, 1988.)

energy prices and supply at the expense of the planet's future economy and ecology.

The vast research and public attention paid to the problem of global warming in the past few years, however, has begun to change this picture. A major scientific study released in 1990 by the United Nations–commissioned Intergovernmental Panel on Climate Change confirmed that a rapid and highly disruptive increase in global temperatures would occur unless emissions are cut. Upon releasing the report, Dr. John Houghton, head of the British Meteorological Service, noted that it represented "remarkable consensus," with fewer than 10 of 200 scientists dissenting. Many governments are now beginning to take global warming more seriously. Although major cuts in carbon dioxide emissions will take decades to accomplish, the targets will be far more difficult to achieve if emissions continue to rise, as has been the case over the past several years (IPPC 1990; Milne 1990; Flavin 1991).

Solutions to Carbon Dioxide Buildup

As it is impossible to remove carbon dioxide from exhaust gases, slowing these trends will require other strategies: increasing energy efficiency, stemming deforestation, and developing technologies based on renewable energy sources.

Tree planting on a scale that will satisfy future demands for fuel wood, lumber, and paper and that will stabilize soils and the hydrological regime would also help restore balance to the carbon cycle by transferring carbon from the atmosphere to terrestrial systems. Planting fast-growing trees on 110 million hectares of land by the year 2000 to provide fuel wood and ecological stability could reduce the carbon emissions of the world's forests by 41%, helping slow the global warming (Postel and Heise 1988).

Checking carbon emissions to ameliorate the warming will require both a considerable increase in energy efficiency and a change in the mix of energy sources. In the next two decades, efficiency will have to play the largest role, since enormous investments can already be justified by conventional economic criteria alone. Later, as the technologies mature, the transition to renewable energy must begin in earnest.

Most official energy projections assume that worldwide energy efficiency will increase by between 0.5% and 1.0% per year. But carbon dioxide buildup is ongoing and cumulative. Even a 1% rate of efficiency improvement would allow an increase in atmospheric carbon dioxide from 348 parts per million in 1987 to about 600 parts per million in 2075, causing unprecedented changes in climate (Mintzer 1987).

An alternative energy scenario based on the global energy model of the Institute for Energy Analysis in the United States found that a successful effort to improve worldwide efficiency by 2% annually would keep carbon dioxide concentrations to 463 parts per million in 2075. The world's climate would still be at least 1°C warmer than today, but the most catastrophic climatic effects would probably be avoided. Irving Mintzer of the World Resources Institute in Washington, D.C., using another model, calculates that a 1.5% annual rate of improvement—together with deliberate policies to limit coal use, to restrict the production of other green-house gases, and to halt deforestation—would reduce the warming projected by 2075 by half (Chandler 1985, 1987; Mintzer 1987).

Improving energy efficiency by 2% per year for several decades is challenging but probably feasible. Over 50 years it would reduce global energy intensity by almost two-thirds. Some industrial market countries have been achieving this mark in the past 15 years. Sustaining this pace through decades of changing conditions and at times when fuel prices are not rising will be a difficult task, however, requiring major institutional changes. Fortunately, many of the needed technologies are already in place, and some countries are showing the way. Buildings worldwide must become as efficient as Sweden's, and industry must become as efficient as Japan's. Automobiles must become as efficient as the best prototype models now found in the engineering facilities of Europe and Japan.

Renewable energy and nuclear power also contribute to carbon reduction by substituting for fossil fuels. For example, fossil-fuel–fired power plants alone now produce one-quarter of the world's fossil-fuel–related carbon emissions. Hydropower effectively displaces 578 million tons of carbon emissions from coal-fired power plants each year, equivalent to over 10% of the 1987 total. Nuclear power displaces somewhat less carbon—about 414 million tons annually (estimates based on Edmonds and Reilly 1983; Rotty 1987; United Nations 1987).

In recent years, nuclear power has grown rapidly, making a significant contribution to climate protection, but its growth will slow in the next few years as nuclear plant completions dwindle. While nuclear power advocates at the International Atomic Energy Agency and elsewhere have argued that this energy source can help protect the climate, its technical, economic, and political problems now appear severe enough to rule out substantial expansion (Flavin 1987).

The use of renewable energy sources, on the other hand, could grow rapidly for many years. A large number of hydropower projects are in the pipeline (although there are ecological costs—see Chapter 7, "Hydro-Development"), and within the next decade major increases in geothermal, wind, and solar electricity are likely. According to a new study by several U.S. government laboratories, renewables could supply the equivalent of 50 to 70% of current U.S. energy use by the year 2030 (Meridian Corporation 1989: DOE, IEA 1990; INEL 1990). Nonetheless, some important countries, including China, India, and the former Soviet Union, are still planning on large increases in coal-fired generating capacity. Greater reliance on renewable energy is essential in the long term if a global warming is to be averted.

Most of the progress so far in slowing the rate of carbon dioxide buildup was stimulated by market forces. Energy markets, however, will have to be greatly improved in order to deal with the problem. Indeed, any significant delay in improving energy efficiency will make it far more difficult to avoid a damaging increase in the Earth's temperature. Avoiding the costs and disruption associated with the warming may well justify expenditures in energy efficiency several times greater than those currently being made.

One question for policymakers is where to find tens of billions of dollars for climate protection at a time when public treasuries are being called on to meet a host of pressing needs, from debt repayment to education. Fortunately, most of the funds for energy efficiency and renewables could be diverted from the billions of dollars spent annually on conventional energy supply projects. The new path would actually be less expensive than the old one.

The investments in climate protection can be made by governments interested in economic competitiveness, by grass-roots organizations wishing to create sustainable fuel wood supplies, by state governments concerned about air pollution, and by corporations worried about the bottom line. In most cases the funds can be raised privately and justified economically, but the end result would be an important step toward climate protection. New policy initiatives can play an important role in encouraging governments and individuals at all levels to increase commitments to energy efficiency, renewable energy, and reforestation.

Regional Air Pollution

Fossil fuels are the main source of the air pollution now choking rural areas and cities. Even in the United States, which has relatively stringent air pollution standards, sulfur dioxide and carbon monoxide emissions are down less than 20% since 1975. Nitrogen oxide emissions are actually up 5%. Many American cities still do not comply with air quality standards laid down more than a decade ago. In the U.S. Congress throughout the 1980s, acid rain bills languished, due in large part to opposition from those representing coal-producing states (U.S. EPA 1987). The Clean Air Act of 1990 finally broke the deadlock and shows promise in reducing air pollution in the longer term if it is enforced.

Nonetheless, after over a decade of control efforts, air pollution remains a growing problem in most cities, and one that improved efficiency can help solve. Improved efficiency has the potential to reduce emissions of most dangerous pollutants, though this depends to some degree on the technologies employed. A 1987 study by the American Council for an Energy-Efficient Economy concluded that increased energy efficiency could cut electricity consumption in the Midwest by 15–25%, making it possible to reduce use of the region's dirtiest coal-fired power plants and so reduce acid rain. Because the efficiency savings are economical in their own right, the funds saved can be used to invest in additional pollution controls (ACEEE 1987).

In most other countries, air pollution continues to increase, particularly in the coal-rich regions of Eastern Europe and China. The incidence of respiratory and cardiac disease is growing rapidly in these areas, as is damage to lakes and forests. Despite strong evidence that air pollution is killing the forests of central Europe, the European Community has been able to agree only on relatively weak standards for emissions reduction, and most East European governments have yet to act at all (Brown and Flavin 1988).

Energy efficiency can be a new weapon in the air pollution wars, complementing flue-gas scrubbers and catalytic converters. Czechoslovakia, Poland, and other East European countries could limit the damage now occurring to their forests by adopting a program to improve industrial energy efficiency. Rome could attack the cause of much of the city's respiratory disease and slow damage to its ancient ruins by doubling the fuel efficiency of its cars. Energy efficiency can provide an immediate first step to a sustainable world economy.

The Untapped Source: The Efficiency Bridge

The post–World War II era is commonly described as the age of oil. Petroleum fueled the engines of industrialization and helped raise living standards around the globe. By similar logic, the current era is the age of energy efficiency. Since 1973, the world has saved far more energy than it has gained from all new sources of supply combined. The energy savings of the industrial market economies since 1973 exceed the combined energy use of Africa, Latin America, and South Asia. Efficiency made it possible for the world to climb out of the severe 1981-82 recession, and led to a 75% decline in the real price of oil between 1981 and 1986 (IEA 1987b).

Energy Efficiency Potential Underestimated

When the Ford Foundation completed a landmark study of the U.S. energy future in 1974, it presented three scenarios. One curve was identified as the "historical growth scenario" and envisioned national energy use doubling between 1970 and 1987. The lowest consumption forecast, termed the *zero growth scenario,* still involved an almost 20% increase in energy use. In fact, over the next thirteen years the U.S. economy expanded by over 35%, but U.S. energy use fell (EPPFF 1974; U.S. DOE Feb. 1987).

While most analysts underestimated the potential for greater efficiency, they overestimated the world's ability to live with the side effects of high levels of energy consumption. It was assumed, for example, that world energy consumption could more than double by the year 2000 without debilitating price increases. The Organization of Petroleum-Exporting Countries (OPEC) was expected to be pumping at least three times as much oil as it now does, unhindered by tanker wars, or a revolutionary regime in Teheran, or Iraq's seizure of Kuwait over disputed oil. Nuclear power was believed capable of supplying at least five times as much energy as it does today, untouched by billion-dollar cost overruns or accidents in Pennsylvania or the Ukraine (Gibbons and Chandler 1981; Flavin 1987, 1992).

As circumstances changed, energy use patterns changed with them. Between 1973 and 1985, the energy intensities of all industrial market countries fell (Table 6.1). The improvement, however, varied widely—from about 6% in Australia and Canada to 18% in West Germany, and 23% in the United States. Japan improved by 31%, remarkable since the nation started with one of the world's most efficient economies. It continued to widen its lead during the eighties (IEA 1987b).

Efficiency Responds to Global Environmental Limits

The Club of Rome's 1972 warning that the world would run out of fuels and raw materials appears contradicted by the fact that the world today faces a glut rather than a shortage of fossil fuels. But that glut is itself a product of temporary shortages and soaring prices. Since the early seventies, the world has encountered a series of limits to growth and has so far shown remarkable adaptability, thanks to dramatic price increases not anticipated by the Club of Rome. But the global envi-

Table 6.1

Energy Intensity (in Megajoules per 1980 Dollar of GNP) of Selected National Economies, 1973–1985

Country	1973	1979	1983	1985	Change, 1973–85 (percent)
Australia	21.6	23.0	22.1	20.3	–6
Canada	38.3	38.8	36.5	36.0	–6
Greece[a]	17.1	18.5	18.9	19.8	+16
Italy	18.5	17.1	15.3	14.9	–19
Japan	18.9	16.7	13.5	13.1	–31
Netherlands	19.8	18.9	15.8	16.2	–18
Turkey	28.4	24.2	25.7	25.2	–11
United Kingdom	19.8	18.0	15.8	15.8	–20
United States	35.6	32.9	28.8	27.5	–23
West Germany	17.1	16.2	14.0	14.0	–18

[a] Energy intensity increased as a result of a move toward energy-intensive industries such as metal processing.

Source: International Energy Agency, *Energy Conservation in IEA Countries,* Paris: Organisation for Economic Co-operation and Development, 1987.

ronmental limits now emerging may turn out to be the most stringent and dangerous of all, sorely testing the resolve of policymakers and citizens (Meadows et al. 1972).

Investment in energy efficiency is the most effective response to those limits, for it simultaneously leads to lowered oil dependence, reduced air pollution, and climate protection. Doubling the fuel efficiency of a typical European car, to 50 miles per gallon, lowers its annual fuel bill by almost $400; generally cuts emissions of nitrogen oxides, hydrocarbons, and carbon monoxide; and reduces carbon emissions by half, or 450 kilograms annually. A similar improvement for the world as a whole would cut carbon emissions by almost 200 million tons annually, a substantial contribution to climate protection.[2]

The Petroleum Quandary

In 1986, oil imports in many countries rose for the first time in almost a decade—by more than 1 million barrels per day in the United States alone. After falling in the 1980s, the Middle East's share of the oil market is expected to increase from 27% in 1990 to nearly 40% in the year 2000, according to the Paris-based International Energy Agency. The reason for this increase is that oil production has begun to decline in much of the rest of the world. The large oil fields

[2] Worldwatch Institute calculations based on data from IEA, International Road Federation, and Rotty (private communication).

discovered in the 1970s and 1980s in Mexico, the North Sea, Alaska, and Siberia are no longer expanding, and in some cases are in sharp retreat. For example, in the United States, the world's number-two producer, production has declined from 10.6 million barrels per day in 1985 to 9 million barrels in 1990. Oil industry analysts expect further declines to 7.4 million barrels per day by the year 2000. The decline is due not only to production costs associated with smaller wells but to real reductions in available supplies. If current trends continue, the United States will be importing more oil than ever by the mid-nineties (Flavin 1992).

Meanwhile, the concentration of remaining petroleum reserves in the Persian Gulf grow ever more pronounced. While the Persian Gulf region had 55% of proven global reserves as recently as 1980, by 1989 that figure had reached 65%. Most of the nations in that area have at least 100 years of proven reserves left at current extraction rates compared with less than 20 years' worth in Europe, North America, and the former Soviet Union (Flavin, 1991). Concern about this supply was a major factor in leading to the Persian Gulf War of 1991. While the war was nominally aimed at liberating Kuwait, it was also about Persian Gulf oil—who controls it and how much it will cost. When Iraq's tanks rumbled into Kuwait in August 1990, the world faced the third oil shock in the space of 17 years. This invasion, which immediately raised Iraq's share of world oil reserves from 10% to 20%, caused a 170% increase in oil prices in 3 months and led to near panic in world financial markets. India, for example, was forced to cut its oil consumption by a remarkable 25% as prices skyrocketed in late 1990. While quick military victory restabilized prices to prewar levels, it did nothing to change the distribution of oil. Iraq and Kuwait will take several years to recover, and Kuwait has permanently lost 10–20% of its long-term production potential as a result of having all its major oil fields set ablaze. Meanwhile, the Gulf region faces overwhelming economic, social, and political problems that will plague the world for years (BPC 1987; DOE EIA May 1987; Flavin et al. 1987; Flavin 1991, 1992).

These growing imbalances in the world oil market jeopardize the energy security of importing nations and the collective security of the world community. Only the fact that the Persian Gulf crisis in 1990 occurred when there was extensive excess capacity—principally in Saudi Arabia—kept prices from soaring toward $100 per barrel. If Middle Eastern oil production reaches 80% of capacity—as it did in 1973 and 1979—it will take only a minor political or military conflict to send prices soaring. The increases that follow could well exceed those of the seventies. To complicate matters, the world's number-one producer, the former Soviet Union, has had a significant drop in oil production over the past few years, and some analysts believe that the fragmentation of the Soviet Union could drive it out of the oil export business altogether by the late-1990s. With world consumption now rising about 1% annually and production plummeting in the United States, the danger zone is likely to be reached in the late-nineties (Flavin et al. 1987; Flavin 1992).

The only realistic means of avoiding another oil crunch is to invest heavily in energy efficiency—largely in transportation. The potential of increased energy efficiency to reduce oil imports is demonstrated by the fact that efficiency largely caused the oil glut of the mid-eighties. One change alone—the increase in the average fuel efficiency of American automobiles from 13.1 miles per gallon (MPG) in 1973 to 17.9 MPG in 1985—cut U.S. gasoline consumption by 20 billion gallons per year, lowering oil imports by 1.3 million barrels per day, two-thirds of the peak production from the rich oil fields of Alaska (DOE EIA April 1987).[3]

Petroleum geologists agree that the United States is unlikely to find another oil field as large as Prudhoe Bay's, but the country could save another 1.9 million barrels per day by the year 2000 by raising new-automobile fuel efficiency to 45 MPG in 1995, according to a study by Deborah Bleviss of the International Institute for Energy Conservation. This may be impractical, but most countries should strive for a minimum 1 MPG annual improvement in the fuel economy of new cars (Flavin et al. 1987; Bleviss 1988).

Energy Efficiency: An International Perspective

Priorities in energy efficiency vary among countries. Industry is the top priority in the Third World and in Eastern Europe and the former Soviet republics, since it is the largest energy user and is a key determinant of both environmental quality and economic competitiveness. In industrial countries, this sector probably needs less government attention than others.

Transportation efficiency is crucial for most countries. Improved automobile fuel economy could be accomplished with a package of consumer incentives for the purchase of more efficient cars, fuel efficiency standards, industry research and development (R&D) programs, and fuel taxes. Automobile fuel efficiencies will eventually reach some practical limits, at which point it will be important to have developed economical alternative fuels, such as ethanol, and be well on the way to more efficiently designed communities that rely on human power and mass transit.

Buildings are the most wasteful energy users in industrial countries, and deserve the greatest attention from government programs. Improvements already made in these nations' buildings spare the atmosphere 225 million tons of carbon emissions annually, but heating, cooling, and lighting those buildings still pumps out over 900 million tons each year—17% of world carbon emissions from fossil fuels. Whereas the energy requirements of automobiles and industry could be halved with available technologies, building energy requirements can be reduced by 75% or more when new buildings are constructed. To sustain that 2% annual rate of improvement over the long run, building efficiency will have to compensate for diminishing returns in industry and transportation.[4]

The investments required to sustain a 2% rate of improvement in energy efficiency in the next decade or two are justified on purely economic grounds, but steps must be taken to make energy markets work much more effectively. The world has achieved over $300 billion worth of annual energy savings since 1973, mostly as a result of private investment decisions. Each additional 2% of savings will reduce bills by about $20 billion annually, at an incremental cost of $5–10 billion (Rosenfeld 1987).

---

[3] U.S. oil savings, Worldwatch estimate based on DOE EIA (April 1987).

[4] Worldwatch Institute estimate based on data from IEA and Rotty (private communication).

In market economies, improved efficiency falls primarily to the private sector, which pays for the big-ticket items such as home weatherization and the modernization of industrial equipment. The world is now spending roughly $20–30 billion annually on improved efficiency, down somewhat from the peak of the early eighties (estimate based on Rosenfeld 1987). Government efficiency research, development, and demonstration programs in the industrial market economies absorb about $600 million annually (IEA 1987a). Government R&D spending in these areas could also be tripled in most countries.

Energy efficiency data for the former centrally planned countries are more difficult to obtain, but available evidence indicates that these nations continue to lag behind in the efficiency revolution. The former Soviet Union, now fragmented into several states, has not bettered its record at all since the early seventies and remains the least energy-efficient industrial economy. Without the discipline of market prices, Soviet industrial managers have in the past looked at energy bills as just another cost to pass on to consumers; many families still use open windows to regulate the temperature in their centrally heated apartments, even in the dead of the Soviet winter (Hewett 1984).

Increasing energy efficiency requires the orchestration of dozens of technologies, policies, and institutions—one reason that countries have differing degrees of success. Indeed, energy efficiency provides an interesting microcosm of economic organization. The former centrally planned economies such as the Soviet Union's have shown virtually no improvement in energy efficiency, whereas market-oriented and highly organized economies such as Sweden's and Japan's have done unusually well.

The United States and the former Soviet Union will have to take leadership roles if the interrelated problems of energy efficiency and climate change are to be taken seriously. The United States is the world's largest energy consumer and led other nations into the profligate energy practices of the postwar period. The Russian republics are among the least energy-efficient in the world, yet their claim on global energy resources is among the fastest growing. Together the United States and the Soviet republics account for 42% of the carbon now entering the atmosphere from fossil fuel use. A joint commitment to improve energy efficiency by both countries would make a major contribution to climate protection and might help mobilize action around the world (Rotty 1987).

The Third World Conundrum

The Third World is also critical to any long-term energy scenario. Indeed, one of the most troubling features of most of the forecasts now in use is the assumption that today's industrial countries will continue to use a disproportionately large share of the world's energy, despite the fact that developing countries have about three-quarters of the world's population. A 1981 study by the International Institute for Applied Systems Analysis was ostensibly attuned to global equity issues, yet it still assumed the Third World would use just 36% of the global energy supply in the year 2020 (IIASA 1981).

Such scenarios imply that while Third World energy use grows, per capita energy use will stagnate, presumably making it impossible for most developing countries to follow the modernization path taken by the newly industrializing nations in recent years. Although consistent with recent experience—many developing nations are burdened with unmanageable foreign debt and have been priced out of the oil market—it is a morally intolerable vision, inconsistent with the articulated goals of the international community. Poverty-induced conservation is not conservation at all. It is just plain poverty.

An energy efficiency gap has also opened among developing nations. Newly industrializing economies, such as Taiwan, South Korea, and Brazil, have begun to incorporate state-of-the-art industrial machines and processes, accompanied by a broad array of energy efficiency standards and financial incentives. But most of the Third World lags further and further behind the industrial world. Many nations are still going through the early, energy-intensive phases of industrialization, and energy intensity is rising rather than falling. Moreover, many countries subsidize energy prices and have not yet implemented effective policies to restrain consumption. For most developing nations, imported oil continues to place a heavy burden on foreign exchange; lagging progress in efficiency will make it increasingly difficult to compete in international markets.

One of the greatest challenges will be to meet the energy needs of the poor without repeating the mistakes of the rich. Only rapid advances in energy efficiency and a decentralized, agriculturally based development path can allow the Third World to fuel improved living standards with limited energy supplies. In the poorer countries of Africa and Latin America, the rapid onset of an energy efficiency revolution is critical. Some Asian countries, including China, with the world's largest coal reserves, have sufficient fossil fuels to last for many decades but face a critical environmental choice. Using energy efficiency to displace coal may be essential to protecting human health as well as the climate.

Energy Efficiency Crucial for Global Economic and Environmental Progress

International economic competitiveness is not just an issue for the Third World. In 1986, the United States used 10% of its gross national product to pay the national fuel bill, but Japan used only 4%. Arthur Rosenfeld of Lawrence Berkeley Laboratory has calculated that, relative to Japan, the United States is effectively paying a $200-billion energy inefficiency tax, leaving less to invest in other areas such as retiring the national debt. Japan is not only richer for its efficiency, it is also in position to dominate the world market for high-efficiency technologies (Rosenfeld 1987).

Another gap is developing between different sectors of society. Efficiency improvements have been most impressive in industry, with buildings and transportation trailing somewhat. Efficiency improvements depend on the structure of the market: buildings, which actually have the most potential for efficiency gains, have progressed slowly due to the fragmented nature of the markets involved and the slow replacement rate of buildings compared with automobiles and industrial equipment. Energy-intensive industries like chemicals have improved remarkably, however. Similarly, the commercial aviation industry has been pushed by punishingly high fuel costs to invest heavily in wide-body aircraft, improved jet turbines, and lighter materials, increasing fuel efficiency by about half (Hirst and O'Hara 1986; Boeing 1987).

To achieve its full potential, energy efficiency must emerge from the obscure corners of energy ministries. It must rise to the top of agendas throughout both government and industry. The very term *energy efficiency* must be transformed from a watchword of specialists to a centerpiece of national and international economic philosophy. Energy efficiency is an essential ingredient of economical and ecological progress; its status should be charted as closely as productivity or inflation. The Commission of the European Communities has suggested the need for such a commitment. At a 1986 meeting, national energy ministers agreed to an ambitious target of a 20% improvement in energy efficiency by 1995 (Directorate-General for Energy 1987).

Data available through mid-1987 indicate that energy efficiency continues to improve in most nations, despite the oil price collapse of 1986. Many of the investments made after the 1979–1981 increases in oil prices are still "in the pipeline." It is only a matter of time, however, before lower oil prices slow the momentum. Around the world, government and private energy efficiency programs are being abandoned, and new investments are being curtailed. Without prompt action, trouble lies ahead (IEA 1987b).

No country has even begun to tap the full potential for further improvements; a range of technologies now coming on the market are more efficient than experts thought feasible just a few years ago. Efficiency gains of at least 50% are available in every sector of the economy. Yet from woodstoves in African villages to office buildings in California, the limiting factor is not technical but rather institutional. If energy efficiency is to realize its full potential to revitalize economies and protect the environment, major institutional reforms will be needed. The most serious current obstacle is what energy conservation professionals call the "payback gap." Energy consumers—from homeowners to factory managers—rarely invest in efficiency measures with payback periods longer than 2 years, while energy producers look to new supply options with payback periods of up to 20 years. If institutional changes can bridge this simple but critical gap, energy efficiency might largely take care of itself.

The key is developing markets for saved energy. Consumers do not want energy; they want the services energy provides. Improved efficiency cuts the energy needed to provide these services. With generating capacity already in place to produce additional power, the energy saved can then be used for other purposes. Firms that find the cheapest ways to conserve energy, while still meeting standards of quality and reliability, should be rewarded. And as more utility systems are opened to competitive bidding, saved energy can compete with independent power.

A large and growing energy conservation industry has developed for commercial buildings and factories in the United States, Canada, and some European countries. The industry will only come into its own, however, when electric and gas utilities are prepared to buy saved energy. If properly rewarded, private energy conservation firms could keep societies on the crest of technical efficiency advances, without unduly straining public coffers.

Energy efficiency improvements are by nature fragmented and unglamorous: thicker insulation and ceramic auto parts are not perhaps as intrinsically captivating as nuclear fusion or orbiting solar collectors. But infatuation with grandiose energy supply options got us into our current predicament; focusing on the mundane may be the only way to get out. Indeed, perhaps no other endeavor is as central to the goal of fostering sustainable societies. Without improved efficiency, it is only a question of which will collapse first: the global economy or its ecological support systems. With greater energy efficiency, we stand at least a fighting chance.

Research and Development Needs

Technological advances are central to the shaping of energy futures, and in recent years technology development has accelerated drastically. Governments have led the way in commercializing a number of large technologies, most notably nuclear power plants, while private industry, particularly since the mid-seventies, has focused on much smaller systems.

The fruits of these efforts are everywhere. Oil drilling is being extended to depths unimaginable two decades ago, coal can be burned with much lower sulfur emissions, nine-watt desk lamps are available that use just $.10 of electricity per month, and some villagers are now generating their power directly from sunlight. Research and development is by nature a hit-and-miss process, and there have been as many failures as successes. The key is to design the R&D process in a way that maximizes the success rate while still pursuing technological options with potentially high payoffs. A review of recent R&D investment confirms the necessity of this approach.

OECD R&D Efforts

The late seventies witnessed a dramatic increase in government spending on energy R&D. The industrial market countries doubled their energy research, development, and demonstration expenditures from $5.1 billion in 1975 to $10.2 billion in 1980. Most have leveled off since then, though budgets have declined precipitously in the United States and West Germany. Today, Italy and Japan lead the members of the Organization for Economic Co-operation and Development (OECD) in energy R&D expenditures as a proportion of gross national product, while the United States has fallen to last place (IEA 1987a).

The most important legacy of the mid-seventies is the creation of new programs to develop renewable energy and efficiency technologies. These grew rapidly until 1981, when they began to level off or fall in most countries. In 1986, the OECD member countries spent $484 million on renewable energy R&D and $622 million on improved energy efficiency R&D (Table 6.2). This is 6.8% and 8.4%, respectively, of total energy R&D expenditures, still a small sum compared to the $4.5 billion spent on nuclear technologies, including breeder reactors and fusion. Spending on renewables and efficiency has been uneven since the early eighties, with greater than 50% cuts in the U.S. and some other programs, and increases in Greece, Italy, and the United Kingdom (IEA 1987a).

Table 6.2

Government Energy R&D Budget in OECD Member Countries, by Energy Source, 1986

Country	Fossil Fuels	Nuclear	Renewables	Efficiency	Total[a]
Japan	310	1,801	99	78	2,311
United States	294	1,134	177	275	2,261
Italy	4	658	30	48	761
West Germany	122	352	66	21	566
United Kingdom	20	271	16	43	378
Canada	138	144	11	34	336
Sweden	9	12	17	29	79
Greece	3	2	10	0	15
Denmark	5	0	3	5	14
Total OECD[a]	990	4,503	484	622	7,133

[a] Total includes minor additional expenditures; excludes France.

Source: International Energy Agency, *Energy Policies and Programmes in IEA Countries, 1986 Review,* Paris: Organisation for Economic Co-operation and Development, 1987.

No simple measures of the overall success of these programs are available, but substantial evidence exists of individual projects that have already led or will soon lead to important new commercial technologies. Some of the clearest successes involve small programs with minimal expenditures. The common ingredients seem to be a well-thought-out development process and successful collaboration between the public and private sectors.

The tiny country of Denmark has led the world in developing intermediate-scale, grid-connected wind turbines, an industry that did not exist before the eighties but reached sales of over $500 million in 1985. Starting in the mid-seventies, Denmark began a program to build on its historical experience with wind power technology. Less than $10 million of annual government funding was used to develop improved blades, rotors, and braking systems. In addition, the government established a national wind turbine test center to support the development of private commercial machines. Turbines had to pass a stringent certification test for tax credit eligibility.[5]

The Danish wind power industry grew slowly until 1982, when wind farming took off in California. Although the Danish machines were generally heavier and less aerodynamically efficient than the American ones, years of painstaking

[5] Value of industry is Worldwatch Institute estimate based on Strategies Unlimited, "International Market Evaluations: Wind Energy Prospects" and International Energy and Trade Policies of California's Export Competitors," California Energy Commission, Sacramento, California, 1987; IEA, *Policies and Programmes, 1986;* IEA, *Renewable Sources of Energy,* Paris: OECD, 1987.

progress made them much more reliable, a key factor in any new technology. Since 1982, the Danish share of California's wind power market has grown steadily, reaching 65% in 1986. As of 1987, Denmark not only dominated the worldwide wind power industry but has helped make wind a reliable, economically competitive energy source (R. Lynette & Associates 1987).

U.S. R&D Efforts

U.S. government-supported energy efficiency programs have also spawned some small but important success stories. In 1976, the Department of Energy and the Lawrence Berkeley Laboratory began a program to develop solid-state ballasts that would make fluorescent lamps 15 to 30% more efficient. By 1980, prototype lamps installed at a federal office building met both reliability and performance standards (Geller et al. 1987).

Several small U.S. manufacturers began producing the new models, taking just 1.4% of the ballast market by 1984. Large electronics companies in Europe and Japan developed a similar technology. In 1984, General Electric, GTE, and all the other major lamp manufacturers entered the market for solid-state ballasts, which are now expected to take half of the U.S. market in a few years. The American Council for an Energy-Efficient Economy found that the 2 million ballasts in use in 1987 were already saving $15 million worth of electricity each year. Over the next 30 years, the new ballasts are expected to save $25 billion worth of electricity—not bad, considering that the government seed money amounted to only $2.7 million (Geller et al. 1987).

The Department of Energy began another program in 1976 to develop special window coatings that allow light to pass unimpeded but retain heat within a building. Federal contractors developed several early generations of the new coatings, and after a few years some were able to raise venture capital to continue the process privately. The technology did not really catch on until 1983, however, when a large window manufacturer adopted it. Just $2 million was spent on the initial federal program to spur the development of the new windows, and private industry later spent over $150 million on facilities to manufacture them. The windows are projected to save $120 million worth of heating fuel annually in the mid-nineties, with cumulative savings reaching $3 billion in the year 2000 (Geller et al. 1987).

It is hard to know just how common such success stories are, but anecdotal evidence indicates that they are widespread. Researchers in dozens of fields report that the state of technology has advanced by huge strides since the late seventies. Most of these advances, however, have not yet reached a stage of commercialization that commands public attention. One example is thin-film photovoltaic cells, which are now used mainly for consumer electronic devices. But the costs are falling so rapidly that household solar power may become practical in little more than a decade.

Government energy research programs should be seen as selective seeding efforts. Not all of the seeds need to sprout in order for the program to be a success. Unfortunately, such programs are put to a yearly political test in many countries, and budgets rise and fall depending on the fashion of the moment and whether a particular project can benefit a powerful constituency. Such uncertainty makes long-term planning difficult and makes investment in renewables less attractive for private investors.

Japan's R&D Commitment: A Model for the United States?

Japan, in contrast to other countries, has maintained a steady commitment to energy research and development programs—from nuclear power to coal gasification and photovoltaics. Government policy states that "it is necessary to offer the appropriate guidance to the private sector's energy conservation efforts, and to prevent the consciousness toward energy conservation or the incentive toward investment in energy conservation from being affected and weakened by short-term trends in the oil markets" (Energy Conservation Center 1987).

Japanese energy R&D spending reflects this commitment. It rose 140% between 1975 and 1980, and then increased another 17% between 1980 and 1986, reaching $2.3 billion. For the first time, Japan leads the world in government energy R&D spending. The Sunshine Program for renewable energy and Moonlight Program for energy efficiency are yielding excellent results, with effective links to the private sector (IEA 1987a).

By contrast, U.S. spending peaked in 1980 at $5.2 billion, then declined sharply to about $2.3 billion in 1986. U.S. budgets were affected by the Reagan Administration's desire to cut all programs with imminent commercial prospects (except some big-ticket nuclear demonstration projects). The administration has also sought to virtually eliminate government support of renewable energy and energy efficiency technologies. Congress has continued to provide funds for these programs, but at less than half the levels of 1980–81. As a result of growth in military expenditures, the Department of Energy has become mainly a nuclear weapons agency (IEA 1987a). With the end of the cold war, however, funding priorities are slowly shifting, with industrial partnerships and environmental research gaining ground.

U.S. energy programs have long been skewed toward long-term nuclear projects with dubious prospects, notwithstanding the small successes mentioned earlier. A 1987 General Accounting Office study found that in most cases where federal support of technologies has been limited to basic research, private industry has not gone forward with commercialization efforts. Overall, U.S. energy R&D programs are marked by an extraordinary degree of confusion and bureaucratic infighting. The Department of Energy's own Inspector General has stated that there is no coherent long-term plan guiding the department's programs (DOE Inspector General 1986; U.S. GAO 1987).

Public/Private and International Cooperation Needed

Recognition is growing in the United States and throughout the world that effective coordination between the public and private sectors is an essential ingredient of success in R&D. Only by working with the thousands of companies that make equipment for producing and consuming energy can the world's vast scientific and engineering expertise be effectively harnessed to develop commercial equipment that will have a meaningful impact.

An effective process must be set up to choose among myriad proposals for research and development. In many countries a built-in bias favors the proposals of large, well-established companies. The better programs, however, are usually those in which new projects are chosen on their merits. Independent engineering panels can be set up to ensure that this occurs, something that has been successful with some of the U.S. energy efficiency programs. Large increases in spending on renewable energy and efficiency will be required in order to right the current imbalance in spending, but these increases will have to proceed gradually, with close technical guidance, if the funds are to be used effectively.

International collaboration in energy R&D should also be made a high priority. The recent mushrooming of energy technology development means that even wealthy nations can no longer afford to have independent research programs in every promising field. So far, the European Economic Community has the largest such joint effort, and many European countries channel most of their energy R&D funds through the program. Total funding is now about $60 million per year for non-nuclear energy R&D, split almost evenly between renewables and efficiency. In addition, the International Energy Agency maintains a cooperative program among OECD member countries that has spent about $600 million during the past decade (IEA 1987a).

It is unfortunate that for the most part such programs have not been broadened to include former centrally planned and developing nations. The Third World has been able to fund only small projects on its own and needs to share in the latest energy technology developments, particularly those involving decentralized technologies. The United Nations Conference on New and Renewable Sources of Energy in 1981 was supposed to lead to such an effort, but it became bogged down in political infighting and bureaucratic inertia. Although some foreign aid programs have effectively promoted the development of indigenous energy resources, these are neither as strong nor as numerous as they could be.

The Market Path

In all nations, the creation of an effective energy policy involves the successful blending of many different elements, any one of which taken in isolation is likely to be inadequate. Perhaps the most important principle is reliance on market forces. Their role is shown by what has happened through their absence in the former Soviet Union and Eastern Europe. Inefficient use of energy, lagging technology development, and choking air pollution have become synonymous with these countries. The former Soviet Union, for example, has pioneered the development of some sophisticated efficiency technologies, but its buildings waste heat due to a lack of simple things like fiberglass insulation and caulking. Many developing countries that rely on state-run enterprises and heavily subsidized energy prices now face similar problems.

The market-based economies have done better, but still not well enough. Even the most capitalistic countries have a long history of inappropriate government intervention, reliance on monopolistic enterprises, and deeply entrenched obstacles to more-efficient buildings and transportation. Large energy supply projects are generally favored over smaller, renewable energy technologies. In short, a real market approach to energy has yet to be given a fair chance.

Governments can unleash the vigor of markets by reforming institutions and reorienting energy subsidies and taxes so that prices reflect marginal costs, including environmental costs. So long as energy is priced as if it were almost free, and so long as consumers have neither the means nor the information to make cost-effective investments in energy efficiency, we will continue to squander our remaining fossil fuels. Reasoned, graduated efficiency standards, though seemingly in contradiction to an emphasis on markets, can play a crucial and consistent role by encouraging manufacturers to develop the technologies that market forces point toward.

The many goals of energy policy—lowered oil dependence, economic competitiveness, environmental protection, and climate preservation—make the formulation of a comprehensive, consistent policy difficult. Any proposed quick solutions to the complex of problems are almost certainly unworkable. The key is to unleash the power of private innovation and individual action, without attempting to force broad government mandates on society.

A goal of reducing national energy expenditures, if pursued rigorously, can lead to a strong emphasis on energy efficiency, improve economic competitiveness, and limit growth of oil dependence. Specific disincentives, however, will be required to ensure that countries do not become too heavily dependent on fossil fuels that threaten life-support systems. With the end of the Cold War and the emergence of market forces around the world, there is a historic opportunity to incorporate energy efficiency worldwide. If all governments work toward this goal, we can immediately embark on a path that will replace creeping disaster with gradual improvement.

References

ACEEE, *Acid Rain and Electricity Conservation*, Washington, D.C., 1987.

Bleviss, D. L., *The New Oil Crisis and Fuel Economy Technologies: Preparing the Light Transportation Industry for the 1990's*, New York: Quorum Press, 1988.

Boeing, Inc., Seattle, Washington, private communications, Sept. 1987.

British Petroleum Company (BP), *BP Statistical Review of World Energy*, London, 1987.

Brown, L. R. and C. Flavin, "The Earth's Vital Signs," *State of the World, 1988*, Chapter 1, W.W. Norton and Company, 1988.

Chandler, W. U., "The Case of China," Worldwatch Institute, Washington, D.C., unpublished, 1987.

Chandler, W. U., Energy Productivity: Key to Environmental Protection and Economic Progress, *Worldwatch Paper 63*, Washington, DC: Worldwatch Institute, January 1985.

Directorate-General For Energy, *Energy in Europe: Energy Policies and Trends in the European Community*, Luxembourg: Commission of the European Communities, 1987.

DOE, EIA, *Monthly Energy Review*, February 1987.

DOE, EIA, *Monthly Energy Review*, April, 1987.

DOE, EIA, *Monthly Energy Review*, May, 1987.

DOE, EIA, *Annual Energy Review 1989,* Washington, DC: 1990.

DOE, Inspector General, "The Coordination of Long-Term Energy Research and Development Planning," Washington, DC, November 1986.

Edmonds, J. A. and J. M. Reilly, "A Long-Term Global Energy-Economic Model of Carbon Dioxide Release from Fossil Fuel Use," *Energy Economics,* April 1983.

Energy Conservation Center, "Energy Conservation in Japan 1986," Tokyo, February 1987.

Energy Policy Project of the Ford Foundation (EPPFF), *A Time to Choose: America's Energy Future,* Cambridge, MA: Ballinger, 1974.

Flavin, C., Reassessing Nuclear Power: The Fallout from Chernobyl, *Worldwatch Paper 75,* Washington, DC: Worldwatch Institute, March 1987.

Flavin, C., "Slowing Global Warming," *State of the World, 1990,* Chapter 2, W.W. Norton and Company, 1990.

Flavin, C., "Designing a Sustainable Energy System," *State of the World, 1991,* Chapter 2, W.W. Norton and Company, 1991.

Flavin, C., "Oil's Shaken Foundation," *World Watch* 5, January-February 1992, pp. 7–9.

Flavin, C., D. Hayes, and J. MacKenzie, "The Oil Rollercoaster," Fund for Renewable Energy and the Environment, Washington, DC, 1987.

Geller, H., J. P. Harris, M. D. Levine, and A. H. Rosenfeld, "The Role of Federal R&D in Advancing Energy Efficiency: A Fifty Billion Dollar Contribution to the U.S. Economy," *In* Annual Reviews, Inc., *Annual Review of Energy,* Vol. 12, Palo Alto, CA, 1987.

Gibbons, J. H. and W. U. Chandler, *Energy: The Conservation Revolution,* New York: Plenum Press, 1981

Hewett, E. A., *Energy, Economics, and Foreign Policy in the Soviet Union,* Washington, DC: Brookings Institution, 1984.

Hirst, E and F. M. O'Hara, *Energy Efficiency in Buildings: Progress and Promise,* Washington, DC: American Council for an Energy-Efficient Economy, 1986.

INEL (Idaho National Engineering Laboratory) *The Potential of Renewable Energy: An Interlaboratory White Paper,* prepared for the Office of Policy, Planning and Analysis, DOE, in support of the National Energy Strategy, Golden, CO: Solar Energy Research Institute (SERI), 1990.

International Energy Agency (IEA), *Energy Policies and Programmes of IEA Countries, 1986 Review,* Paris: Organisation for Economic Co-operation and Development (OECD), 1987a.

International Energy Agency (IEA), *Energy Conservation in IEA Countries,* Paris: Organization for Economic Co-operation and Development (OECD), 1987b.

International Institute for Applied Systems Analysis, (IIASA), *Energy in a Finite World: Paths to a Sustainable Future,* Cambridge, MA: Ballinger, 1981.

IPPC, "Policymakers' Summary of the Scientific Assessment of Climate Change," Report from Working Group I, June 1990.

Meadows, D. H., D. L. Meadows, J. Randers, and W. W. Behrens III, *The Limits to Growth,* New York: Universe Books, 1972.

Meridian Corporation, "Characterization of U.S. Energy Resources and Reserves," prepared for Deputy Assistant Secretary for Renewable Energy, DOE, Alexandria, VA., June 1989.

Milne, R., "Pressure Grows for US to Act on Global Warming," *New Scientist,* June 2, 1990.

Mintzer, I. M., *A Matter of Degrees: The Potential for Controlling the Greenhouse Effect,* Washington, DC: World Resources Institute, 1987.

Postel, S. and L. Heise, "Reforesting the Earth," *State of the World, 1988,* Chapter 5, W.W. Norton and Company, 1988.

R. Lynette & Associates, "The Lessons of the California Wind Farm: How Developing Countries Can Learn From the American Experience," Redmond, WA, May 1987.

Rosenfeld, A. H., "Conservation and Competitiveness," Testimony in Hearings on Economic Growth Opportunities in Energy Conservation Research, Task Force on Community and Natural Resources, Budget Committee, U.S. House of Representatives, July 15, 1987.

Rotty, R., University of New Orleans (formerly of Institute for Energy Analysis, Oak Ridge Associated Universities, Oak Ridge TN.), private communication on carbon emissions, June 16, 1987.

United Nations, Department of Information, Economics, and Social Affairs, Statistical Office, *1985 Energy Statistics Yearbook,* New York: United Nations, 1987.

U.S. Department of Energy (DOE), Energy Information Administration (EIA), *Monthly Energy Review,* February, 1987.

U.S. Environmental Protection Agency, "National Air Pollutant Emission Estimates, 1940–1985," Washington, DC, January 1987.

U.S. General Accounting Office, "Energy R&D: Changes in Federal Funding Criteria and Industry Response," Washington, DC, February 1987.

Chapter 7

Hydro-Development

Eric F. Karlin, Howard Horowitz, and William J. Makofske

The less developed countries (LDCs) are looking to further their industrial development, to improve their agriculture, and generally to raise their standard of living. A vital component of such development is the availability of energy. The major options for energy development may be classified into two types: nonrenewable energy, such as fossil fuels and nuclear power; and renewable energy, consisting of hydroelectric power, wind power, biomass (including the use of wood and crop residues), and direct solar. Many of the LDCs, however, are notably lacking in terms of indigenous nonrenewable energy resources. Given the poverty and often large external debt of these countries, it is unlikely that they will be able to import significant amounts of fossil fuels. Nuclear power provides even greater difficulties; it is presently the most expensive form of energy, and most LDCs do not have the technical expertise and infrastructure to maintain and repair nuclear power plants. In most cases, however, LDCs do have indigenous renewable energy sources, particularly hydro and biomass. These energy sources are renewable resources in the sense that they are available indefinitely, provided that they are utilized in a sustainable way. Not surprisingly, many LDCs are focusing on renewable energy as the primary way to meet their energy needs (Hayes 1977; Deudney 1981; Flavin 1986; Shea 1988).

Hydro-development is particularly appealing as an energy source since the technology is mature and well known. Hydropower, an established technology that works well, has been in use for centuries. Many of the LDCs have extensive river systems and mountainous landscapes, both prerequisites for successful hydro-development. If developed and managed properly, hydro facilities should last for a relatively long time. Such facilities also provide energy, via a relatively pollution-free energy technology, that is a prerequisite for further development. Electrical energy is very important for the development of industry and the creation of a wide diversity of jobs. Small electrical motors allow many possibilities for different kinds of product manufacturing and also increase productivity. In addition, there are other advantages of hydro-development besides the generation of electrical power. Reservoirs are created, which store water for irrigation, public water works, recreational use, and flood control. In spite of these additional advantages, hydro-development is still looked at primarily in the context of energy development. Hydro-development also represents a highly visible sign of progress (Figure 7.1), and is a technology that has been readily financed by the industrial nations.

Hydro-development has had major positive effects in the more developed countries (MDCs). In the United States, the Tennessee Valley Authority dams triggered development that had a favorable impact on the impoverished Appalachian region, beginning in the 1930s (Deudney 1981). In the Pacific Northwest, where over 90% of the electricity is produced by hydropower, the lower cost of power has been an incentive for industry relocation and growth.

Given all of the benefits associated with hydro-development, it is all too easy to overlook the fact that there are negative impacts derived from such technology. Indeed, if all the impacts of hydro-development were only positive, there would be little need to do anything except advocate it. Unfortunately, this technology does have negative impacts and these must be weighed against its more obvious benefits (Alexis 1984; Goldsmith and Hildyard 1984a). These negative effects include social, economic, environmental, and health impacts that may, in some cases, outweigh the benefits. There is little doubt that a great deal of balancing must be done. Lack of attention to these impacts has caused many hydro-development projects to fail to achieve their expected beneficial potential. In this chapter, we take a broad perspective on hydro-development and outline the areas that must be investigated thoroughly in order to plan for the mitigation of negative impacts.

Hydropower Growth and Potential

About 14% of the electricity generated in the United States today comes from hydropower, an amount that is comparable to that generated by nuclear power (Sawyer

Figure 7.1. A large hydroelectric dam (Hoover Dam). (From Bureau of Reclamation, photo by E. E. Herzog. *In* Turk, J., *Introduction to Environmental Studies,* Philadelphia: Saunders College Publishing, 1980.)

1986). On a worldwide basis, however, about one-quarter of all electricity is produced by hydropower (Deudney 1981). Indeed, there are over 35 countries where over two-thirds of the electricity produced comes from hydropower; many of these are industrialized nations including Norway, Switzerland, Austria, and New Zealand (Deudney 1981). Developing countries have a particularly heavy dependence on hydropower. For example, Zaire, Zimbabwe, Kenya, Nepal, Brazil, Bolivia, and Columbia obtain more than half of their electricity from hydropower (Flavin 1986).

The amount of energy that can be obtained from hydro-development depends upon, first of all, the amount of water present in the rivers of a country. This, in turn, is related to the amount of precipitation and the percentage of the precipitation that becomes runoff or groundwater. It is water from these sources that is being harnessed. Secondly, there is the extent of elevational change. The greater the distance that water falls, the more energy that it releases. If a country has mountains, it can harness more hydropower than a country having a similar amount of water but a level landscape. Once such information is known, it is possible to predict the maximum amount of energy available from hydropower for any given area. It is a relatively simple matter to calculate what percentage of that energy is already being utilized. For instance, Europe has harnessed about 59% of all the hydropower that is potentially available. North America has harnessed about 36% of its potential (Deudney 1981).

In the mid-1980s, the total electrical generating capacity of the LDCs was about 450,000 MW (megawatts) or about 120 watts per person (compared to the 688,000 MW and 2900 watts per person for the United States) (Flavin 1986). This is in spite of the nearly sixfold increase in LDC electrical power since the 1960s. In 1980, hydropower provided 38% of the electricity in LDCs compared to 30% from coal and 26% from oil. Yet only 5–10% of the energy potentially available from hydropower has been harnessed in Africa, Asia, and Latin America (Deudney 1981). The most extensive new development of hydropower is thus most likely to occur in those developing countries that have a large untapped hydropower potential.

Hydro-Technology

The scientific principles behind hydroelectric technology are very simple. Basically, when water is situated above sea level, the pull of gravity forces it to flow to successively lower elevations until it reaches sea level. Energy is transformed from potential energy to kinetic energy as the water drops in elevation. The power available in this process is directly proportional to the volume flow rate of water and the height difference over which it falls. Hydroelectric power is based upon transforming this kinetic energy of water into the turning of electric generators, thus converting the water's kinetic energy to electrical energy. This is accomplished by building a dam and impoundment area or reservoir, allowing the water to flow through a pipe or penstock from the top of the dam onto a water turbine at the bottom. The water flowing down the pipe turns the water turbine, which is mechanically connected to a generator (Figure 7.2). The process is a straightforward and reliable method of generating electricity (Kleinbach and Salvagin 1986).

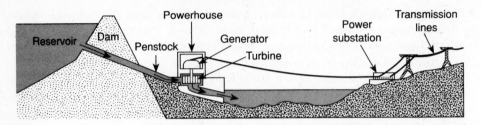

Figure 7.2. Diagram of a typical hydroelectric facility. (From Kleinbach, M. H. and C. E. Salvagin, *Energy Technologies and Conversion Systems,* Englewood Cliffs, NJ: Prentice-Hall. Original source: The Encyclopedia Americana, Danbury, CT.: Grolier, 1982).

Appropriate site selection is critical to the success of hydro facilities. While the theoretical power from large volume flows with low vertical height drop (hydraulic head) might be equivalent to smaller volume flows with larger vertical drops, practically they are not the same. Because high dams and large pipes are expensive, a large natural height drop is by far the more important variable. The ideal topography for hydro-development is mountainous terrain where the banks are high and the river is narrow, dropping quickly in height (see Figure 7.1).

Because rivers do not have constant flow rates throughout the year, and the head is so important, it is necessary to build both a dam and reservoir. This provides water storage for the dry season and allows the hydro plant to produce electrical power on a more constant basis. The height of the dam will depend on the amount of water that needs to be stored, the head needed, and the surrounding topography. In relatively flat regions, the width of the dam and the size of the impoundment area may be substantial, leading to increased costs and greater land use impacts.

Environmental Impacts

There is an increasing awareness that hydro-development may have serious environmental impacts. These include the loss of agricultural land, wilderness, and forests; loss of biodiversity; increased sedimentation; reduction of fisheries; and the potential creation of earthquakes. In many cases, these impacts are linked together. Fortunately, many but not all of these impacts may be minimized by careful placement, design, construction, and maintenance of the hydro facility. Unfortunately, many of these impacts are more severe in tropical environments, where extensive hydro-development is taking place, and sufficient efforts have not been undertaken to lessen or mitigate them.

Loss of Land

The amount of land area used for hydro-development depends greatly on the terrain. In mountainous regions, where 60 to 170 kilowatts may be produced per hectare of reservoir area, economically significant amounts of power can be obtained from relatively small reservoirs. Regions having a flat topography, however, require large reservoirs, as only 1 to 30 kilowatts per hectare of reservoir area are generated in such situations. Some prime examples of the massive sacrifice of land for hydro-development can be seen in the Amazon basin of Brazil (Shea 1988).

Throughout the twentieth century, Brazil has tried to develop the seemingly limitless potential of its natural resources. The Amazon and its tributaries have by far the greatest volume of water of any river system in the world, but the main stem and many tributaries have too flat a gradient to be suitable for traditional hydroelectric dam construction. Until the 1960s Brazil made little effort to exploit the basin's water power; most of the large-scale natural resources development projects, such as American-owned rubber plantations at Fordlandia and Balterra, were basically extractive and did not require large inputs of electrical energy. They failed, as did so many subsequent Amazonian development schemes, and by the late 1960s the Brazilian government decided to change its approach to emphasize the harnessing of hydroelectric potential. This, it was believed, would stimulate the rapid growth of the region's urban areas (Manaus, Belem, and Santarem) by attracting heavy industry, such as aluminum smelting, with the inducement of abundant cheap electricity. Assessments of total hydroelectric potential in the Amazon basin ranged up to 50,000 megawatts (Sternberg 1985). Some of the north-flowing tributaries have sufficient gradient for large-scale power generation, and the first of these large dam-building projects was completed in 1984 at Tucurui on the Tocantins River. The reservoir behind Tucurui Dam flooded 243,000 hectares of land (over 900 square miles). Its power turbines generated 8,000 megawatts of firm power, which was far in excess of the amount needed locally in the sparsely populated region. Tucurui has, indeed, attracted much heavy industry to the region, but this in turn has imposed additional direct costs, such as a complex infrastructure of transmission lines, highways, river locks, railroads, and housing facilities (Sternberg 1985). All of these things have transformed the region's natural environment, but they are consistent with Brazil's intention to "occupy" the Amazonian interior more fully.

Balbina Dam, completed in 1988 on the Uatama River about 100 miles north of Manaus, has flooded an even larger area of tropical rainforest than Tucurui: 254,000 hectares (almost 1,000 square miles) (Sternberg 1985). Unlike Tucurui, Balbina generates very little electric power. The Uatama River basin is so flat that Balbina provides a maximum output of only 250 megawatts, less than 1 kilowatt per hectare of flooded land (Sternberg 1985). This is less than half of Manaus's electrical needs, and comes at a very high price; if the city is to expand in the manner hoped for by the government, additional electrical supplies will still be required. The negative impacts of the dam have been dramatic and have generated much controversy within the region.

Unfortunately, Balbina is not by any means a unique example of a dam that flooded a large area with little potential for electrical power in return; Curua Una, on the river by the same name near Santarem, Brazil, is a smaller project than Balbina but with a similar litany of troubles. In Ghana, the Akosombo Dam has a reservoir area that covers 850,000 hectares (over 3,200 square miles), yet it only generates 833 megawatts of electricity at full power (Sternberg 1985). In practice, Akosombo generates less power than that, and floods less area, because the annual stream flow in the Volta River has been inadequate to ever actually fill the reservoir to capacity.

Although the above examples have focused on the loss of wilderness to hydro-development in tropical regions in LDCs, large-scale hydro-development is also affecting the wilderness areas of MDC's. The James Bay Hydroelectric Project is intended to be the world's largest hydroelectric network, and will be carved out of the wilderness of northern Quebec, Canada (Borowitz 1990; Picard 1990; Linden 1991; Thurston 1991). The first phase of this massive undertaking has been completed and consists of a series of reservoirs flooding 967,500 hectares (over 3,700 square miles) and having an installed capacity of 14,800 megawatts (Berkes 1990; Picard 1990). If completed as planned, the total project would flood more than 2,590,000 hectares (over 10,000 square miles), an area the size of Lake Erie, and supply 27,000 megawatts (Picard 1990; Thurston 1991).

Work on another massive hydro-development project has recently begun in China: the Three Gorges Dam on the Yangtze River. When completed the dam will form a 563 km (350 mile) reservoir that will not only inundate many natural areas, including some of China's finest scenery, but will also displace some 1.2 million people. As the Yangtze River carries a high load of eroded soil material, there is a major concern that the reservoir may silt in very quickly. In addition, the project may be too expensive for the Chinese to undertake at this time. Supporters of the dam point out that the project is needed for the electrical power that will be generated (17,680 megawatts) and also for flood control. Critics argue that it would be simpler and less costly to build several smaller hydro-electric dams and question the need for such a large-scale flood control system (Kristof 1993; Burton 1994).

As is the case for the Three Gorges Dam, the lands being inundated by dams are often located in the flood plains of the associated rivers. Such sites typically are very fertile and the creation of a reservoir represents a loss of productive farmland. Many times there are significant populations living in the areas to be flooded, and resettlement may involve having to relocate up to 100,000 people, far more than that number in the case of the Three Gorges Dam (see above) or the proposed Tehri Dam in India. These people are resettled onto higher ground, where the soils are often not as fertile as those associated with the flood plains of rivers. This can be a very serious problem for the people who are being resettled. Although others may benefit, the resettled population often does not. Two small Indian tribes were displaced from the reservoir area of Balbina Dam; they had been self-sufficient, but were resettled elsewhere after their efforts to prevent the dam were defeated, and have since become sedentary and dependent on government welfare (Balbina: Master of Destruction 1989). In that respect, the indigenous people shared much the same fate as the indigenous animals: some were relocated, some were displaced, some just disappeared, and a few (such as a rare species of turtle) were provided with "artificial housing" by the state electrical company.

Displacement of people occurs not only in tropical regions but also in temperate regions as well. Recently, American Indians in North Dakota sued the Federal government for their loss of land along the Missouri River that was flooded to make Lake Saqaquoia. They feel that their uprooting was very costly in terms of lost fertile farmlands as well as the loss of their homes and cultural ways. The James Bay Hydroelectric Project has also displaced many Cree and Inuit communities, and has the potential of radically disrupting the life-styles of thousands of these indigenous people. Legal action by the Cree has resulted in the requirement of a full environmental review of the entire James Bay project (Newman 1991; Thurston 1991).

As noted above, land lost to hydro-development is often wilderness and other natural areas. Unfortunately, government and industry often consider such land to be wasteland. Consequently, its loss is considered to have a minimal impact. In reality, the ecological value of such areas is tremendous, especially wilderness areas. Natural ecosystems play a most significant role in regulating regional hydrology and climate, minimizing soil erosion, providing natural products (wood, game species, medicinal plants), and maintaining biodiversity. Many in the environmental field consider the loss of wilderness, especially in the tropics, to be one of the most serious environmental problems at the present time (see Chapter 8, "The Loss of Biodiversity").

Dam Failure and Seismic Activity

Another negative aspect is the risk of catastrophic dam failure. Theoretically, if they are engineered correctly dams should not collapse, but large dams have been known to fail. The record of dam failure is really quite substantial. In the twentieth century we have seen several major dam collapses: Malpasset Dam in France (1959), Teton Dam in the United States (1975), and elsewhere. Some of these involved substantial loss of life to people who lived downstream.

Also, it is now known that dams, under certain circumstances, can induce earthquake activity in the geologic substructure below them (Bolt 1978). Since the weight of the water behind a large dam represents an enormous additional stress on the Earth's crust, the crust has some flexible ways of responding. One of those is called seismic inducement. There have been areas, which although having no previous earthquake history, began to have earthquakes after large dams were built. These include the Koyna Dam in India (magnitude 6.5 on the Richter scale), the huge Kariba Dam in Zimbabwe (magnitude 5.8), and the Vouglans Dam in France (Bolt 1978). This has happened often enough to make it a major concern. Today, much more thorough engineering studies are undertaken to assure that a proposed site will be stable (Goodland 1986).

Decline in Water Quality

In the past, major reservoirs were usually created in desert, savanna, or agricultural regions. These ecosystems have relatively little biomass and thus only small amounts of biotic material were submerged by impoundment. Recently, however, a number of large-scale reservoirs have been created in areas having a large biomass (such as tropical wet forests). If such ecosystems are flooded without first removing most of the existing biomass, significant problems in water quality will develop (Goodland 1977). This is particularly the case in the tropics, where the biomass can be quite substantial and where year-round warm temperatures allow for the rapid decay of biotic materials. The decomposition of submerged biomass depletes oxygen from the water, often to the point that an extensive anaerobic (oxygen-free) zone develops in the reservoir. Such a condition causes fish kills and extensive loss of other aquatic organisms. Very few of the millions of trees within the reservoir area of the Balbina Dam could be salvaged; instead they decomposed in the stagnant waters and produced massive fish kills, stench, and anoxia. The anoxia at Curua Una was so severe that, in addition to the algae blooms and fish kills, there were multiple underwater explosions caused by the large-scale formation of sulfuric acid! Mercury contamination can also result from decaying biomass submerged in reservoirs. The bacterial population associated with the decay of submerged trees in the James Bay reservoirs has transformed inorganic mercury that had lain dormant in the soil and rock into methyl mercury (Picard 1990). This form of mercury can, and has, entered the food chain, causing the closure of the local fisheries (Berkes 1990; Linden 1991). Hydrogen sulfide would also be produced by anaerobic decay processes, representing an additional environmental problem. Although problems caused by submerged biomass should disappear once the material has been fully decayed, it may take 50 years for that to happen (Linden 1991; Thurston 1991).

Obviously, most of the biomass should be removed prior to impoundment in order to avoid these problems. Unfortunately, such a large-scale removal of vegetation presents another problem. Not only does the biomass have to be removed, it requires disposal as well. A staggering amount of material would have to be cleared from several hundred square miles of tropical wet forest, and there are not always ready markets for the timber. So even though it will cause water problems, much of the vegetation is often left on the reservoir site.

Sedimentation

Dams also impede the flow of sediment (sand, silt, and clay) that rivers carry. Normally rivers carry sediment downstream towards the ocean and in the process the sediment plays important roles in the overall ecosystem. The fertility of river bottom lands, for example, comes from sediments. Often, the sediments carry nutrients to the estuaries and bays, supporting the aquatic life in these areas.

When a dam is constructed, a region of quiet water is created in the reservoir. When water stops moving, the sediments that are normally carried by the flow of the stream settle out, and thus the reservoir begins to fill up. In the long run, all reservoirs will become filled with sediment. That might take hundreds or thousands of years under optimal circumstances. Unfortunately, it usually occurs much more quickly, particularly in areas where there has been deforestation and other disturbances in the upstream area that cause increased soil erosion. Also, because the water released by a dam has lost its sediment load, its erosive capacity may have increased, leading to a higher rate of river bank erosion downstream. The Aswan Dam in Egypt is a classic example. The downstream area has been eroded much more rapidly since completion of the dam, and the Nile delta, which once was expanding in area, began to decrease in extent (White 1988).

Loss of Fisheries and Decreased Land Fertility Downstream

The sediments also carry nutrients, and as the sediments settle in the reservoir, the nutrients associated with them are no longer available to downstream areas. Thus, soils downstream from a dam that had once received enrichment from the sediments deposited on them now become progressively less fertile. Fisheries downstream are also affected because the fish populations are dependent on the nutrients that the sediment carried. A major contributing factor to the collapse of the sardine fishery in the Mediterranean was the loss of nutrients brought about by the retention of sediments by the Aswan Dam (White 1988).

Fish migrations are also blocked by dams and in some cases this has caused major declines in the populations of the affected fish species. The salmon population in the Columbia River system of Washington and Idaho is only 15 to 25% of what it once was, and this reduction is primarily attributed to the presence of many dams (Muckleston 1990). Not only are the adult salmon blocked from returning to their spawning grounds, but the smolt suffer a high mortality as they attempt to migrate to the oceans (from passing through turbines, having delayed migrations because of lowered water velocity, and increased disease because of higher water temperatures). Although a number of mitigating options can be implemented (fish ladders, fish hatcheries, and spill ways), they are often not effective enough to allow for the fish populations to increase back to previous levels (Stowell 1987; Eley and Watkins 1991). In fact, the United States National Marine Fisheries Service may soon recommend that the once abundant coho salmon (*Oncorhynchus kisutch*) be listed under the Endangered Species Act (Satchell 1994).

Health Impacts

The damming of rivers and the associated creation of reservoirs and extensive irrigation systems have exacted a heavy toll in terms of human health. In particular, both malaria and schistosomiasis (also known as bilharzia) have significantly increased in many areas having large-scale water projects (The Social and Environmental Effects of Large Dams 1984; Goldsmith and Hildyard 1984a; Goodland 1986). In the case of malaria, the presence of extensive permanent irrigation systems provides an excellent breeding environment for the *Anopheles* mosquito, which transmits the malaria pathogen (the protozoan *Plasmodium*). This, coupled with the ability of the mosquitos to develop resistance to the biocides currently being used against them, has allowed for a major resurgence of malaria in many countries (The Social and Environmental Effects of Large Dams 1984).

Schistosomiasis: A Case Study

Although schistosomiasis is one of the major parasitic diseases in the tropics and subtropics, many people in the MDCs (which are primarily in temperate regions) are generally not even aware of its existence. Schistosomiasis is caused by a parasitic flatworm called the blood fluke (*Schistosoma,* phylum Platyhelminthes) (Cheng 1973). There are three *Schistosoma* species that can cause the disease, and they primarily occur in tropical and subtropical regions. Schistosomiasis is known from southern Asia and Japan, Africa, the Middle East, northern South America, and the Caribbean (Cheng 1973).

Interestingly, blood flukes require two hosts. Part of their life cycle is spent in a snail, and part of their life is spent in a human; both are required for the completion of the blood fluke life cycle. The snails become infected when exposed to larvae that hatch from eggs found in untreated human body waste. In turn, humans are infected when their skin comes into contact with water contaminated with immature blood flukes that have been asexually produced in the snails (Figure 7.3) (Cheng 1973; Stein 1977a).

Many hydro projects have been associated with an outbreak of this disease. One example is the impact of the Aswan Dam in Egypt. Schistosomiasis, which had previously not been prevalent in the area around the dam, began to proliferate after the dam was constructed. The infection rate reached 100% in some communities (Goldsmith and Hildyard 1984a). Another example is Ghana's Volta Lake project, where schistosomiasis increased from 2% of the population prior to hydro-development to over 80% afterwards (Goodland 1986). Prior to the presence of a reservoir and associated irrigation systems, snails were not abundant because of limited habitat availability. Snails grow best in relatively calm, stagnant waters, where there is an abundance of aquatic vascular plants (macrophytes) that the snails use as food (Stein 1977a). Before the dams were developed, such habitats were not common in many of the affected areas. Hydro-development resulted in the widespread creation of favorable habitats for the snails. The resulting increase in snail populations allowed for a corresponding outbreak of schistosomiasis.

An increasing snail population, however, is only one aspect of the problem. Schistosomiasis could be effectively controlled by the use of proper sewage treatment technology. Unfortunately, sewage treatment facilities in many LDCs are often primitive or nonexistent, especially in the undeveloped parts of countries where many dams would likely be built (Stein 1977b). But even if a sewage treatment system were to be built, there would also have to be a change in cultural habits to successfully utilize it. A fisherman out fishing for the day may not be inclined to head back to shore to use a toilet, but rather use the lake as a repository for his wastes, as he always had done before. So not only does hydro-development provide a good environment for snails, but the codevelopment of proper sewage treatment systems and cultural adjustment programs is often lacking (Stein 1977b). In the previously described Brazilian reservoirs of Tucurui, Balbina, and Curua Una, schistosomiasis became so prevalent that authorities considered establishing a 2-km-wide "cordon sanitaire" around the waters, to separate them from human contact. This proved to be quite impossible, as hunters, fishermen, and others are attracted to water and the banks are too extensive to patrol (Sternberg 1985).

Several conclusions result from the study of schistosomiasis. One is that technology that has been developed and successfully implemented in temperate parts of the world may not be suited for other regions, such as the tropics or the polar tundras. These areas have dramatically different ecosystems with a whole spectrum of different environmental parameters. For instance, planners, engineers, and physicians in the United States generally do not have to deal with schistosomiasis. It was not a factor that had to be considered when we built dams.

Secondly, before any technology is introduced, especially technology new to a particular region, a complete impact study needs to be done, not only on environmental impacts, but on all other impacts as well. Again, this is very important because it cannot be assumed that our experience in Europe or the United States will provide answers for what we will find in the Amazon River basin of South America.

Finally, it should be recognized that any technology introduced into an area, especially in undeveloped regions, will require extensive secondary projects. Hydro-development involves much more than just the technology needed to construct a dam, reservoir, and irrigation systems. It also involves changing a way of life for the local people and the creation of an infrastructure required to mitigate the potential for an epidemic outbreak of schistosomiasis, or other diseases.

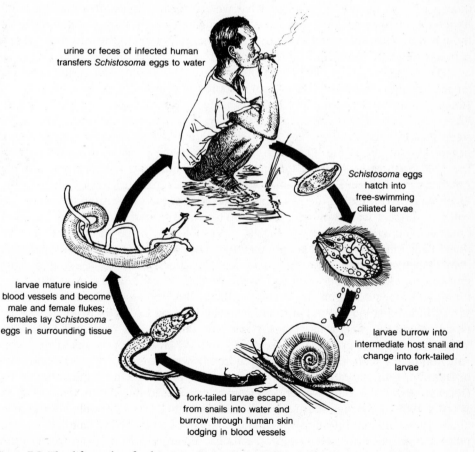

urine or feces of infected human transfers *Schistosoma* eggs to water

Schistosoma eggs hatch into free-swimming ciliated larvae

larvae mature inside blood vessels and become male and female flukes; females lay *Schistosoma* eggs in surrounding tissue

larvae burrow into intermediate host snail and change into fork-tailed larvae

fork-tailed larvae escape from snails into water and burrow through human skin lodging in blood vessels

Figure 7.3. The life cycle of schistosomiasis. (From Miller, G. T., *Environmental Science,* Belmont, CA: Wadsworth.)

Social Impacts

In addition to environmental and health impacts, hydro-development may also have other negative effects on the society and culture. These may be broadly classified as social impacts. These often involve subgroups of the population that are adversely affected by the hydro-development project. In some cases, a hydro project may be so ill conceived that there are negative effects on the society as a whole. Previously, we have mentioned that populations often are moved from along fertile river banks to less fertile land at higher elevations. Resettlement of native people from their traditional lands may affect the culture and their livelihood. In most instances, people forced to move because of dam and reservoir construction view the change as negative. Other less obvious negative changes, however, often result.

In addition to resettlement, many valuable archaeological sites are lost in building a dam and creating a reservoir. The Aswan Dam is a classic example. In that case, there was an effort to move some structures to higher ground but some were lost. Many dams are not built in such thoroughly studied locations, and in those cases we do not even know what we are losing. Some sites that are being flooded have spectacular natural aesthetic qualities, such as major waterfalls. Although waterfalls inherently indicate a high energy potential, their aesthetic aspect may have equally significant value.

The battle over the construction of Hetch-Hetchy Reservoir was literally a watershed event in the history of natural resource development in the United States. Hetch-Hetchy valley—the spectacular glacier-carved canyon of the Tuolumne River—was one of two great scenic valleys in Yosemite National Park. The river's steep gradient and the canyon's high walls made it an ideal water supply for the not-too-distant city of San Francisco, which was in the process of rebuilding from the 1906 earthquake. Gifford Pinchot and John Muir, two leading figures in the American conservation movement, fought bitterly over the issue of Hetch-Hetchy: Pinchot favored building the reservoir, while Muir opposed it. After substantial debate, the project was approved, and the dam was finished in 1912. San Francisco obtained a superb water supply that still serves it today, while part of Yosemite National Park was lost to subsequent generations of park visitors. This battle between dam construction and preservation of free-flowing rivers was re-enacted many times in the decades following the Hetch-Hetchy decision, and dam construction triumphed repeatedly. On the Columbia River, the Bonneville Dam drowned the sacred Indian fishing ground at Celilo Falls, and on the Colorado River, Lake Powell inundated the numerous glades, pools, and falls of Glen Canyon (Porter 1963; Wilkinson and Conner 1987). In the 1960s and 1970s, the opponents of dam construction finally won some battles, saving free-flowing stretches of the Yampa River in Dinosaur National Monument, the Colorado River in the Grand Canyon, and the Delaware River above the Water Gap from the proposed inundation (Stegner 1955; Worster 1985). Elsewhere in the world, a similar pattern may be seen: after generations of unchallenged dominance by the builders of large hydroelectric dams, significant opposition to specified proposals has emerged in Europe (regarding the Danube River), in China (regarding the enormous Three Forks Dam proposed for the Yangtse River), in India (regarding Tehri Dam proposed in the earthquake-prone headwaters region of the Ganges River), and in Australia (regarding wild rivers on the island of Tasmania) (Steffen 1984).

Another major concern is how the energy will be used. In many cases, the electricity being generated is not needed locally in such large amounts. Often, many of the people are very poor and cannot afford electricity of any type. In fact, the rural population usually benefits very little from this development. Less than one-third of the people living in rural areas in developing countries actually have access to electricity (Flavin 1986). Although this number is expected to increase over the next 15 to 20 years, one is still talking about a relatively small percentage of people that are going to have access to electricity. The benefits of hydro-development will not be spread evenly throughout the society, with the more well-to-do parts of society benefitting disproportionately. In the case of the James Bay Hydroelectric Project, a large percentage of the electrical power that is generated in northern Quebec will be used by the northeastern United States and not by Canada (Thurston 1991).

Between 90% and 95% of electricity in LDCs goes to large cities and industries, bypassing the rural areas where the electricity is being generated (Flavin 1986). That may exacerbate a major trend, the movement of the population from the rural areas to the cities. It would actually be better to discourage the explosive growth in cities. Yet a pattern where the electricity produced by hydro-development is made available to cities and not to rural areas supports and encourages a trend towards increasing urbanization. This distorts governmental policies towards development, provides more potential for social unrest, and creates increasingly concentrated pollution problems.

Hydropower has also tended to attract large-scale industrial complexes, usually owned by multinational corporations, which require large amounts of electrical energy. In order to encourage this, the governments often price the power cheaply, masking the fact that a large debt was accrued in building the hydro-development project. The multinational corporations use the energy to produce energy-intensive products like aluminum, which are for the most part exported to the industrialized countries (Deudney 1981; Sawyer 1986). If that happens, the energy provided by hydropower is not entirely beneficial to the host country and may, in fact, have a significant negative impact as relatively few local jobs are created. Even the monies that LDCs borrowed to build the dams often return directly to the industrialized countries because that is where the generators, turbines, transmission lines, and the engineering all come from (Deudney 1981). In a sense, the LDCs are being loaned money so that they will then buy the products of the MDCs. This does little to enhance the social and economic equity in these countries. Tucurui, in the Brazilian Amazon, exemplifies all of these problems, and is part of the reason why Brazil is burdened with such massive foreign debt. Similarly, Akosombo has created far more debt than benefits for Ghana (Freeman 1974).

Economic Impacts

There are compelling short-term reasons to build dams. First of all, many jobs are created in the country involved. There will be more irrigated land and more hydropower, which can be translated into exports and other economic benefits. Dam building is a very profitable business for the large multinational companies that do this work. So it is quite compelling in the short run to build the dams. However, the long-term economic analysis has often been very inadequate.

Investment in new electrical capacity has been a significant factor in the LDC debt crisis. With the oil crisis of the 1970s and early 1980s, hydropower development became much more popular; indeed, hydro capacity of the LDCs doubled in the decade 1980 to 1990. The initial projected costs for these projects are typically in the range of $2000–$3000 per kilowatt, a sum that is very expensive (Flavin 1986). As is typical with large-scale projects, however, construction costs escalated and the time for completion expanded. The resulting interest and foreign exchange payments were often more than the utility could handle. Because utilities in most LDCs countries are government monopolies, the utility debt became the national debt. About one-fifth of the LDC debt, or $180 billion, is now estimated to be due to utility debt (Flavin 1986).

The rapid expansion of the electrical supply coupled with the debt crisis has created severe problems for the utility industry. In a number of LDCs, financial pressures have forced cutbacks in maintenance, leading to reliability and power loss problems. The $60 billion investment per year that the World Bank projects that LDCs will need to make to keep up with a demand growth of 7% per year will simply not be attainable. In addition, in many cases, poor management decisions, inefficiency, and scarce technical expertise have allowed power systems to deteriorate (Flavin 1986).

Of course, the costs associated with hydro-development are not only related to problems of financing and operation. For example, nobody expected the Aswan Dam in Egypt to devastate the sardine fishery in the Mediterranean, although in fact it did do that. It could have been predicted if a proper environmental analysis had been done. The process of laying out all the potential problems ahead of time, thinking them through, trying to quantify the benefits and risks in a realistic manner, and really analyzing the situation was not done in the case of the Aswan Dam, where the former Soviet Union was a prime builder. Each of the developed countries of the world, however, including the United States, the former Soviet Union, and various European nations, have had their share of disastrous dam projects. While not all dam projects are disastrous, the list of disastrous ones is, unfortunately, very long, and many times the problems developed because there had not been an adequate analysis. In some cases, political considerations may overwhelm the decision-making process (Goldsmith and Hildyard 1984b).

Mitigation: Balancing Benefits and Costs

We have tended to focus on the negative impacts of hydro-development. Despite the litany of woes, we must not forget that there are significant benefits provided by dams. Trade-offs exist, with valid and vital concerns on both sides. Since hydropower is one of the most readily available energy sources, hydro-development will probably take place in the LDCs. What can be done to minimize the potential problems associated with it?

Most of the developments in the tropical regions have been caused or financed by banks based in MDCs. Several banks have recently come under heavy criticism because of their role in financing extensive development projects without having looked at the full range of impacts: health impacts, ecological impacts, economic impacts, and social impacts. For many decades, the World Bank was notorious for promoting very large-scale hydro-development with very little environmental assessment (see Chapter 9, "Tropical Forests: Problems and Prospects"). While the governments themselves were participants in the process, many of these governments were hardly representative in the sense that we are familiar with in the United States. Unfortunately, many of the governments in developing countries have tended to be military dictatorships that have a fairly short life span. When they are overthrown and replaced by another, a question arises as to who was really the local beneficiary of the hydro-development. Was it the populace and country as a whole or was it the government officials and a few individuals associated with the project? In the final analysis, however, the country gets saddled with the resulting debt, regardless of who benefited.

To its credit, the World Bank has, in the last several years, made an effort to reconsider some of its almost automatic support for dam projects, and has been working on doing a more thorough analysis and exploring what can be done to mitigate some of the negative impacts (Goodland 1985, 1986). One of the things presently being done by the World Bank is to acquire land adjacent to dams and set it aside as a preserve, so some land is saved to compensate for the land lost. It is not necessarily a one-to-one relationship, however, as the land lost is usually flood plain and the land saved may be at a higher elevation and thus less fertile. Such action is also in the best interest of hydro-development. By keeping the surrounding watershed intact, erosion will be minimized and the life of the reservoir maximized.

One criteria for looking at whether a site is suitable for hydro-development is to look at how much power is going to be generated in relation to the amount of land that will be flooded. This will provide a ratio of the number of kilowatts generated per acre flooded. In areas of relatively level terrain, such analysis usually yields an unfavorable ratio (Goodland 1977).

There are some critics of dams who argue that few, if any, dams could adequately satisfy a full set of requirements that includes: an adequate environmental assessment; benefits to a large segment of the population; enhancement of local food productivity; protection of public health and safety; protection of heritage lands or unique wild areas; siltation protection; avoidance of salinization and waterlogging; avoidance of displacement of indigenous populations; incorporation of design and engineering safeguards; minimal damage to fisheries; and protection of water quality and quantity downstream. Indeed, some of these impacts may be unavoidable regardless of the mitigation measures taken (Goldsmith and Hildyard 1984a).

Alternatives to Hydro-Development

In projecting the costs and benefits of large scale hydro-development, it may be useful to consider the major goal (which should be sustainable development) and to ask whether there are other options that can attain the goal better. Ultimately, it is a question of whether some other investment might provide more overall benefits for the country. Indeed, development specialists are beginning to realize that some alternative investments could have much greater benefits than large-scale hydro-development.

One alternative is to develop small-scale hydro (generally defined as less than 15 MW capacity; also called micro-hydro) as opposed to large-scale hydro. Dams can be of any size. In some countries, for example, the emphasis has been on building many smaller dams that collectively have less impact than a few large dams would have. The best example of such a policy is in China today. Starting about 1968, Chinese leaders decided to build some large dams, but also many smaller dams. They have constructed about 76,000 hydro plants since 1970, most of them village based (Flavin 1986). They are developed, planned, and built by the rural community where they exist. This approach has some powerful advantages, especially since the hydropower is used locally, so that the villagers do benefit from the dam. A secondary benefit is that water is also available for local irrigation during the dry periods. They have also used the resulting ponds and reservoirs for aquaculture, the growing of fish. These systems are thus viewed as integrated systems that need to be balanced in terms of power production, irrigation, and fisheries. China still is, of course, building large-scale hydro-development projects. But nonetheless small-scale hydro has been quite successful. A major stumbling block, however, has been the limited availability of financing for small-scale hydro projects (Flavin 1986). Poor funding of small-scale hydro may be attributed to lack of knowledge of its potential by planners, to lack of technical expertise, and a bias toward large-scale projects by multinational firms.

In spite of these problems, the potential growth of small-scale hydro looks promising. In a number of countries, such as Peru, Costa Rica, Guatemala, Nepal, Guinea, and Madagascar, the small-scale hydro potential exceeds the existing total installed electrical generating capacity. Between 1983 and 1991, the small-scale hydro capacity of LDCs was projected to increase from 10,000 MW to about 29,000 MW. The worldwide potential may well be more than 100,000 MW (Flavin 1986). In addition, small-scale hydro installations may often be developed without drastically altering the local environment (Figure 7.4).

There is an analogy to the United States that is interesting. We have discovered in the years since the energy crisis of 1973 that electricity is a very high quality, and thus expensive, form of energy. Using electricity to heat homes or to heat water is one of the most expensive ways to get these jobs done. The LDCs are also starting to realize that. For example, increasing efficiency and conservation may be a very important thing to do concurrently with hydro-development so that one does not waste this very expensive, high quality energy, which is very limited in these countries.

In order to promote worthwhile alternatives, realistic studies must be done to identify what the needs are for electrical energy. Ways that the electricity could be best utilized to promote the general welfare of the people of the country must be given priority over uses that may have very little benefit to the country. For example, there were studies done in Brazil that showed that projected electrical consumption in selected areas such as lighting, refrigeration, and motors could be cut 30% by implementing energy efficiency by the year 2000. While $10 billion would be needed to implement such a program, it would eliminate the need for 22,000 MW of installed capacity and save $44 billion in construction and other costs (Flavin 1986). So conservation could play a major role in meeting the energy needs of a country in a very cost-effective manner; a concept that, unfortunately, has not really been fully appreciated in the United States. But the LDCs have to give it serious consideration. They cannot afford to be in a position to needlessly squander their limited supply of energy, and their large foreign debt must be reduced.

Figure 7.4. A hydroelectric power plant using a head of 1193 feet on the Pit River in California. The water descends from the reservoir on the Mc-Cloud River through 5500 feet of steel penstock. This arrangement minimizes the environmental impact compared to the development of a large dam facility. (From Dorf, R. C., *Energy, Resources & Policy,* Reading, PA: Addison-Wesley, 1978. Courtesy of Pacific Gas and Electric Company.)

Conclusions

There is no doubt that further hydro-development should and will occur. The question is how to minimize the negative ecological and social impacts that go along with hydro-development. Micro-hydro may be one of the ways to minimize some of these effects. Efforts need to be continually made to make the governments and the population aware of the importance of natural systems in maintaining the ecological balance of regions. Some free-flowing rivers should be left intact in order to preserve ecosystems that are dependent upon "untamed" river systems. Additional studies are also needed in order to generate predictions on what the impacts of hydro-development will be, and how such impacts can be minimized. In some cases it may be determined that development should not occur in a particular region because the impacts cannot be mitigated. Increasingly, banks lending money need to assess potential impacts from proposed development in far more depth and detail. These are all very important considerations. If they are seriously pursued, we will see more ecologically based (and economically sound) hydro-development in the future.

References

Alexis, L., "The Damnation of Paradise, Sri Lanka's Mahaweli Scheme," *The Ecologist* 14(5–6), 1984, pp. 206–215.

"Balbina: Master of Destruction," Videodocumentary for Brazilian TV, available through L.A.V.A. Archives, 1989.

Berkes, F., "The James Bay Hydroelectric Project," *Alternatives* 17(3), 1990, p. 20.

Bolt, B. A., *Earthquakes: A Primer,* San Francisco: W. H. Freeman & Co., 1978, pp. 121–130.

Borowitz, S., "How to Drown a Wilderness: James Bay, Canada," *Sierra* 75 (May–June), 1990, pp. 82–83.

Burton, S. "Taming the Wild River," *Time* 144(25), 1994, pp. 62–64.

Cheng, T. C., *General Parasitology.* New York: Academic Press, 1973.

Deudney, D., *Rivers of Energy: The Hydropower Potential, Worldwatch Paper 44,* Washington, DC: Worldwatch Institute, 1981.

Eley, J. and T. H. Watkins, "In a Sea of Trouble: The Uncertain Fate of the Pacific Salmon," *Wilderness* 55, Fall 1991, pp. 18–26.

Flavin, C., *Electricity for a Developing World: New Directions, Worldwatch Paper 70,* Washington, DC: Worldwatch Institute, 1986.

Freeman, P. H., *The Environmental Impact of Tropical Dams: Volta Hydroproject Case Study,* Washington, DC: Smithsonian Institute, 1974.

Goldsmith, E. and N. Hildyard, "The Myth of the Benign Super Dam," *The Ecologist* 14 (5-6), 1984a, pp. 217–220.

Goldsmith, E. and N. Hildyard, "The Politics of Damming," *The Ecologist* 14 (5-6), 1984b, pp. 221–231.

Goodland, R., "Environmental Optimization in Hydro-Development of Tropical Forest Regions," *In* R. S. Panday (ed.), *Proceedings of the Symposium on Man-Made Lakes and Human Health,* University of Suriname, Paramaribo, Suriname, October 23–25, 1977.

Goodland, R., "Environmental Aspects of Hydroelectric Power and Water Storage Projects," *In Environmental Impact Assessment of Water Resources Projects,* Volume III, UNESCO/UNEP, Roorkee, UP, India 1985, pp. 1–30.

Goodland, R., "Hydro and the Environment: Evaluating the Tradeoffs," *Water Power and Dam Construction* (Nov.), 1986, pp. 25–29.

Hayes, D., *Energy for Development: Third World Options, Worldwatch Paper 15,* Washington, DC: Worldwatch Institute, 1977.

Kleinbach, M. H. and C. E. Salvagin, *Energy Technologies and Conversion Systems,* Englewood Cliffs, NJ: Prentice-Hall, 1986.

Kristof, N. D., "China Breaks Ground for World's Largest Dam," *New York Times,* 22 June 1993.

Linden, E., "Bury My Heart at James Bay," *Time* (July 15), 1991, pp. 60–61.

Muckleston, K. W., "Salmon vs. Hydropower: Striking a Balance in the Pacific Northwest," *Environment,* 32, 1990, pp. 10–15, 32–35.

Newman, P. C., "The Beaching of a Great Whale," *Macleans* 104 (Sept. 16), 1991, p. 38.

Picard, A., "James Bay II," *Amicus Journal* 12(4), 1990, pp. 10–16.

Porter, E., *The Place No One Knew,* San Francisco, CA: Sierra Club Books, 1963.

Satchell, M. "Fish with Nowhere to Run," *U. S. News & World Report* 117(24), 1994, pp. 34–36.

Sawyer, S. W., *Renewable Energy: Progress, Prospects,* Washington, DC: Association of American Geographers, 1986, p. 30.

Shea, C. P., *Renewable Energy: Today's Contribution, Tomorrow's Promise, Worldwatch Paper 81,* Washington, DC: Worldwatch Institute, 1988.

Steffen, W., "Furor over the Franklin," *Sierra* 69, 1984, pp. 43–49.

Stegner, W., *This Is Dinosaur.* New York: Alfred A. Knopf, 1955.

Stein, J., "Water for the Wealthy," *Environment* 19(4), 1977a, pp. 6–14.

Stein, J., "Fumbled Help at the Well," *Environment* 19(5), 1977b, pp. 14–17, 41–45.

Sternberg, R., "Hydroelectric Energy: An Agent of Change in Amazonia (Northern Brazil)," *In* F. J. Calzonetti and B. D. Solamon (eds.), *Geographical Dimensions of Energy,* Norwell, MA: Kluwer Academic Publishers, 1985.

Stowell, C. D., "Salmon: Continuity for a Culture," *In* High Country News Staff (eds.), *Western Water Made Simple,* Washington DC: Island Press, 1987, pp. 71–80.

"The Social and Environmental Effects of Large Dams," Briefing Document, *The Ecologist,* Worthyvale Manor, Camelford, UK, 1984

Thurston, H., "Power in a Land of Remembrance," *Audubon* (Nov.–Dec.), 1991, pp. 52–59.

Wilkinson, C. F. and D. K. Conner, "A Great Loneliness of the Spirit," *In* High Country News Staff (eds,), *Western Water Made Simple,* Washington DC: Island Press, 1987, pp. 54–64.

White, G. F., "The Environmental Effects of the High Dam at Aswan," *Environment* 30(7), 1988, pp. 5–11, 34–40.

Worster, D., *Rivers of Empire: Water, Aridity and Growth of the American West,* New York: Pantheon Books, 1985.

Chapter 8

The Loss of Biodiversity

Howard Horowitz and Eric F. Karlin

In the next few decades, human activities are likely to result in the extinction of perhaps one-half of all the species on the planet (Wilson 1988). Scientists have identified some 1.5 million species so far, but estimates of the total number of living species range from 3 to 50 million (May 1988, 1992). The large majority of species have not been scientifically identified and are located in the tropics, where comparatively little biological and ecological research has taken place. This astounding loss of biodiversity is being caused by many activities, including deforestation, desertification, large-scale hydro-development, agriculture, pollution, fishing, hunting, and poaching. Decreasing biodiversity will result in severe ecological, scientific, and economic loss, and has aesthetic, recreational, and ethical ramifications as well (Lubchenko et al. 1991).

The problems we encounter trying to understand and protect biodiversity are as staggering as that of diversity itself. Our comprehension of the normal functioning of natural ecosystems is still superficial, although the more we learn about them, the more wondrous and complex they appear to be. Critical interactions between plants, animals, and microorganisms are poorly understood, yet such understandings are necessary for assessing the impacts of ecosystem disturbance. Some examples discussed later will illustrate this complexity, and the risks of our ignorance of it. Differences between natural extinction and human-caused extinctions are also examined, along with attempts to protect biodiversity through species preservation and ecosystem preservation. Finally, the potential impacts of greenhouse warming, acid deposition, and other major environmental changes on biodiversity are considered.

The Evolution of Life

Early Species Development

One of the remarkable properties of life is that species have the capacity to change, via the process of evolution, into new species. When life first appeared on Earth it may have been represented by only one, or perhaps, a handful of very simple one-celled organisms. Today there are millions of species, exhibiting a bewildering array of bodies and life cycles. Clearly, the vessels that carry the spark of life have become greatly diversified in number and structure since life first appeared on Earth several billion years ago. Evolutionary change is thought to have been very slow at first, but as time passed, more effective genetic and reproductive systems were developed, and this greatly facilitated the process and rate of evolution.[1]

New species evolved, each with a different capacity to utilize the environmental resources present. With time, environmental space not previously occupied by life became occupied. For instance, prior to the development of a stratospheric ozone layer capable of absorbing a significant percentage of the more damaging wavelengths of ultraviolet radiation, life was primarily limited to aquatic environments. The terrestrial habitat was a vast unoccupied environmental space. Once an effective ozone layer had developed (about 450–500 million years ago), life was able to survive on land (see Chapter 13, "Stratospheric Ozone Depletion"). At first, the invasion was limited to wetlands and other moist terrestrial environments. Eventually, however, life developed the capacity to occupy almost all of the environmental space that was available on land—even the polar regions.

The Diversity of Life: Speciation and Extinction

The fossil record indicates that not only have new species evolved, but that there has also been a loss of species. No species lasts forever. So the diversity of life on Earth is not simply a function of speciation, it is actually a balance between speciation and extinction. Given the richness of life currently inhabiting the planet, it is obvious that speciation has clearly outpaced extinction, at least for a significant proportion of the Earth's history (Wilson 1989, 1992).

There have been times, however, when species extinction was far greater than speciation. Episodes of mass extinction punctuate the fossil record periodically. Two of the most dramatic marked the end of the Permian Period (225 million years ago) and the end of the Cretaceous Period (65 million years ago) (Raup and Sepkoski 1982; Wilson 1989). The latter event, which included the extinction of dinosaurs and ammonites, has engendered much controversy regarding the possible causes of such large-scale extinctions. Explanations include global climate changes, sea level changes, intense volcanic activity, outbreaks of disease, competition with other species, and large asteroids striking the Earth (Bakker 1986; Alvarez 1987; Gore 1989; Alvarez and Asaro 1990; Courtillot 1990). The geologic evidence that massive asteroids impact the Earth periodically is irrefutable, but whether or not they were actually responsible for mass extinctions is not at all certain. Many scientists remain skeptical of this explanation and interpret the fossil record to indicate

[1]Among the more significant evolutionary developments was the eukaryotic cell, which first appeared in the fossil record about 1.4 to 1.9 billion years ago (Knoll 1992). Although single celled, the first eukaryotic organisms represented a major evolutionary advance. They possessed a nucleus, mitochondria, and chloroplasts. They also eventually developed the process of sexual reproduction, the most effective means of promoting genetic recombination yet evolved. With sexual reproduction, the process of evolution could take place far more quickly. Rates of speciation exploded and, in addition, far more intricate and complex biological structures evolved. For instance, it was the eukaryotic line that gave rise to all multicellular organisms (Knoll 1992). Two of the most diverse phyla today are the Arthropoda (insects, spiders, and crustaceans) and the Magnoliophyta (flowering plants) (May 1988; Wilson 1988).

that previous mass extinctions occurred over periods of hundreds of thousands of years or more.

Following all of the previous mass extinctions, however, there was an accelerated rate of speciation. It is thought that the surviving species were able to utilize the extensive amounts of environmental space formerly occupied by the species that had become extinct. For instance, mammalian evolution was quite rapid and extensive after the extinction of the dinosaurs (Raup 1988). This was in large part because mammals were occupying the space formerly held by the dinosaurs. Thus, although there was a massive loss of life, there was also a multitude of opportunities for the surviving species to utilize. Many species were lost, but new ones replaced them. The evolution of new species, however, took place over periods of millions of years, not overnight. For instance, it took 5 to 10 million years for the Earth's biodiversity to be restored after the Cretaceous extinction episode (Wilson 1989).

Humans and the Threat of Extinctions

The human species (*Homo sapiens*) is a relatively young species, being in the vicinity of 250,000 years old (Campbell 1985; Stringer 1990). Unlike any other species, however, human use of environmental space is not limited by the biological constraints of the body. Our technology transcends our biological limitations and allows us to occupy almost any environmental space we choose. Although humans are unable to obtain energy by eating wood, we can release its energy by burning it. While a forest may not provide much food for human needs, we can destroy the forest and replace it with a field of wheat. Even though humans are not biologically equipped to live in the Arctic, our technology (clothing, fire, buildings) allows us to do so. As our technological capacity advances, so does our capacity to utilize environmental space.

Early human populations were small and had a relatively minimal technology, thus human impact on other species was slight. Species losing environmental space to humans could usually find haven in the many extensive areas that humans did not yet occupy. Even so, it is believed that early humans may have played a significant role in the extinction of many species that disappeared between 7,000 and 20,000 years ago (i.e., the ground sloth, mammoth, cave bear). In particular, the migration of people across the Bering Strait, and their rapid diffusion through the American continents appears to correspond to a significant extinction of large mammals, perhaps because of overhunting (Martin 1973; Vermeij 1986). As the human population increased and as technology became more sophisticated, humans occupied an ever increasing percentage of the Earth's environmental space. Although rates of extinction increased dramatically, many species could still survive in wilderness areas up until relatively recent times. At the present time, however, the human population is so large and our technology is so powerful, that virtually all parts of the globe are being impacted, including the Arctic tundra (Watkins 1990) and the oceans (Culotta 1994). Humans have grasped all of the readily available environmental space and are now making serious efforts to occupy even the relatively inaccessible environmental space. For many species, there are no longer any "safe" areas left for their survival. Lacking a refuge, hundreds of thousands, if not millions, of species are facing imminent extinction. It is estimated that up to 50% of all extant species may be driven to extinction in the next few decades, with the majority of these from the tropical regions (May 1988; Wilson 1988, 1992; see also Chapter 9, "Tropical Forests: Problems and Perspectives"). But widespread extinction is also taking place in other regions as well (Franklin 1988; Eley and Watkins 1991; Ryan 1992; Wilson 1992).

This current mass extinction event is quite unique and differs dramatically from prior mass extinctions. First, it is caused by a biotic source—a single species at that (humans). Our activities, both direct (i.e., habitat destruction, hunting) and indirect (i.e., acid deposition, stratospheric ozone depletion, pesticides) are the sole cause of this current wave of extinctions. Secondly, all ecosystems and all biological groups are threatened, whereas previous mass extinctions usually affected certain ecosystems or species groups (Myers 1988). Third, the speed of the extinction process is alarming. We are currently looking at an extinction process that is at least 10,000 times greater than before human intervention (Wilson 1989, 1992). If the projected loss of half all extant species within the next 25 to 50 year period comes to pass, the rate of extinction caused by humans will be even higher (May 1988). There is also one other significant aspect. If we assume that large numbers of humans survive for the foreseeable future, there will be little unoccupied environmental space after the mass extinctions. Humans will occupy most of the environmental space, and human ecosystems are relatively uniform throughout the world. Thus, unlike the situation following previous mass extinctions, surviving species will have very little opportunity to diversify by utilizing unoccupied environmental space.

Understanding the Biodiversity Crisis

If our knowledge of "healthy" ecosystem functions is fuzzy, then our knowledge of the effects of human-caused disturbances on these systems is perhaps even fuzzier. The disappearance of a large predator or a common bird may be obvious; the disappearance of an obscure rodent, plant, or insect may be quite unnoticed. The full consequences of the loss of any species, however—whether large mammal or microscopic algae—are most uncertain. The empty niche is likely to be filled by other species, but the entire ecosystem is affected. Using laboratory based ecosystems, Naeem et al. (1994) have demonstrated that a decline in biodiversity may result in a drop in primary production. Significant ecosystem disruption may occur if the species that is extirpated played a significant role in ecosystem processes. Even if major disruptions are not associated with the loss of a species, ecosystems may become less resilient to future perturbations as their biological diversity declines.

Usurping Net Primary Production

The extent to which humans have appropriated the world's ecosystems for our own usage has been estimated by studies of global net primary production (NPP). People now consume roughly 3% of the global NPP through direct consumption, such as direct food consumption and livestock feeding. When our indirect consumption is included, such as the NPP in farm fields and pastures, and the biomass consumed in forest burning and clearing, the human portion of the global NPP may be as high as 19%. Human appropriation of land-based NPP may be as high as 31% (Vitousek et al. 1986)! Furthermore, these farms, pastures, and clearings often have less potential NPP than the more diverse and productive natural ecosystems that would otherwise have occupied these areas. Overall, this is an enormous share of the world's natural production to be taken over by one species.

The price for this takeover by humans and their domesticated supporters is that all the rest of the plant and animal species just have to get by on the leftovers. As the human portion of the global NPP gets progressively larger, the other species' portion gets progressively smaller. As human population increases further, and the farms, pastures, and tree farms reach further into marginal lands, the human share of the global NPP could surpass 50%. Even the oceans are not immune to this takeover; drift nets 35 miles long are making a big dent in the ocean's resource production. Projected expansions of the squid and krill fisheries would have an even greater impact by reaching further down into the food webs of the ocean. This uncontrolled growth of the human share of the global NPP cannot be sustained forever. It is truly a form of "biomass piracy" that can only lead to devastation for the wild forms of life that are left without sufficient habitat on land and without sufficient food supplies in the ocean. Large-scale extinctions and impoverished ecosystems will be the ultimate outcome of this trend, even for species that are not directly hunted or harvested by humans (see Chapter 2, "Population Growth and the Global Environment: An Ecological Perspective").

Loss of Habitat

Overhunting has certainly wiped out various large mammals, flightless birds, and other vulnerable targets, but most species loss is due to reduction of suitable habitat. The destruction of extensive areas of natural ecosystems is affecting the planet's biodiversity significantly, with tropical deforestation having the most devastating impact (see Chapter 9, "Tropical Forests: Problems and Prospects"). But extinction can also occur when ecosystems are not completely destroyed. If the once-continuous habitat is badly fragmented, populations will become isolated from each other and may become too small to be sustainable (Harris 1984; Wilcove et al. 1986). This is an especially big risk if each breeding pair of a species requires a large home range; predators such as the northern spotted owl (*Strix occidentalis* ssp. *caurina*) and the cougar (*Felis concolor*) are at particular risk from this kind of fragmentation.

The rapid conversion of the Great Plains—from vast prairies with migrating buffalo herds to cow pastures fenced with barbed wire—caused the black-footed ferret (*Mustela nigripes*) population to drop precipitously close to extinction. This small predator feeds almost exclusively on prairie dogs (*Cynomys* spp.), but the great prairie dog "towns" that once dotted the plains were systematically decimated by cattle ranchers. They did this partly out of fear that the burrows and holes could cause economically significant levels of injury to their stock, and partly out of unwillingness to share the grass with another herbivore (although the "towns" supported a different but rich vegetation community that the buffalo herds sought out). Today's prairie dog population is just a small remnant of what it once was (as the buffalo population is just a small remnant of what it once was) (Clark et al. 1989). Both are largely confined to national parks and a few roadside tourist attractions. Although the prairie dog is not in itself in danger of extinction, its predator, the black-footed ferret, is now endangered; without greater areas of prairie dog habitat, the ferret's prospects for survival are grim (Biggins and Crete 1989).

The black-footed ferret, then, is an unintended victim of the prevailing economic activity that has come to dominance throughout nearly all of its formerly extensive range. This example is hardly exceptional; numerous other endangered species, from the red-cockaded woodpeckers (*Dendrocopos borealis*) that require mature southern pine forest, to the desert pupfish (*Cyprinodon* spp.) whose isolated warm pools are drying up due to the lowering of water tables by well-water withdrawal, are accidental victims of "resource economics." Habitat loss is a consequence of activities that form the economic foundation of many regions. Reconciling the economic determinism of "managed landscapes" with the biotic requirements of "natural landscapes" will be a challenge we must try to meet, for the sake of species survival.

Other Causes of Biodiversity Loss

Overexploitation of Species

While habitat loss is the primary problem leading to extinctions, it is not the only one. Commercial, subsistence, and sport hunting have in the past, and will in the future, threaten the extinction of many species. Today, Bengal tigers, rhinoceros, elephants, and certain whales are among the numerous species that are at risk of extinction. In the past, the American bison (*Bison bison*) and the snowy egret (*Leucophoyx thula*) were brought to the edge of extinction by overexploitation. The passenger pigeon (*Ectopistes migratorius*) did not survive its exploitation; the most populous bird in North America literally disappeared in a few decades. Sometimes exploitation of one species for harvesting can threaten another species; such is the case with some methods of commercial tuna harvesting and dolphins.

Plants are also threatened by overexploitation. Although partially protected by law, populations of the saguaro cactus (*Carnegiea gigantea*) are threatened by cactus rustling. The venus fly trap (*Dionaea muscipula*), naturally occurring only in wetlands of North and South Carolina, is threatened both by overcollecting for horticultural purposes and also by habitat loss, as many sites where it occurs are being drained for forest production. Many species of orchids have been so overcollected that they are no longer found in areas where they were once common (Koopowitz and Kaye 1983).

Competition from Introduced Species

Because of their isolation from mainland contacts, many unique and distinctive species and ecosystems have evolved on islands. These "endemic" ecological communities lack the overall biodiversity of mainland areas. Their unique, but simplified, communities are generally quite vulnerable to catastrophic loss of their endemic species after the introduction of more aggressive species from distant mainland areas. The introduction of rats, cats, and other mainland species has invariably been disastrous to the native ecosystems.

An excellent example of this loss of diversity may be found in the Hawaiian Islands. In the past 200 years, out of 43 species of land birds, 15 have become extinct and 19 are threatened or endangered. In the same time period, 10% of the 1,250 species of native flowering plants have become extinct, while an additional 40% are threatened or endangered. An additional 45 species of land birds that were present 1,500 years ago when the Polynesians first arrived had become extinct prior to the discovery of the Hawaiian Islands by Western Civilization in 1778 (Raven 1985).

Although extinction due to species introductions is less common on continents than on islands, mainland ecosystems often experience extensive disruption from the invasion of introduced species. One classic example is the precipitous decline of the American chestnut (*Castanea dentata*) in the early 1900s, which was brought about by the introduction of chestnut blight (*Endothecia parasitica*) from the Orient (Koopowitz and Kaye 1983). The American chestnut was once a dominant species in the deciduous forests of eastern North America. Its wood was highly valued for lumber and its nuts were a major source of food for wildlife. Today the species has been nearly exterminated by chestnut blight. Although the American chestnut is still extant, its ecosystem function has largely been eliminated.

There are many other examples of the impacts of introduced species. Purple loosestrife (*Lythrum salicaria*) has taken over many wetlands in eastern North America. In addition to replacing native plants such as cattails (*Typha* spp.), purple loosestrife provides little food for wildlife. Other introduced plant species that have greatly impacted mainland ecosystems include prickly pear cacti (*Opuntia* spp.) in Australia and South Africa and kudzu (*Pueraria lobata*) in the southeastern United States (Koopowitz and Kaye 1983). The Argenitine fire ant (*Solenopsis invicta*) was first observed in the United States (at Mobile, Alabama) in 1942. Today it is found throughout the southeastern United States and is significantly altering many of the ecosystems in which it occurs, especially the super-colony (polygyne) form. The number of arthropod species (particularly ants) present in an ecosystem often decline significantly when fire ants invade it (Mann 1994). Aquatic ecosystems have also been affected by introduced species. The introduction of the sea lamprey (*Pteromyzon marinus* ssp. *dorsatus*) to the Great Lakes of North America significantly contributed to the major decline in the native fish populations of these lakes. The zebra mussel (*Dreissena polymorpha*), although only recently introduced to North America (in the mid-1980s), is already causing extensive damage to aquatic ecosystems. It is even directly affecting humans by clogging water pipes and growing on boat hulls (Ross 1994; see Chapter 11, "Coastal Ocean Resources and Pollution: A Tragedy of the Commons").

Pollution and Extinction

It is well known that chemical pollutants can play a significant role in the extinction of a species. The threat of chemical pollutants depends on many factors: longevity in the environment, dispersal to sensitive areas, and accumulation in organisms. Perhaps the most famous case is that of DDT, a pesticide widely used after World War II that decimated the populations of a number of fish-eating and predatory birds. It was the increasing concentration of the pesticide up the food chain, called *biological amplification,* that ultimately reduced the calcium in the shells of the bird eggs, causing many of them to break prematurely. After banning DDT in 1972, most of the affected bird populations recovered somewhat. Tragically DDT residues are still found in many species. This is because many birds still carry pesticides obtained from past exposures, residual DDT products (like DDE) are still abundant in many ecosystems, and DDT is still in use (Niles et al. 1989). For instance, the pesticide Dicofol contained 10% DDT as an "inert ingredient" (Dicofol 1984). The combination of various organochlorine pesticide residues has been linked to a recent decline in bald eagle populations along the Columbia River (Frost 1990).

Of course, DDT represents only one of many possible pollutants. Excess heat released to waterways during electrical power plant operation has severely affected some local species of fish. The significant depletion of the stratospheric ozone layer, leading to enhanced ultraviolet radiation in the Antarctic region, may have serious consequences for life in polar and subpolar regions. Based on the ever-increasing pollution levels around the world, pollution, particularly regional and global pollution, is likely to be a much greater threat to biodiversity in the future (see subsequent section, *Global Threats to Maintaining Biodiversity*).

Industrial Agriculture and Monoculture

The greatest threats to diversity are not just confined to natural ecosystems and wild species. Changes in agriculture over the last half-century have accelerated a tremendous reduction in the intraspecies genetic diversity of our domesticated plants and animals. Indigenous cultivators developed and maintained hundreds of distinctive local varieties of virtually every domesticated crop plant, and also maintained many diverse forms of domesticated animals. The shift to modern agribusiness has resulted in the widespread dissemination of a very few high-yielding varieties, while hundreds of locally adapted varieties are neglected and at risk of being lost.

In traditional agriculture, the planting of multiple varieties and species of crops in distinctive spatial and temporal patterns—based on long-standing agroecological systems—is a proven hedge against risk, and provides for stability of both culture and agriculture. The high-yielding varieties of modern agribusiness increase both yield and risk, and are dependent on the supports provided by the petrochemical industries—fertilizers, pesticides, and mechanized equipment. These high-yielding varieties may be adopted by farmers in many parts of the world with great apparent success, at least as measured by our rather short time frame of experience. They are most likely to succeed on farmland with better-than-average soils; in remote uplands and on marginal soils, the commercial high-yielding varieties are less likely to be adopted and rarely succeed even when tried. These marginal places, then, have become the reservoirs of genetic diversity for the world's agricultural crops. For example, when botanist Hugh Iltis circulated his Christmas card with an old illustration of teosinte—the wild perennial ancestor of corn, long lost and thought to be probably extinct—it triggered a remarkable chain of events leading to its rediscovery on a 6-acre patch of the Sierra de Manantlan in Mexico. This perennial ancestor to today's annual maize species is hardly just a botanical curiosity; it is remarkably virus resistant and has multimillion dollar value as a source of genetic material for the development of more vigorous varieties of maize (Iltis et al. 1979; Iltis 1983, 1988).

Not only is biological diversity threatened by the transformation of complex ecosystems into monocultural cash crops, the monocultures are doomed to failure, in the long run, because they cannot provide the biodiversity required to sustain the soils that nourish them. Soils are perhaps the least understood and least appreciated component of terrestrial ecosystems, but they are also perhaps the most remarkable. The productivity of the site, whether it is "wild" or cultivated, depends on the health of the soil ecosystem: its numerous varieties of bacteria and fungi, animals of many kinds, photosynthate from plant roots, and the cycling of nutrients. Complex interactions between all of these things must be maintained, or else the soil will suffer breakdowns in both fertility and physical structure. Two examples of the hazards of treating ecosystems as if they were simplistic commodity production units can be drawn from the Douglas-fir (*Pseudotsuga menziesii*) forests of the Pacific Northwest.

Misuse of Herbicides to Kill Pioneer Species

In order to shorten the time needed for lumber production, there is widespread aerial application of herbicides to kill "unwanted" shrub and hardwood species, such as alder (*Alnus* spp.) and ceanothus (*Ceanothus* spp.). These species are natural pioneers after a disturbance on these sites, and clear-cutting the old forest and then burning the logging residue is indeed such a disturbance. They are targets of herbicide spray because they are seen by the foresters as "competition" for the available sunlight, moisture, and nutrients. In fact, the species interactions are far more complex than that: these so-called competitors host a vast network of soil organisms that benefit the entire forest community in many ways. For example, the nitrogen-fixing bacteria found in the root nodules of these "target species" provide the largest source of usable nitrogen to the site over the entire life cycle of the forest (Cromack et al. 1979). Hundreds of years after the alder and ceanothus have gone, the nitrogen thus fixed would still be utilized by the Douglas-fir trees that naturally succeeded them. Yet by poisoning these "competitors," the nitrogen will not be fixed, and future conifer growth on the site is likely to be limited by a shortage of nitrogen. (The high-tech response of the foresters is to suggest several aerial applications of nitrogen fertilizer, starting about ten years after the aerial herbicide application). The alder and ceanothus also provide many other site benefits; these range from the amelioration of harsh sites to the support of soil organisms that kill specific pathogens that can be fatal to the Douglas-fir (Rose 1979). Finally, field measurements of the growth interactions between crop trees and target species revealed that the "competition" was greatly exaggerated. On many sites, there was no reduction in Douglas-fir growth at all that could be attributed to the target shrubs; the trees were growing vigorously right through the ceanothus (Horowitz 1982).

Consequences of Unwitting Interference with Small Mammal Mycophagy

One rodent, the mountain beaver (*Aplodontia rufa,* not a true beaver), has acquired a taste for planted Douglas-fir seedlings and inflicts damage to young plantations. Another, the deer mouse (*Peromyscus maniculatus*), consumes seed and can also damage the regeneration. In an effort to reduce this damage, foresters applied poisoned seed onto the ground in these areas. A closer investigation of the small mammal populations that inhabited these ecosystems, conducted by a team of soil scientists and wildlife biologists, discovered that things were not as simple as they had appeared, and that the poison seed was having unintended negative consequences for the future forests (Maser et al. 1978a,b). The seed was eaten not only by mountain beavers, but also by a wide spectrum of small rodents, including voles (*Clethrionomys* spp., *Microtus* spp.), northern flying squirrels (*Glaucomys sabrinus*), and Townsend's chipmunks (*Eutamias townsendii*). Some of these rodents were also mycophagous—that is, fungus-eaters—and among the fungi they ate were mycorrhizal fungi. One of the most exciting areas of scientific discovery in recent years has involved the realization of the importance of mycorrhizal fungi in soil ecology: there are thousands of species of fungi associated with plant roots in mycorrhizal relationships. Each fungal species is dependent upon the root systems of particular host plant species for photosynthate, and their presence is also essential for the well-being of the host plants. They live in the soil, become attached to the roots of their host species, and greatly enhance the host plant's uptake of soil nutrients by drawing them into the roots through long, fine filaments called *hyphae*. Their utilization is critical for the successful reclamation of many degraded ecosystems (Perry and Amaranthus 1990).

The study of the young Douglas-fir plantation revealed that the dispersion of mycorrhizal fungi around the soil occurs primarily by way of spores transported in the fecal pellets of fungus-eating rodents. Besides functioning as agents of dispersion for mycorrhizal fungi, these small rodents make important contributions both to soil structure and to nutrient cycling. A sharp reduction in their numbers, which was an incidental outcome of the poison seed treatment, was a great loss to the long-term health (and therefore to the productivity) of the ecosystem.

These examples indicate that in order to manage a forest (or any other ecosystem) for long-term productivity, we must respect the biological diversity it encompasses. The small "noncommercial" species may play important roles that we do not always fully understand. As Maser said,

> Our world is losing both species and habitats because we are not sensitive to how and why they are functionally interconnected . . . If we do not understand the organism and its function within the habitat, how can we understand the results of unexpected changes in the habitat when the organism is removed? Dare we assign even a tentative negative value to any organism in a forest if we do not have the slightest idea of its diversity of functions? (Maser 1988).

Preserving Species and Ecosystems

The Anthropocentric Perspective

Despite the dramatic losses of North American wildlife species over the last century, conscious efforts to protect biotic diversity were not formulated and implemented until the last quarter-century. We emphasize that "conscious efforts" are recent: although some protection was provided to certain ecosystems through the establishment of national parks a century ago, their intent was primarily to provide "pleasuring grounds" for the public enjoyment of our most spectacular natural scenery. The parks were primarily anthropocentric—people-centered. They have never been representative of the diversity of our nation's ecosystems. High mountains and deep canyons are well represented, but there are few tallgrass prairies or low-elevation forests. The rapidity with which major habitats were destroyed in the late nineteenth and early twentieth centuries was not a great concern to policymakers, despite the shocking elimination of once-common bird and mammal species, because policy was geared to promote the clearing and settlement of the land. In this light, the passing of the passenger pigeon was viewed with some sad nostalgia, but such feelings were totally disembodied from any ecological context. The elimination of predators such as wolves was deliberate and ruthless, and was pursued with dispatch within the national parks as well as on private and state-owned lands. This reflected American society's profound hostility toward predators other than ourselves; this hatred of predators is not shared by all cultures, but can be traced back across the Atlantic Ocean to deep-rooted Medieval mythology and biblical iconography (Lopez 1984). This hatred of predators persists despite extensive knowledge of their behavior, their long-term ecosystem functions, and the changes wrought by their elimination.

Of course, much of the damage to biodiversity is done "inadvertently" by human use of land and water resources. The construction of human habitat (cities and suburbs), mining, plowing, and lumbering are a few of the activities that impact on the ability of natural communities to find food, to migrate, and to breed. The drainage of fresh water wetlands, for example, was conducted on a massive scale throughout the United States during the nineteenth and early twentieth centuries. A local example of this was the reclamation of the "drowned lands," a former Pleistocene lake bed that had become a great swamp, with extensive cedar stands and wet meadows, located around the Walkill River in Orange County, New York. These wetlands were cleared, cut, and drained in the mid-nineteenth century by the massive hand labor of Irish immigrants, and have subsequently been maintained as prime agricultural land through the labor of later generations of Polish and German immigrants. Thousands of hours were spent digging ditches, dams, and canals to drain these wetlands; the resulting "black dirt" farms now provide rich yields of onions and sod. Although this reclamation project is still paying rich dividends a century later, it was very controversial at the time. In fact, it triggered the Beaver-Muskrat War, a series of guerrilla battles fought between 1835 and 1871 between proponents and opponents of the conversion. Before drainage, the site had a remarkable abundance of fur-bearers, game birds, and deer; many of the local people made their living as trappers. These "Muskrats" opposed the drainage efforts, and would make nocturnal raids to destroy the dams and canals of the "Beavers." In the end, the great majority of the land was drained, but the bitter feelings lasted for over a generation (Hull 1976; BOCES 1983).

Although the anthropocentric perspective does not recognize intrinsic values in nonhuman species, it does acknowledge the value of other life to humans (Norton 1987). The function of other species in providing ecosystem processes like decomposing wastes, detoxifying poisonous wastes, regulating water supplies, and moderating climate have real economic importance. Many wild species of plants have the potential to provide genetic material for new crop strains and for future biological engineering. Materials extracted from many plants are used today in medicines, and new ones may provide cures for serious diseases. Animals are extensively used in testing drugs, vaccines, and chemicals to further human understanding of disease. Preserving the diversity of life purely for beauty and recreational pleasure are also important. But all of this is based on the value of other species in meeting human needs. With an anthropocentric viewpoint, the reasons for preserving species are purely utilitarian. Anthropocentrism, however, as the only basis for decision-making about human activities as they affect other species may not be adequate for preserving biodiversity.

Other Perspectives

What perhaps was needed for policymakers to fully recognize the biodiversity crisis was the emergence of another vision that could challenge the assumptions of anthropocentrism. The articulation of the nonanthropocentric perspective (Leopold 1949; Taylor 1981; Callicott 1986) and the weakly anthropocentric perspective (Norton 1987) have provided this alternative vision. Nonanthropocentrism (which includes biocentric and ecocentric perspectives) recognizes that species and ecosystems have intrinsic value. With this perspective, nature is viewed as having a value independent of its value to humans. The weakly anthropocentric perspective does not recognize that nonhuman species and ecosystems have intrinsic value, but one of its central tenets is that nonhuman species and ecosystems have both "demand value" and "transformative value" for humans (Norton 1987). The inclusion of transformative values allows for another dimension to be added to species and ecosystems: the value of their contribution to the formation of human ideals (Norton 1987). Indeed, the biophilia hypothesis states that contact and interaction with the natural environment is essential for the full development and well being of humans (Wilson 1993). The political muscle of these alternative perspectives is still not strong when pitted directly against the anthropocentric power of market economics, but nonetheless they have made some dramatic changes in public policy over the last several decades.

Actions to Preserve Species Diversity

Preserving Species

One approach to the preservation of endangered species is to determine which species are in danger of extinction and then to mount programs to protect them (Figure 8.1). The passage of the Endangered Species Act in 1973 was a landmark in this approach to protecting biological diversity. The law requires the U.S. Fish and Wildlife Service to establish lists of *endangered* and *threatened* species. Each listing follows the completion of appropriate scientific studies. A species is *endangered* if it is facing imminent extinction, and *threatened* if it is likely to become endangered in the foreseeable future (Bartel 1987).

(a)

(b)

Figure 8.1. The Haleakala silversword (*Argyroxiphium sandwicense*) is endemic to the Hawaiian Islands (growing in the craters and on the summits of volcanic mountains). It is recognized by the United States as being an endangered species. (a) Close-up of the plant in its rosette stage. (b) Habitat of the Haleakala silversword: Haleakala Crater, Haleakala National Park, Maui.

The Endangered Species Act is limited by a number of factors. First, its focus is on individual species that have already been reduced to a level that threatens their survival. Some species may face inevitable extinction by the time they are depleted sufficiently to get listed. In some cases the genetic variation among the surviving individuals may be insufficient to maintain species vigor. This kind of population bottleneck, with subsequent inbreeding and reduction of fertility, is exemplified by the current status of the cheetah (*Acinonyx jubatus*), which has a very limited genetic variability (Obrien et al. 1983). Second, only a small number of species have sufficient databases at the present time to allow for a determination of endangered status. So in most cases a several year study is required to determine and document if a particular species actually meets the criteria established for being endangered. There have been cases where a species became extinct while this investigation was being undertaken! Third, those species that we know the most about are for the most part the popular ones (i.e., blue whale (*Balaenoptera musculus*), giant panda (*Ailuropoda melanoleuca*), and bald eagle (*Haliaeetus leucocephalus*). There is far less public support for the less "exciting" species (i.e., mosses, beetles, fungi, and nematodes) (Kellert 1986; Randall 1986; Norton 1987). For instance, animals received protection long before plants did in many states; several states have yet to pass legislation that would protect endangered plant species at the state level (McMahan 1987). Thus, a popularity contest of sorts may dictate which species receive protection. This is unfortunate, because every species has an ecological value and deserves protection. Some of the less popular species may actually have far more ecological significance than the more popular ones.

The Endangered Species Act has other limitations as well. Once a list of endangered species is generated, the decision must be made as to which species have higher priority, especially since there is not enough funding to protect all of the species at risk. One could argue that those at highest risk should receive higher priority. Should time and money, however, be spent on species at the very brink of extinction, or would it be more effective to work with less critically endangered species, which have a better chance of survival?

The Endangered Species Act does mandate federal efforts and expenditures for critical habitat preservation, if such actions are required to save a listed species. In some cases, the critical habitat is already protected in parks or refuges, or else the additional areas that need to be protected for this purpose are relatively modest; these set-asides can be accomplished with a minimum of controversy. Certain species, however, require habitat protection that can be established only at great expense, and in the face of opposition from powerful economic and political interests. These situations pose the hardest tests for legislative efforts to save biodiversity; the economic activities that are responsible for habitat destruction may prove to be stronger than the protection offered by the Endangered Species Act.

One major conflict of this kind involved the snail darter (*Percina tanasi*), a small fish native to a few stretches of free-flowing water in upper tributaries of the Tennessee River. Part of its very limited habitat was jeopardized by the proposed construction of Tellico Dam. After a heated debate, Congress granted an exemption from the "do not harm" provision of the Endangered Species Act and authorized the building of the dam.

A similar, but larger, conflict is unfolding over the fate of the northern spotted owl (*Strix occidentalis* ssp. *caurina*) in the old-growth Douglas-fir forests of the Pacific Northwest. These ancient forests have tremendous economic value as lumber, and 90% of them have been logged over since the beginning of the twentieth century. The remaining 10% consists of many fragmented patches of various sizes, mostly located on national forest lands; the majority of these old-growth fragments are included in timber harvest plans over the next several decades. The ancient forests, however, are a unique and disappearing habitat for a number of rare plant and animal species, including several birds, bats, salamanders, lichens, and probably many upper-canopy insects; very few of these species has been extensively studied (Harris 1984). One that has been studied sufficiently for federal listing (as "threatened") is the northern spotted owl (Ehrlich et al. 1992). Each breeding pair requires approximately 800 hectares of old growth forest (Carey et al. 1990). One of three spotted owl subspecies, the northern spotted owl was originally believed to occur only in the old growth forests of Oregon and Washington. There is recent evidence, however, that the northern spotted owl may also be found in a variety of forest types in northern California (Esterbrook 1994).

A Fish and Wildlife Service study, led by biologist Jack W. Thomas, proposed a protection plan for the northern spotted owl that would set aside a pattern of old-growth "blocks," reducing allowable timber harvest considerably. This triggered a storm of protest from the lumber industry, which projected widespread job losses and mill closures if the few remaining old-growth areas were to be protected from logging (Sampson 1990). In the midst of this controversy, Judge W. L. Dwyer (Federal District Court in Seattle) halted sales of new logging concessions on 24 million hectares of national forests in the Pacific Northwest in 1991, determining that such activity was in violation of the Endangered Species Act. In 1993, the Clinton Administration proposed a new protection plan: old-growth forests would be protected on federal lands while restrictions on logging in the vicinity of northern spotted owl nesting sites on state and private lands would be reduced. Although this proposal would provide some protection for a large percentage of the northern spotted owls presently known to occur in the United States (Ehrlich et al. 1992), many scientists believe that the plan falls short of what is needed to ensure the survival of the northern spotted owl (Cushman 1994). In June 1994, Judge Dwyer rescinded the 1991 injunction banning the sale of logging concessions. Hearings on the legality of the Clinton Administration's plan were scheduled to begin in September 1994 (Tokar 1994). The final outcome of this story is still in doubt, although it is clear that unless a significant percentage of old-growth forests are protected, some of the species that are endemic to them will face extinction (Franklin et al. 1981).

There are other treaties and laws that take a species approach. The 1975 Convention on International Trade in Endangered Species of Wild Flora and Fauna (CITES) is administered by the UN Environment Program. It limits certain species (about 675) from being commercially traded as live specimens or wildlife products because they are endangered or threatened. Unfortunately, it is signed by only half the world's nations, and it is not strictly enforced by all of these (Campbell 1987). One of CITES most effective actions has been the 1989 ban on trade in ivory, which has greatly reduced the poaching of elephants. There are some drawbacks, however, to this action. Some nations have populations of elephants that are large enough for "sustainable utilization." Such a program would allow for the generation of revenues to help support nature preserves in Africa and be an effective way to prevent elephant herds from exceeding carrying capacity. Because of the ban, however, any ivory that is obtained from sustainable utilization cannot be sold. An unsuccessful attempt was made to have the ban repealed in 1992 (Begley 1993; Bonner 1993a, b; Kaufman and Franz 1993).

Another approach involves the setting up of wildlife refuges. Started by President Theodore Roosevelt in 1903, the United States had some 437 refuges by 1988. The majority of these are wetlands set aside for migratory waterfowl, however, and over 85% of all the refuge area is found in Alaska. Thus only a limited diversity of habitats and species have been protected by this approach. Also, there are few guidelines for what can occur in the National Wildlife Refuge System. More than half are open to hunting and fishing. Activities such as lumbering, grazing, farming, and mineral exploration and development are common. Many of the large refuge areas in Alaska are threatened by extensive oil and gas development (Watkins 1990), particularly in the wake of the Persian Gulf crisis.

In some cases, protecting a species may involve removing it from its habitat and providing artificial protection (ex situ conservation programs). Zoos, aquariums, and botanical gardens are examples of this approach. Unfortunately, relatively few species are able to be protected by this method, and sometimes the breeding and maintenance programs fail. Gene banks, storage of seed, semen, or even embryos of endangered species are also used. In agriculture, many known and potentially useful varieties of a few crops have been saved in this way. Unfortunately, again only a relatively small number of species are saved. And the success of ex situ conservation programs is dependent upon continued interest and expensive long-term support (Campbell 1987; Ashton 1988; Conway 1988; Dresser 1988).

Although they may still exist in human-made environments, species that have been extirpated from natural ecosystems, or almost extirpated from them, are ecologically extinct. Because they are no longer found in nature, or are present in such small numbers, their ecological function has been lost. The impact of ecological extinction may be quite significant. A single plant species often supports a large number of animals and other organisms. Its disappearance from an ecosystem, or a drastic reduction in its population size, would lead to a major impact on the species that utilized it. Likewise, the extirpation of a large predator such as the wolf from a region represents the loss of a species that had previously helped to control the population size of large herbivores such as deer. Although a species can be saved in a biological sense, it may in fact no longer exist in an ecological sense. Simply saving a species from biological extinction does not necessarily mean that ecosystems are spared from species loss and ecological disruption. Clearly, working only at the species level of protection has its limits, especially if one were to factor in the problem of human greed, corruption, and ignorance (Bonner 1993a, b; Schaller 1993).

Preserving Ecosystems

Many researchers have found that a parallel approach to species preservation may be more effective. This is to focus on the preservation of representatives of all ecosystem types that occur in a given region (Norton 1987; Tear et al. 1993). By such action, it is hoped that living space is preserved for most of the species that naturally occur there. Particular attention is given to the preservation of the rare types of ecosystems, as they provide the unique habitats required by many rare species. Thus, by preserving the rare ecosystems, there is a high probability of protecting a significant proportion of those species that are rare and endangered in that particular region. Using this approach, it would not be necessary to undertake the lengthy and cumbersome process of determining the endangered status of each of the hundreds to thousands of species that occur in each ecosystem (Figure 8.2).

The United Nations Educational, Social, and Cultural Organization (UNESCO) has developed a global program to create parks and reserves in order to protect representative samples of ecological zones. Set up through UNESCO's Man and the Biosphere Program, over 300 "biosphere reserves" had been established in 76 countries by the early 1990s. The New Jersey Pinelands region is one such biosphere reserve (Figure 8.3). The biosphere concept combines both conservation and sustainable development; different zones within a reserve satisfy the goals of habitat protection and sustainable extraction of resources. Unfortunately, the land area set aside as biosphere reserves would have to be at least tripled in order to preserve most of the representative samples of the Earth's ecosystems (Wolf 1988).

Simply establishing biological reserves is only part of the answer. The management of these preserves needs to be addressed. What level of public access will be allowed and how will this be controlled? What ecological management techniques are needed to maintain the ecosystem (fire, mowing, flooding)? How will migrating organisms be able to move from one reserve to another? Where will the funding come from to manage the reserve? Even if a successful management system is developed and implemented, there are other problems that would be difficult to deal with. Contamination by pesticides and other agrochemicals applied to adjacent areas would be hard to avoid. Invasion of the reserve by introduced species is also a

Figure 8.2. Entrance to the Waikamoi Nature Preserve, Maui, a natural area owned and maintained by the Nature Conservancy. Located adjacent to Haleakala National Park, Waikamoi includes remnants of high elevation wet forests (see Fig. 9.3), which provide habitat for many native bird species.

Figure 8.3. A southern white cedar (*Chamaecyparis thyoides*) swamp and associated water-lily (*Nuphar variegata*) wetland in the New Jersey Pine Barrens. Located in southern New Jersey, the New Jersey Pine Barrens region is one of the biosphere reserves recognized by the UNESCO Man and the Biosphere program.

possibility. Water flowing into the area from outside areas may be diverted, or polluted, and thus cause significant degradation. The Everglades of Florida is suffering from this problem (Ehrenfeld 1972; Holloway 1994). Obviously, ecosystems that have been spared from development or other forms of direct human impact have not necessarily escaped the danger of being degraded or destroyed in other ways (Janzen 1986; Campbell 1987; Ryan 1992).

Several other problems would also have to be addressed. One would be the problem of ecosystem isolation—the island effect. Large land masses support greater species diversity than small ones. Studies show that the number of species found in island systems increases approximately to the fourth root of the land area (MacArthur and Wilson 1967). The land area-species relationships found on real islands are useful for understanding the limitations to biodiversity imposed by habitat fragmentation in large mainland ecosystems. For example, the remaining fragments of a once-large tropical rain forest that are now surrounded by a "sea" of pastureland or banana plantations are limited in their potential diversity by their smaller size and isolation from other natural areas, just as "real" islands in the ocean are so limited. Therefore, fragmentation of such rain forests (and any contiguous large ecosystems) poses a grave threat to their biological diversity and ecosystem integrity (Lovejoy et al. 1986; Wilcove et al. 1986). It is not enough to preserve "representative samples" of such natural areas; the preserves must be large enough to allow for the maintenance of the entire ecological community

(Lewin 1984). The critical size and shape required for this purpose varies from one ecosystem to another, but it is generally much larger than was originally thought when parks were set aside simply for "recreational" purposes (Mann and Plummer 1993). For example, many of the national parks in the western United States have suffered significant losses in mammalian diversity since their establishment a century ago. The extent of these losses appears correlated to park size, with "small" parks of 250 to 1,250 km² suffering greater losses than parks of 2,500 to 4,000 km². The largest parks, such as Yellowstone (8,984 km²) suffered the smallest losses; only the giant Jasper-Banff park complex (17,519 km²) in the Canadian Rockies showed no loss of species (Gleick 1986).

One of the major issues at the United Nations Conference on Environment and Development (UNCED) held in Rio de Janeiro in June 1992 was the Convention on Biological Diversity. The primary goal of this treaty was to generate an international agreement to protect the Earth's endangered species and ecosystems. The United States was one of the few nations not to sign the convention. Then-President Bush declined to sign because certain provisions in the treaty mandated that technology relating to biodiversity (particularly biotechnology) be made available to those LDCs that provide the genetic resources. The United States claimed that this would undermine the patent laws that protect its biotechnology industry (Burk et al. 1993). Although still having reservations about certain aspects of the convention, the Clinton Administration signed the Convention on Biological Diversity on 4 June 1993.

Restoring Ecosystems

Given the large losses of forested regions that have already occurred, and the fact that almost half of the Amazon Basin forest destroyed for grazing and cropland is now abandoned, ecological restoration of degraded areas may be a necessary strategy in maintaining biodiversity (Berger 1990). We know from biogeographic theory that smaller isolated areas lose biodiversity quicker than larger areas. The hope is that if fragmented areas between abandoned lands can be reconnected into larger areas, some extinctions may be averted. Research into this approach is promising, but whether regeneration can occur at a rate matching degeneration is an unanswered question (Wolf 1988). Unfortunately, ecosystem restoration is in its infancy, and many question whether humans are currently capable of duplicating functional ecosystems (Holloway 1994).

Among the most highly damaged ecosystems in the world are the tallgrass prairies of the United States, which have been practically eliminated. Most of the tallgrass prairie present today has been developed by restoration programs. Prairie restoration, begun by Aldo Leopold in 1934, has now been accomplished in many places in the midwest, including several hundred acres at the Fermi National Accelerator Laboratory in Batavia, Illinois. The process is not a simple one, however, in that native species must be introduced in such a way as to mimic natural succession (Wolf 1988). Research on prairie restoration may be tested in a proposed National Tallgrass Prairie Preserve in Oklahoma. This 12,145 hectare prairie is now managed by the Nature Conservancy, using prescribed burning and the reintroduction of bison (Hudson 1994).

Managed Nature?

Natural ecosystems did not appear overnight, but have been evolving over a several billion year period. The complex relationships and interactions between producers, consumers, and decomposers represent co-evolution on a major scale. Throughout this time, ecosystem development and maintenance was controlled by environmental constraints. Amazingly, ecosystems developed the capacity to achieve a balance between the needs of life and environmental limitations; long-term sustainable systems were developed.

This situation is now changing. Large scale wilderness areas may soon be largely destroyed, and most remaining natural areas will likely exist in environments where natural conditions do not control their fate. Instead, the surviving natural ecosystems are likely to be preserves, existing as islands in a human-dominated landscape. Under such conditions, it will be humans who largely decide their fate, not natural processes. Our decisions and actions will be major variables in determining how these remnant ecosystems survive and function. As our knowledge about ecosystems and the processes that maintain them is so incomplete, it is unlikely that we will be able to manage ecosystems as well as natural processes have. Thus the more dependent ecosystems are on human management for their survival, the greater the chance that they will experience degradation. Hopefully, this situation will be minimized by setting aside ecosystems large enough to function primarily under natural conditions, and thus require a minimum amount of human management. The preservation of species and ecosystems is pointless if the end result is to simply have token, dysfunctional examples of what once was.

Global Threats to Maintaining Biodiversity

There are several global level environmental problems that could also significantly impact upon the preservation of endangered species and ecosystems. As these problems are beyond the local level of control, even the best local management system would be relatively powerless to deal with them.

One is the impact of acid deposition and other forms of air pollution. Many ecosystems may already be severely damaged by these agents (spruce-fir forests in New England and the Southern Appalachians; northern hardwood forests in Canada), and the only way to prevent the damage is to stop the source of pollution (see Chapter 12, "World Air Pollution: Sources, Impacts, and Solutions"). Another is the reduction of stratospheric ozone, which allows for increasing amounts of ultraviolet light to reach the biosphere. Elevated levels of ultraviolet light would not only be harmful to humans, but would also have the potential to disrupt natural ecosystems (see Chapter 13, "Stratospheric Ozone Depletion"). A third is the impact of climatic changes brought by a major global warming trend (see Chapter 14, "The Challenge of Global Climate Change").

Although this is only a potential threat at the present, global warming could prove to be one of the most significant problems to species preservation if it comes to pass. The mixed deciduous forests of the Southern Appalachians, for example, are not likely to be able to move northward rapidly enough to keep pace with its appropriate climatic zone, which could have shifted up to Ontario, Canada, by the next century. Although these forests have migrated both northward and southward due to past changes in climate, this migration would be more difficult because the change is likely to occur over a very short period of time, according to some climate models (Abrahamson 1989; Peters and Darling 1989; Kolata 1989). The natural mechanisms of forest dispersal, such as wind-blown seed or animal-carried seed, are not likely to be fast enough. In fact, one of the best seed dispersal agents in the entire region is no longer around to help this time: the extinct passenger pigeon. It traveled in huge flocks, dispersing literally billions of seeds in great migratory flights north and south over the continent, before being driven to extinction in the late 1800s by massive overhunting and habitat destruction (Schorger 1973). To meet this challenge, people may have to try a bolder reforestation program than we have ever attempted: the conscious transplanting of an entire forest ecosystem, rather than the usual one- or two-species monocultural plantations. It could possibly be accomplished, but at present we are poorly prepared, in both theory and practice.

Conclusions

Even as our awareness of the biodiversity issue is increasing, the threats continue to multiply. The problems already described are becoming worse: (1) loss of natural habitat due to human uses; (2) loss of locally adapted agricultural systems and their replacement by global-controlled monocultures; (3) the introduction of aggressive exotic species that decimate vulnerable endemic species; (4) species loss due to overhunting; and (5) species loss and ecosystem degradation by pollution. In addition to these, we are facing a new threat: the probable rise of global temperatures due to the increased emission of greenhouse gases into the atmosphere. If global warming proceeds as rapidly as some scientists project, it will lead to major disruptions of most of the Earth's ecosystems.

While solutions can be envisioned that would help to reduce the loss of biodiversity, their effectiveness is tied to a number of global trends, including population growth, development of the less developed countries (LDCs), pressures to increase agricultural output, increasing resource use, and numerous forms of pollution. Unless these issues are resolved, today's stream of extinctions will soon become a mighty river with serious consequences for all life on the planet. In the final analysis, a conscious decision and a significant effort by all nations is needed to preserve the diverse array of life presently found on our planet. Nothing short of this will be effective.

References

Abrahamson, D. (ed.), *The Challenge of Global Warming,* Washington, DC: Island Press, 1989.

Alvarez, L. W., "Mass Extinctions Caused by Large Bolide Impacts," *Physics Today* 40, 1987, pp. 24–33.

Alvarez, L. W. and F. Asaro, "What Caused the Mass Extinction? An Extraterrestrial Impact," *Scientific American* 263(4), 1990, pp. 78–84.

Ashton, P. S., "Conservation of Biological Diversity in Botanical Gardens," *In* E. O. Wilson (ed.), *Biodiversity*, Washington, DC: National Academy Press, 1988, pp. 269–278.

Bakker, R. T., *The Dinosaur Heresies: New Theories Unlocking the Mysteries of the Dinosaurs and their Extinction*, New York: William Morrow & Co., 1986.

Bartel, J. A., "The Federal Listing of Rare and Endangered Plants: What Is Involved and What Does It Mean?," *In* T. S. Elias (ed.), *Conservation and Management of Rare and Endangered Plants*. Sacramento, CA: California Native Plant Society, 1987, pp. 15–22.

Begley, S., "Killed by Kindness," *Newsweek* 12 April 1993, pp. 50–54.

Berger, J., *Environmental Restoration*, Washington, DC: Island Press, 1990.

Biggins, D. and R. Crete, "Black-footed Ferret Recovery," *In* U.S. Forest Service Gen. Tech. Rep. RM-171, 1989, pp. 59–63.

BOCES, *Orange County: A Journey Through Time*, Goshen, NY: Orange-Ulster BOCES Services, 1983.

Bonner, R., "Crying Wolf Over Elephants," *New York Times Magazine*, 7 February 1993a, pp. 16–19 30, 52–53.

Bonner, R., *At the Hand of Man: Peril and Hope for Africa's Wildlife*, New York: Knopf, 1993b.

Burk, D. L., K. Barovsky, and G. H. Monroy, "Biodiversity and Biotechnology," *Science* 260, 1993, pp. 1900–1901.

Callicott, J. B., "On the Intrinsic Value of Nonhuman Species," *In* B. G. Norton (ed.), *The Preservation of Species*, Princeton, NJ: Princeton University Press, 1986, pp. 138–172.

Carey A. B., J. A., Reid, and S. P. Horton, "Spotted Owl Home Range and Habitat Use in Southern Oregon Coast Range," *Journal Wildlife Management* 54, 1990, pp. 11–17.

Campbell, B., *Human Evolution, 3rd ed.*, New York: Aldine, 1985.

Campbell, F. T., "The Potential for Permanent Plant Protection," *In* T. S. Elias (ed.), *Conservation and Management of Rare and Endangered Plants*, Sacramento, CA: California Native Plant Society, 1987, pp. 7–13.

Clark, T. W., D. Hinckley, and T. Rich (eds.), *The Prairie-dog Ecosystem: Managing for Biological Diversity*, Montana BLM Wildlife Tech. Bull. #2, 1989.

Conway, W., "Can Technology Aid Species Preservation?," *In* E. O. Wilson (ed.), *Biodiversity*, Washington, DC: National Academy Press, 1988, pp. 263–268.

Courtillot, V. E., "What Caused the Mass Extinction? A Volcanic Eruption," *Scientific American* 263(4), 1990, pp. 85–92.

Cromack, K., C. C. Delwiche, and D. H. McNabb, "Prospects and Problems of Nitrogen Management Using Symbiotic Nitrogen Fixers," *In* J. Gordon et al. (ed.), *Symbiotic Nitrogen Fixation in the Management of Temperate Forests*, Forest Research Lab, Oregon State Univ., Corvallis, Oregon, 1979.

Cullota, E., "Is Marine Diversity at Risk?," *Science* 263, 1994, pp. 918–920.

Cushman, J. H., Jr., "Owl Issue Tests Reliance on Consensus in Environmentalism," *New York Times*, 6 March 1994, Sec 1 p. 14(N) p. 28(L).

Dicofol: Special Review of Pesticide Products Containing Dicofol, Environmental Protection Agency, Federal Register, March 2, 1984, pp. 10569–10572.

Dresser, B. L., "Cryobiology, Embryo Transfer and Artificial Insemination in *Ex Situ* Animal Conservation Programs," *In* E. O. Wilson (ed.), *Biodiversity,* Washington, DC: National Academy Press, 1988, pp. 296–308.

Easterbrook, G., "The Birds—the Spotted Owl: An Environmental Parable," *The New Republic* 210, 28 March 1994, pp. 22–27.

Ehrenfeld, D. W., *Conserving Life on Earth,* New York: Oxford University Press, 1972.

Ehrlich, P. H., D. S. Dobkin, and D. Wheye, *Birds in Jeopardy,* Stanford, CA: Stanford University Press. 1992.

Eley, T. J. and T. H. Watkins, "In a Sea of Trouble," Wilderness 55, 1991, pp. 18–26.

Franklin, J. F., K. Cromack, W. Denison, A. McKee, C. Maser, J. Sedell, F. Swanson, and G. Juday, *Ecological Characteristics of Old-Growth Douglas-Fir Forests,* USDA Forest Service General Technical Report PNW-118, Forest Service, Pacific Northwest Forest and Range Experiment Station, USDA, Portland, Oregon, 1981.

Franklin, J. F., "Structural and Functional Diversity in Temperate Forests," *In* E. O. Wilson (ed.), *Biodiversity,* Washington, DC: National Academy Press, 1988, pp. 166–175.

Frost, P., "Bald Eagles on the Columbia River: Threats from Persistent Organochlorines," *Journal of Pesticide Reform* 10, 1990, pp. 5–9.

Gleick, J., "Species Vanishing from Many Parks," *New York Times,* 3 February 1986.

Gore, R., "Extinctions," *National Geographic* 175, 1989, pp. 662–699.

Harris, L. D., *The Fragmented Forest,* Chicago: University of Chicago Press, 1984.

Holloway, M., "Nurturing Nature," *Scientific American* 270(4), 1994, pp. 98–108.

Horowitz, H., *Conifer-Shade Growth Interactions on Proposed Brush Control Sites in the Oregon Cascades,* Ph.D. thesis, University of Oregon, 1982.

Hudson, E., "Remnant Grassland Survives in Oklahoma," *High Country News,* 7 March 1994 p. 4.

Hull, R., *People of the Valley: A History of the Town of Warwick, 1700–1976,* Warwick, New York, 1976.

Iltis, H. H., "From Teosinte to Maize: The Catastrophic Sexual Transmutation," *Science* 222, 1983, pp. 886–894.

Iltis, H. H., "Serendipity in the Exploration of Biodiversity," *In* E. O. Wilson (ed.), *Biodiversity,* Washington DC: National Academy Press, 1988, pp. 98–105.

Iltis, H. H., J. F. Doebly, R. Guzman, M. and B. Pazy, "*Zea diploperennis* (Gramineae): A New Teosinte from Mexico," *Science* 203, 1979, pp. 186–188.

Janzen, D. H., "The Eternal External Threat," *In* M. E. Soule (ed.), *Conservation Biology,* Sunderland, MA: Sinauer Associates, 1986, pp. 286–303.

Kaufman, D. G. and C. M. Franz, *Biosphere 2000: Protecting Our Global Environment,* New York: HarperCollins, 1993.

Kellert, S. R., "Social and Perceptual Factors in the Preservation of Animal Species," *In* B. G. Norton (ed.), *The Preservation of Species,* Princeton, NJ: Princeton University Press, 1986, pp. 50–73.

Knoll, A. H., "The Early Evolution of Eukaryotes: A Geologic Perspective," *Science* 256, 1992, pp. 622–627.

Kolata, C., "How Fast Can Trees Migrate?," *Science,* Feb. 10, 1989, pp. 735–737.

Koopowitz, H. and H. Kaye, *Plant Extinction: A Global Crises.* Washington, DC: Stone Wall Press, 1983.

Leopold, A., *A Sand County Almanac,* London: Oxford University Press, 1949.

Lewin, R., "How Big Is Big Enough?," *Science* 225, 1984, pp. 611–612.

Lopez, B., *Of Wolves and Men,* New York: Scribners & Sons, 1984.

Lovejoy, T. E., R. O. Bierregaard, A. B. Rylands, J. R. Malcom, C. E. Quintella, L. H. Harper, K. S. Brown, A. H. Powell, G. V. N. Powell, H. O. R. Shubart, and M. B. Hays, "Edge and Other Effects of Isolation on Amazon Forest Fragments," *In* M. E. Soule (ed.), *Conservation Biology,* Sunderland MA: Sinauer Associates, 1986, pp. 257–285.

Lubchenco, J., A. M. Olson, L. B. Brubaker, S. R. Carpenter, M. M. Holland, S. P. Hubbell, S. A. Levin, J. A. MacMahon, P. A. Matson, J. M. Melillo, H. A. Mooney, C. H. Peterson, H. R. Pulliam, L. A. Real, P. J. Regal, and P. G. Risser, "The Sustainable Biosphere Initiative: An Ecological Research Agenda," *Ecology* 72, 1991, pp. 371–412.

MacArthur, R. H. and E. O. Wilson, *The Theory of Island Biogeography,* Princeton, NJ: Princeton University Press, 1967.

Mann, C. C., "Fire Ants Parlay Their Queens Into a Threat to Biodiversity," *Science* 263, 1994, pp. 1560–1561.

Mann, C. C. and M. L. Plummer, "The High Cost of Biodiversity," *Science* 260, 1993, pp. 1868–1871.

Martin, P. S., "The Discovery of America," *Science* 179, 1973, pp. 969–974.

Maser, C., *The Redesigned Forest,* San Pedro, CA: R. and E. Miles, 1988.

Maser, C., Trappe, J., and R. Nussbaum, "Fungal–Small Mammal Interrelationships with Emphasis on Oregon Coniferous Forests," *Ecology* 59, 1978a, pp. 799–809.

Maser, C., Trappe, J., and R. Nussbaum, *Implications of Small Mammal Mycophagy to the Management of Western Coniferous Forests,* Washington, DC: Wildlife Management Institute, 1978b.

May, R. M., "How Many Species Are There on Earth?," *Science* 241, 1988, pp. 1441–1449.

May, R. M., "How Many Species Inhabit the Earth?," *Scientific American* 267(4), 1992, pp. 42–48.

McMahan, L. R., "Rare Plant Conservation by State Government: Case Studies from the Western United States," *In* T. S. Elias (ed.), *Conservation and Management of Rare and Endangered Plants.* Sacramento, CA: California Native Plant Society, 1987, pp. 23–31.

Myers, N., "Tropical Forests and their Species: Going, Going,. . . ?," *In* E. O. Wilson (ed.), *Biodiversity,* Washington, DC: National Academy Press, 1988, pp. 28–35.

Naeem, S., L. J. Thompson, S. P. Lawler, J. H. Lawton, and R. M. Woodfin, "Declining Biodiversity Can Alter the Performance of Ecosystems, *Nature* 368, 1994, pp. 734–737.

Niles, L. J., K. E. Clark, and S. Paturzo, "Status, Protection and Future Needs of Birds of New Jersey," *In* E. F. Karlin (ed.), *New Jersey's Rare and Endangered Plants and Animals,* Mahwah, NJ: Institute for Environmental Studies, 1989, pp. 69–103.

Norton, B. G., *Why Preserve Natural Variety?,* Princeton, NJ: Princeton University Press, 1987.

Obrien, S., D. E. Wildt, R. Goldman, C. R. Merril, and M. Bush, "The Cheetah is Depauperate in Genetic Variation," *Science* 221, 1983, pp. 459–461.

Perry, D. and M. Amaranthus, "The Plant-Soil Bootstrap: Microorganisms and the Reclamation of Degraded Ecosystems," *In* J. Berger (ed.), *Environmental Restoration.* Washington, DC: Island Press, 1990.

Peters, R. and J. Darling, "The Green House Effect and Nature Reserves," *Bioscience* 35, 1989, pp. 707–717.

Randall, A., "Human Preferences, Economics, and the Preservation of Species," *In* B. G. Norton (ed.), *The Preservation of Species,* Princeton, NJ: Princeton University Press, 1986, pp. 79–109.

Raven, P. H., "Disappearing Species: A Global Tragedy," *Futurist,* October 1985, pp. 8–14.

Raup, D. M., "Diversity Crises in the Geological Past," *In* E. O. Wilson (ed.), *Biodiversity,* Washington, DC: National Academy Press, 1988, pp. 51–57.

Raup, D. M. and J. J. Sepkoski, "Mass Extinctions in the Marine Fossil Record," *Science* 215, 1982, pp. 1501–1503.

Rose, S., "A Streptomycete Antagonist to *Phellinus weirei, Fomes annosis,* and *Phytopthera cinnamoni,*" *Canadian Journal of Microbiology* 26, 1979, pp. 583–587.

Ross, J., "An Aquatic Invader is Running Amok in United States Waterways," *Smithsonian* 24(11), 1994, pp. 40–50.

Ryan, J. C., "Conserving Biological Diversity," *In* L. R. Brown et al. (eds.), *State of the World 1992,* New York: W. W. Norton & Co., 1992, pp. 9–26.

Sampson, N., "Updating the Old-Growth Wars," *American Forests,* December 1990, pp. 17–20.

Schaller, G., *The Last Panda,* Chicago: University of Chicago Press, 1993.

Schorger, A. W., *The Passenger Pigeon,* Norman, OK: University of Oklahoma Press, 1973.

Stringer, C. B., "The Emergence of Modern Humans," *Scientific American* 263(6), 1990, pp. 98–104.

Taylor, P. W., "The Ethics of Respect for Nature," *Environmental Ethics* 3, 1981, pp. 197–218.

Tear, T. H., J. M. Scott, P. H. Hayward, and B. Griffith, "Status and Prospects for Success of the Endangered Species Act: A Look at Restoration Plans," *Science* 262, 1993, pp. 976–977.

Tokar, B., "Between the Loggers and the Owls: The Clinton Northwest Forest Plan," *The Ecologist* 24(4), 1994, pp. 149–153.

Vermeij, G. J., "The Biology of Human-Caused Extinction," *In* B. G. Norton (ed.), *The Preservation of Species,* Princeton, NJ: Princeton University Press, 1986, pp. 28–49.

Vitousek, P. M., P. R. Ehrlich, A. H. Ehrlich, and P. A. Matson, "Human Appropriation of the Products of Photosynthesis," *Bioscience* 36, 1986, pp. 368–373.

Watkins, T. H., "The Perils of Experience," *Wilderness* 54, 1990, pp. 22–84.

Wilcove, D. S., C. H. McLellan, and A. P. Dobson, "Habitat Fragmentation in the Temperate Zone," *In* M. E. Soule (ed.), *Conservation Biology,* Sunderland, MA: Sinauer Associates, 1986, pp. 237–256.

Wilson, E. O., "The Current State of Biological Diversity," *In* E. O. Wilson (ed.), *Biodiversity,* Washington, DC: National Academy Press, 1988, pp. 3-18.

Wilson, E. O., "Threats to Biodiversity," *Scientific American,* 261(3), 1989, pp. 108–116.

Wilson, E. O., *The Diversity of Life,* Cambridge, MA: Belknap Press of Harvard University Press, 1992.

Wilson, E. O., "Biophilia and the Conservation Ethic," *In* S. R. Kellert and E. O. Wilson (eds.), *The Biophilia Hypothesis,* Washington. D.C.: Island Press, 1993, pp. 31–41.

Wolf, E. C., "Avoiding a Mass Extinction of Species," *In* L. R. Brown et al. (eds.), *State of the World 1988,* New York: W. W. Norton and Company, 1988, pp. 101–117.

Chapter 9

Tropical Forests: Problems and Prospects

Howard Horowitz, Eric F. Karlin, and Barbara Bramble

Tropical forests represent a major component of the ecological systems of our planet, having a surprisingly high concentration of biological activity. Until relatively recent times, tropical forests covered 16% of the Earth's land surface, contributed 29% of the total global net primary production, and stored 56% of the world's biomass (Whittaker 1975). Foremost among the diverse array of forest types that are found in the tropics are the wet forests, which include the rain forests (Figure 9.1). These broadleaf evergreen forests occur in areas where rain is plentiful throughout the year and are the world's most diverse ecosystems, perhaps containing over 50% of all species on Earth (Myers 1986; Wilson 1989, 1992). Seasonal forests, where deciduous trees are dominant, are also prominent, occurring in regions that experience annual dry seasons (Figure 9.2). Other tropical forest types include cloud forests (Figure 9.3) and thornwoods (Kricher 1989).

Unfortunately, recent human activities have led to a 50% reduction in tropical forests—an extraordinary percentage (Figures 9.4 and 9.5a) (Wilson 1989; Terborgh 1992a). Estimates of the rate of tropical deforestation vary widely, in part because remote sensing technology is not sufficiently advanced to differentiate between various types of similar ecosystems. The United Nations Food and Agricultural Organization (FAO) estimates that an average of 154,000 km² (38 million acres) of tropical forests of all types were destroyed annually during the 1980s, 36% greater than the 113,000 km² (27.9 million acres) lost each year during the 1970s (Aldhous 1993). Currently, between 140,000 and 200,000 km² (35–50 million acres) of wet tropical forest are being destroyed annually, representing a deforestation rate of 1.8 to 2.5% a year (Terborgh 1992a). This rate is at least twice the rate of deforestation of tropical wet forest that existed in 1980 (Myers 1986; Wilson 1989; Terborgh 1992a). Given the current trends, it is expected that 90% of the remaining primary tropical wet forests will be destroyed by the year 2025 (Terborgh 1992a). Remaining tropical wet forests are concentrated in the Amazon Basin, primarily in Brazil, in central Africa around the Congo River Basin, in Southeastern Asia, and in Indonesia (Myers 1986). Central America, Malaysia, and the Philippines

(a)   (b)

Figure 9.1. Tropical wet forests. (a) Lowland wet forest in Corcovado National Park, Costa Rica; (b) Montane wet forest in Volcanoes National Park, Hawaii. Note the extensive subcanopy formed by tree ferns (*Cibotium* sp.).

have lost most of their forests, and only 1% of the once immense Atlantic coast forest of Brazil now remains (Wilson 1988). Even forested areas that are not logged are impacted, with fragmentation and edge effects causing extensive ecosystem degradation (Skole and Tucker 1993).

Deforestation was one of the major issues discussed at the United Nations Conference on Environment and Development (UNCED) held in Rio de Janeiro in June 1992. It was hoped that a legally binding treaty could be agreed upon that would limit unnecessary destruction of forests. The less developed countries (LDCs) argued that the treaty was focused on tropical forests and that it did not adequately address the deforestation that was taking place in the more developed countries (MDCs), particularly the loss of old-growth forests in Canada and the United States (Figure 9.5b). Unfortunately, a compromise could not be reached, and in the end all that was accomplished was the signing of a weak statement that was not legally binding (Raven et al. 1993).

One of the most impressive features about tropical wet forests is their high species diversity and complex structure (Figures 9.1–9.3). Many plants grow on the trunks and branches of trees as epiphytes, never coming into contact with the ground (Figure 9.6), and many animals spend most of their lives in the habitat pro-

Figure 9.2. Tropical seasonal forest at Santa Rosa, Costa Rica. The picture was taken in early May at the end of the dry season.

vided by the extensive tree canopies. These complex ecosystems are difficult, if not impossible, to fully comprehend and study. For instance, there are up to 300 different tree species per hectare in a tropical wet forest (Wilson 1989), as opposed to the 10 to 15 tree species per hectare typically found in temperate deciduous forests. Instead of many individuals of the same species growing in proximity to one another, which is common in temperate forests, individuals of many tropical tree species tend to be highly dispersed. Often only a few individuals of any particular tree species will occur per hectare. Another complicating aspect is that only a small fraction of the species living in tropical forests have been identified; most are unknown to science. Literally millions of undiscovered species (mostly in the phylum Arthropoda) are believed to exist in tropical forests (Myers 1986, 1988; May 1988; Wilson 1989).

Another difference between temperate and tropical forests is the nutrient status of the soil. The soils of many tropical forests are very infertile: the bulk of the mineral nutrients are found in the biomass component of the ecosystem and not in the soil. A very finely tuned system of mineral recycling has evolved in these forests, where minerals are picked up by the roots of plants and by fungi almost as quickly as they are released by decomposing tissues. In most temperate forests the situation is reversed, with the soils containing most of the mineral nutrients. When trees are removed from tropical wet forests in timbering operations, a large proportion of the ecosystem's nutrient minerals are lost. A large percentage of the minerals may also be lost when extensive tracts of forest are cut and burned. This is because of the intense leaching that occurs in the absence of a healthy ecosystem, which would normally capture and retain the released minerals in the bodies of the plants and animals. Deforestation in the

Figure 9.3. Tropical cloud forest in the Nature Conservancy's Waikamoi Nature Preserve, Maui, Hawaii. Note the stunted and gnarled appearance of the trees.

tropics has the potential for much greater impacts on the ecosystems, especially their ability to grow back, than is the case for temperate regions.

Activities Affecting Deforestation

The causes of this deforestation are rooted in the history of development and poverty in the Third World. The most destructive agent in tropical deforestation is the small-scale farmer, who accounts for about half of the forest destruction and degradation (Myers 1986). Commercial logging is the second most significant factor, accounting for almost one-fourth. Nonsustainable fuelwood gathering, cattle ranching, and hydro-development also play a major role (Myers 1986; see also Chapter 7, "Hydro-Development"). Often the causes of deforestation are intimately linked with social, political, and economic conditions within a country.

Population growth is contributing mightily to the loss of tropical forests; an ever-increasing number of people have the same desperate need to find a place to live and grow food. As land becomes degraded, these people move on to new forest areas and begin the cycle of degradation again. In many developing countries, fuel for cooking and heating is primarily fulfilled by wood. Wood gathering for home fuel use can cause rapid deforestation, especially in areas with seasonal forests.

In much of Central America, the Philippines, and elsewhere, good farmland, comprising a relatively small percentage of all the land, is in the hands of a very few wealthy people. Much of this land is underutilized—either held for speculation

(a)

(b)

Figure 9.4. The loss of tropical forests in the Talamanca region of southern Costa Rica. (a) Intact forests. (b) Extensive deforested areas.

or used for cattle ranching. This use is unproductive in the long run and results in low employment benefits. Some of the best land is used for export crops that mostly benefit the rich landowners. These arable lands are not available for people

(a)

(b)

Figure 9.5. Deforestation in the United States and Canada. (a) Deforested lower slopes of Haleakala Volcano, Maui, Hawaii. (b) Extensive clear cutting in the Pacific Northwest forests of British Columbia, Canada.

to grow food and live on, so the poor are driven into the less fertile forest lands instead. In many countries, it is the combination of population growth and the badly distributed land tenure that is causing significant deforestation.

Clearing of forest land in order to grow drug plants has also contributed to deforestation, especially in the montane forests of the Andes Mountains of South America. Ten percent of the deforestation that has occurred in Peru this century was done to make fields for growing coca, and it appears that the opium poppy is now also being grown there (Goodman 1993).

Unfortunately, badly conceived attempts to improve conditions are leading to further deforestation. Large projects such as dams, mines, and colonization of forest regions are funded by outside interests such as banks and aid agencies. There are also local initiatives, including commercial timber operations and cattle ranching schemes, which are aided by subsidies and tax incentives within a number of countries. Often these activities cause significant deforestation.

Consequences of Deforestation

The loss of tropical forests has many devastating impacts, and some of them are at the global level. The most significant impacts are the loss of biodiversity, loss of soil and soil productivity, flooding, and potential climate change.

Biodiversity Loss

The high biodiversity of tropical wet forests is not only important ecologically, it also has great potential to benefit humans. The potential for the discovery and development of new crop plants or new, more resistant varieties of existing crop species is quite high. Gums, oils, and dyes from tropical plants presently have many specific applications in industry. How many more are yet to be discovered? Some species may have chemicals that could serve as the base for a new generation of natural pesticides. The potential for the discovery of new medicines is also quite high. One promising example is epibatidine, a chemical obtained from a poison arrow frog (*Epipedobates tricolor*) native to Ecuador. This recently discovered chemical has been found to be 200 times more effective than morphine at blocking pain. Although too powerful for use on humans, it is hoped that it will serve as a model for the development of a new class of pain killers (Bradley 1993). Many drugs have already been obtained from tropical plants, including quinine (for malaria), diosgenin (used in the

Figure 9.6. Bromeliads (epiphytes in the pineapple family) growing on a tree limb at Lankaster Garden, Costa Rica.

synthesis of progesterone, testosterone, and cortisone), reserpine (for hypertension), and anti-leukemia drugs (Koopowitz and Kaye 1983). Unfortunately, the rapid destruction of tropical forests and the accompanying extinction of the associated species threatens this potential source of new medicines. Close to one half of the species on this planet may be driven to extinction or become highly endangered if all, or even most, of the tropical forests are destroyed (Raven 1988). This catastrophic loss of species represents one of the most significant environmental problems confronting humanity (see Chapter 8, "The Loss of Biodiversity").

Besides the local loss of biodiversity, the destruction of tropical forests also has an impact on biodiversity outside the tropics. Over the past several decades there has been a major decline in the populations of many migratory birds, with some species being threatened with extinction. For instance, Rock Creek Park (Washington, DC) has experienced an almost 90% drop (between 1940 and the 1970s) in the number of long distance migrant species that nest there. Another numbing statistic is that the number of migrating flights over the Gulf Coast of the United States has dropped 50% in the last 30 years. Two major variables that affect the populations of these birds is the loss of breeding/nesting habitats in temperate regions and the loss of winter habitats in the tropical forests. Over 250 species of birds that breed in the United States over winter in tropical and subtropical regions. Clearly what happens in the tropics will have a dramatic impact on the migratory bird populations in North America (Terborgh 1992b).

Loss of Soils and Resulting Damage

On a more regional basis, the loss of forests leads to soil erosion, resulting in reduced productivity and ultimately in the abandonment of farmland. In some cases, hydroelectric power and irrigation projects located in forested areas have attracted migrant farmers, who have followed the development roads and settled in the watershed areas, clearing them for subsistence crops. The water projects then become rapidly filled in by sediments eroded off the adjacent denuded hillsides (Goldsmith and Hildyard 1986). It is a real economic disaster for a country to lose the benefits of these expensive power investments because of failure to protect adjacent ecosystems. Coastal wetlands and critical fish nursery habitats are also being choked by soil washing down from the upland streams. Excessive sedimentation also results in a deterioration of the coral reef habitats offshore. For many countries, healthy coastal wetlands and coral reefs could be a tremendous source of tourist income and fisheries production. Yet, upland deforestation is endangering this potential.

Loss of trees also exacerbates wind-induced soil erosion, especially in areas that experience seasonal drought. This, in turn, has an immediate and harmful impact on agriculture in regions that have been deforested (Dover and Talbot 1987; see also Chapter 10, "Desertification"). In such areas, one of the best ways to improve agricultural yield is to plant trees. This may be counterintuitive to a farmer, who wants open land for crops. Although the planting of trees to minimize soil erosion takes some of the land out of crop production, this loss is more than offset by the increased yields made possible by intact soils.

Effects on the Hydrologic Cycle and Local Climate

Deforestation affects regional hydrology in many ways, especially in tropical regions where heavy rainfall is seasonal (Nepstad et al. 1994). Floods are often more frequent and devastating in deforested areas, especially in the wet season. When deforestation results in increased amounts of runoff, a smaller percentage of the precipitation enters the groundwater system. The decline in groundwater recharge, in turn, results in less groundwater being available for use during the dry season.

Forests have also long been recognized for their ability to moderate the climate on a local and regional basis. Forests transpire vast quantities of water and provide natural air conditioning. Regional rainfall patterns are often directly related to the ability of forests to provide moisture to the atmosphere. Some of the drought patterns, particularly in Africa, may in fact be caused by deforestation (Brown and Wolf 1985; Timberlake 1985; see also Chapter 15, "Predicting Climate Change: The State of the Art").

Forest Loss and Global Climate Change

Finally, what are the consequences of tropical forest loss on atmospheric carbon dioxide? Although a large percentage of carbon dioxide that is added to the atmosphere comes from the burning of fossil fuels, it is becoming clear that the second most important source of carbon dioxide increase in the atmosphere is the burning of tropical forests (Flavin 1989). In the temperate regions, total forest biomass is more or less stable. Deforestation is in balance with reforestation. We may not like the mix of trees that are being replanted, or particular regional trends, such as the destruction of the ancient forests of the Pacific Northwest. But overall forest cover is relatively stable in Europe, the Soviet Union, Canada, and the United States. By contrast, deforestation significantly outpaces reforestation in the tropics, where the extent of tropical forests is currently decreasing by about 2% each year (Terborgh 1992a). This massive deforestation results in a net flux of carbon out of tropical ecosystems and into the atmosphere, causing a significant increase in atmospheric carbon dioxide (Houghton et al. 1983; Dixon et al. 1994). This, in turn, contributes to the greenhouse effect.

Alternatives to Deforestation

What can be done to change destructive patterns of tropical forest use? We are beginning to see major changes because of the interest of environmental groups on these topics in the United States. For example, while the National Wildlife Federation has been a leader on a number of domestic issues, only in the last several years has it started to pay attention to the consequences of tropical rain forest destruction. The organization began by looking at why forests are being lost, and the actions and needs of people that are creating this destruction. One of the items that has been discussed among a number of governments and conservation groups is alternatives for timber production. There is clearly a demand for wood; the United

States is one of the biggest importers of tropical hardwoods. But there are ways of producing timber on lands that have already been degraded without having continuous incursions into the primary tropical forests, which are so important for biological diversity. Natural forest management, replanting native tree plantations in degraded areas, and setting up "extractive reserves" are all potentially viable alternatives to current destructive forestry practices. In addition, rethinking the need and rationale for other damaging activities, such as construction of hydroelectric dams, cattle ranching, and colonization projects, can help to limit the destruction to tropical forest ecosystems.

Extractive Reserves

There are other related alternatives that encompass shifting cultivation. We are learning about some very interesting alternatives from people who actually live in tropical forests and harvest products produced by those forests on a sustainable basis. One of the very exciting proposals being implemented is a new land use method called an *extractive reserve* (Allegretti 1990; Holloway 1993; see Chapter 5, "Appropriate Technology for Sustainable Development"). These reserves are like national parks or national forests, except that people live in and extract products from the forest. These products, which include palms, nuts, rubber, ornamental leaves and flowers, and seeds of exotic plants, can be harvested without harming the forest if environmentally sensitive methods are used. Because the people who live on and work these lands know how much they depend on the forest, they have a strong incentive to ensure that the forest remains intact. The extractive reserves will be established by law and should be based upon management plans agreed to by the local people, who will have long-term "use concessions" but will not own the land.

One of the most fully developed plans for sustainable development of harvestable resources within the context of a tropical rain forest is the proposal for extractive reserves in the State of Acre in the Brazilian Amazon. The area has been the scene of prolonged conflict between four groups of people—indigenous cultures, rubber tappers, impoverished migrant settlers, and cattle barons—for several decades; each of these groups has had a different agenda for the region's future. With major improvements in the road network between adjacent Rondonia and Acre, there has been an influx of settlers, and the physical domain of the cattle ranches has increased dramatically. These activities threaten both the native peoples and the independent community of rubber tappers (*seringueiros*). The *seringueiros* have been engaged in extractive commodity production on forest lands where they have been living for several generations, without having legal title to the land they live and work on. They formed a trade union in the 1970s, and became politically active in opposition to forest clearance by the cattle ranchers. In 1985 the rubber tappers, led by Chico Mendes, proposed a detailed plan for the operation of extractive reserves in Acre. This would provide legal protection to the sustainable harvest of rubber, Brazil nuts, rattan, and other forest products (Mendes 1990). The rubber tappers have overcome a historical tradition of hostile relations with the indigenous communities to establish a strategic alliance intended to protect the rain forest from massive conversion to cattle ranches. The explosive nature of these conflicts culminated in the assassination of Mendes by ranchers in December 1989, after four previous attempts had failed. This tragedy brought international attention to the social conflicts within the region, but has not stopped the rubber tappers union from pressing forward with its plan for extractive reserves. Indeed, the Chico Mendes extractive reserve was recently established by the Brazilian government

Another example of extractive reserves, drawn from another continent, involves the traditional utilization of the rain forests around Mount Oku in Cameroon. These forests provide a variety of goods to the villagers who inhabit the slopes around the base of the mountain, including cash from the marketing of medicinal bark, honey, and specialty woodcraft products. These activities conserve the forest canopy and do not threaten the rare birds and mammals that are found only in the rain forest. The key economic activity is the sale of tree bark, from *Pygaeum africanum,* used to produce a drug marketed by international pharmaceutical firms in the treatment of prostate enlargement. Bark collection used to be destructive to the trees, but now is regulated to require techniques that do not kill the tree (by leaving vertical strips of bark intact on opposite sides of the trunk). A rapidly growing village cooperative gathers and markets honey, and also sells a variety of locally produced wooden craft items. The Mount Oku region has been threatened by road construction projects and an influx of forest-clearing migrants, and the efforts to protect the extractive reserve are now receiving support from the Missouri Botanical Garden, World Wildlife Fund, and US-AID (Gradwohl 1988).

Unfortunately, while extractive reserves may offer some solutions to tropical deforestation, they also have many limitations: in many cases, they may not be economically or ecologically viable on a large scale (Terborgh 1992a; Alper 1993). Economically, the gathering life style is a marginal one and involves large areas of forest to support small family groups. Countries with rapidly expanding populations may find it hard to justify the use of land in this way. In addition, many tropical forest products have a very limited market, often limited to the country or region of origin. They are not attractive for sale at the international level. If a significant market were developed for a particular product, then a number of issues would have to be confronted: (1) the need to balance increased harvesting/availability of a product with the price people are willing to pay for it, (2) the need to limit overharvesting as forest products gain in value, and (3) the need to ensure that these remote areas will not be damaged or further developed as accessibility is increased by road construction and other forms of transportation.

Medicinal Plants

Although fortunes have been made from the many medicinal chemicals that have been derived from tropical plants, only a small fraction of these profits have been returned to the countries where these plants naturally occur. A recent agreement between Merck and the Costa Rican government represents one way to help rectify this situation. Merck will pay 2 million dollars over a 2 year period to the

Costa Rican National Institute of Biodiversity (INBio) for the right to explore for new medicinal plants in Costa Rica. If medicinal plants are discovered and marketed, Merck will also pay INBio a share of the royalties. Natural ecosystems also gain from this arrangement, as 10% of the 2 million dollars and 50% of its share of the royalties must be used by Costa Rica to support biological conservation (Roberts 1992).

Natural Forest Management

The notion of extractive reserves is not readily applicable to many tropical forest systems around the world, because they do not all have a sufficient number and variety of extractive products for which markets exist. Virtually all tropical forests, however, could be managed sustainably for harvests of wood, as well as for so-called minor forest products (some of which are not so minor). Many conservationists are working for the implementation of "natural forest management," the goal of which is to harvest timber in a sustainable manner from naturally occurring forests (Terborgh 1992a). Natural forest management techniques vary according to the dictates of each specific ecosystem, but all are preferable to the prevailing method of managing tropical forests: harvesting the most valuable species, damaging many of the unwanted trees in the process, and then abandoning the land without replanting. With standard logging methods, about 50% of the tree canopy is destroyed in order to harvest 10% of the trees from a given area of tropical wet forest, with another 10 to 20% of the canopy being removed to make way for roads and skid trails (Alper 1993). In many cases, the damaged forests are soon completely destroyed by conversion of the land to agricultural uses, cattle ranches, or monocultures of exotic tree species.

One way to minimize damage to tropical forests is to sever the lianas that bind the tree canopies together before harvesting the trees. This way when the tree falls it is less likely to pull down other trees. By also preplanning where to place roads and skid trails, damage to the forest canopy can be reduced by 50% (Alper 1993). Another method being tested is called *strip cutting* (Hartshorn 1990). It attempts to mimic the size of cut that a natural tree-fall event might cause. A large tree in a tropical forest is hooked into many other trees, vines, and interspersed growth. When a large tree falls, it causes a fairly sizable gap in the forest; up to 50 meters in diameter, perhaps more in some cases. If one cuts a strip of forest along a contour line no more than 50 meters wide, one has an area that is very conducive to reforestation in a natural way. The seeds that fall from trees adjacent to the opening can reach and reforest even the center of the strip. Thus, the cutover area does not need to be limited to a circular spot. Narrow strips are expected to revegetate themselves within about 15 years. This never produces the dried out, eroded, and gullied bare ground that is the typical legacy of large clear-cuts. Well designed strip cutting protects the fragile soils, leaving the potential for growth of a productive forest again in the near future.

Gradwohl (1988) and Terborgh (1992a) discuss a number of natural forest management systems that have been implemented in the tropics. These include the Malaysian Uniform System, which is based on the biology of the major family of tree species present in Southeast Asia (Dipterocarpaceae), the tropical shelterwood system developed for West Africa by British foresters, the CELOS Silvicultural System developed by Dutch foresters for New World tropical forests in Suriname, and the Palcazu Management System developed for Peru. Although based on extensive planning and research, these natural forest management systems have all failed (Terborgh 1992a). In some cases the failure was due to technical deficiencies, but the fundamental weakness has been the instability of the political, economic, and social conditions of the countries in which the projects were located. The activities of the Shining Path guerrillas forced the USAID mission to abandon the Palcazu project in Peru; a military coup led to the termination of the CELOS project in Suriname; plantations of rubber trees and oil palms proved to be more economically attractive, at least in the short term, than natural forest management in Malayasia (Terborgh 1992a). It is tragic that not one of these systems lasted through one harvest cycle, especially because it implies that long-term planning relating to sustainable forestry practices may not easily be achieved in tropical countries. Yet natural forest management is central to the long-term preservation of tropical forests. Obviously it is not enough to simply focus on the scientific answers to environmental problems, but the social and political aspects must be addressed as well.

Natural forest management may be more viable in the tropics when it is based at the grass-roots level (Alper 1993). The indigenous people living in tropical forests around the world find themselves in imminent peril from the expansion of agriculture and logging activities. In literally hundreds of places throughout the tropics, they are responding to this threat by establishing formal organizations to promote sustained-yield management programs that provide legal protection to their lands, as well as income to their communities. In a book published by Cultural Survival, Inc., an organization devoted to documenting the situations of indigenous peoples around the world, Clay (1988) identified many sustainable-forestry programs. A typical example is CRIC, an organization of 56 Indian community councils in Columbia that has developed reforestation and timber management programs based entirely on the maintenance and replanting of native tree species.

Reforestation: Exotic Species Versus Native Species

A potential use for degraded tropical regions is to reforest damaged areas by developing ecologically based tree plantations (Nepstad et al. 1990). Focusing on native species of trees for wood, on fruit trees for cash income, and on leguminous plants for soil enrichment may well provide the basis for sustainable tree plantations. There is clearly a great deal to be learned, however, about how to grow trees for continuous harvest on short-term cycles, especially in the tropics. Multinational companies are trying to achieve this, but most of the studies show extremely low yields, probably because of the infertile soil. High rainfall in many tropical regions contributes to nutrient loss, as the minerals are quickly dissolved and washed out of the soils. Without investment in expensive fertilizer, the first cycle of trees grows well, the second less well, and the third hardly grows at all.

Although foresters have long engaged in planting monocultures of fast-growing exotic trees, there are many problems with the practice. To begin with, they are alien to the ecosystem; they have not co-evolved with associated plants, animals, mycorrhizae, and other soil organisms; and they have no natural enemies or symbionts. Exotic monoculture plantations are biologically rather sterile in comparison to the vibrant diversity of native forest ecosystems. Furthermore, the rapid growth of lumber in these "industrial" plantations may not be dependable in the long run; success for one or two rotations could be followed by catastrophic blight in the future.

For these reasons, ecologists are reluctant to endorse the widespread planting of exotic species. In fact, recent federal policy calls for their eradication from the National Parks within the United States. Industrial forestry in tropical areas, however, is still based primarily on a narrow range of "high-yielding" exotic species. The Tropical Forest Action Plan, endorsed in 1985 by an international task force convened by the World Resources Institute, the World Bank, and the U.N. Development Program, includes a major section that calls for a worldwide expansion of industrial forestry based on the planting of exotic monocultures (W.R.I. etc. 1985).

Despite the widespread preference for exotic species with already established markets, there are many valuable lumber trees growing among the hundreds of native species found locally in each rain forest region. There are also rapidly growing fuelwood and forage trees that are specific to each region that could be utilized in preference to large-scale monocultures of exotic species. One of the most important tasks of sustainable forestry research is to accumulate the knowledge about how to identify and grow the most valuable native species in each region, and how to develop markets for their products.

In Costa Rica, for example, nearly all of the tree species that are subsidized for the purpose of encouraging reforestation are exotics, including eucalyptus, teak, pine, and gmelina. These tax policies result in the "reforestation" of timber stands that are totally lacking in the unique characteristics of Costa Rica's native tropical forests. There is some good news, however; the Organization of Tropical Studies (OTS), with financial backing from the MacArthur Foundation, has established a series of replicated-block planting trials at their nursery site in the La Selva Reserve to help determine which native tree species have a good potential for lumber or for agroforestry uses such as fuelwood. The trials have involved 94 species to date, and about 10 of them are regarded as promising. Several of these species, from the genus *Vochysia,* may provide significant additional benefits besides fast-growing, high-quality hardwood. They have the ability to draw aluminum out of the soil in substantial quantities and incorporate it into the tree's biomass. Since excessive aluminum is a major problem in many tropical soils, this feature may enhance the soil quality for the benefit of other trees as well. It is to be hoped that future Costa Rican forest tax policy will change to encourage more widespread use of beneficial native species.

Other Uses of Degraded Land

While plantation management with native tree species offers considerable promise in regenerating the forests, there are other potential uses of degraded lands that may be desirable (Holloway 1993). Two such uses, small-scale sustainable agriculture and community reforestation, come to mind. Unfortunately, only 5% of the tropical forest soils that have been surveyed in Indonesia and Brazil are fertile enough for sustainable agriculture. Most of the soils are quite barren once the trees are removed. They offer little promise for agriculture unless a lot of money is available for fertilizer, which is rarely the case in a developing country's agricultural scheme.

Developing community forests may be a more promising activity. In several countries there are specific community forests that are being planted to meet the needs for wood fuel, timber, fodder, fruits and nuts, small poles to build houses, and other uses (Case Studies . . . 1987; Sanwal 1988). The advantage of the community forest is that the people who are going to use and care for the forest actually control and receive income from the land, and so have incentive for good stewardship.

Ecological Reserves

Representative areas of primary tropical forest should also be set aside as protected natural areas where human activities would be highly limited. Such reserves would allow for the preservation of these natural ecosystems and the species that live in them. The formal designation of forest reserves, however, even if done with fanfare by centralized governments, does not by itself guarantee that the reserve areas are actually protected from exploitation and development (see Chapter 8, "The Loss of Biodiversity"). Far too often, areas protected on maps do not show much evidence of protection when visited in the field.

A prime example of this is found in the highlands of Chiapas, Mexico, where the pressure of too many people on too little land, combined with contradictory policies of different branches of government, is undermining the integrity of the El Triunfo Reserve. One hundred and twenty square miles (30,000 hectares) of magnificent mist-shrouded cloud forest, encompassing the crest of the Sierra Madre of Chiapas, was set aside by the government in 1971. Little money was provided for actually managing or even guarding the preserve, however, and so its lower slope margins have become a patchwork of slash-and-burn clearings by migrant settlers. A different government agency subsequently divided portions of the land within the boundary of El Triunfo for *ejidos*—land grants for permanent use by peasant agricultural communities. Las Palmas, for example, is an ejido of 400 families located entirely within the boundaries of the reserve. In addition, $100 million of World Bank-supported projects—including bridges, roads, and flood-control structures—are planned for the Pacific-facing slopes below the forest reserve, with no provisions for buffer zones or other ecological considerations. The core area of El Triunfo is still a vibrant cloud forest, with jaguars, toucans, and tree ferns. In the absence of an integrated watershed plan that recognizes the needs of both people and ecosystem, however, the future of El Triunfo Reserve is far from secure (Cloud 1988).

Ecotourism

With growing popular awareness of the complexity and beauty of tropical forest ecosystems, they are an increasingly attractive destination for visitors from the United States, Europe, and other affluent temperate places. The governments of tropical nations are eager to encourage such travel, as a means to derive economic benefits from the preservation of wet forests and wildlife species (Miller and Tangley 1991). The nations that have led the way in promoting "ecotourism" have been Costa Rica in Latin America and Kenya in Africa, although other nations such as Belize, Rwanda, and Nepal have also made efforts to attract ecotourists.

The establishment of national parks and wildlife preserves is the first step in ecotourism development. The money generated by the tourists can be used to pay for the management of the parks, as well as to provide income for the local population. The parks alone are not enough for successful ecotourism; two additional prerequisites are an adequate infrastructure to accommodate the visitors with at least a modest degree of comfort, and sufficient political stability to allay fears of violence that would otherwise keep travelers away. The necessary infrastructure is likely to include a mixture of public and private facilities, ranging from field stations for scientists and students to hotels, shops, and restaurants for tourists to sleep in and spend their money in. Costa Rica's outstanding success with ecotourism is attributable to its strength in all of these areas: it has superb national parks, a good infrastructure of roads and facilities to meet tourist needs, and a long tradition of political stability. By contrast, outbreaks of political violence, such as those that occurred in southern Mexico and in Rwanda in early 1994, are devastating to efforts to build an ecotourism industry.

Even where ecotourism thrives, it is not free of drawbacks. Increased human usage of natural areas may lead to degradation, both directly (via trampling of plants and disturbance of wildlife) and indirectly (more garbage, sewage, and forest cutting to provide wood for the cooking and heating needs of tourists). In some cases, a large percentage of the money generated from ecotourism ends up in the hands of foreign investors, with very little cash trickling down to the local population.

Despite these problems, ecotourism can be a tremendous economic asset as well as a force for ecosystem preservation. Many tropical nations, such as Bolivia and Jamaica, are now trying to establish ambitious national park systems. Few of them, however, have yet developed the infrastructure needed to generate large-scale success with ecotourism (Eyre 1990).

Protecting Tropical Forests from Unsuitable Activities

Hydropower has been considered to be a major option for the production of energy in many LDCs (see Chapter 7, "Hydro-Development"). Many proposed hydro-development sites, however, are located in regions of flat terrain. If developed, they would flood very large areas and yet produce only a small amount of power. This was the case with Brazil's recently completed Balbina Dam. This project has proven very uneconomical because the water flow and power produced are not as high as expected (Sternberg 1985). If hydropower is not rational in some areas, what are the alternatives? Ironically, energy conservation and investments in end-use efficiency are, at the moment, the very best alternatives from an economic point of view. Goldemberg et al. (1987) and Geller et al. (1988) both found that Brazil, which is planning to invest billions of dollars in hydroelectric power over the next 20 years, could provide the needed power by investing in efficiency instead. Indeed, they could save $44 billion worth of investment in energy facilities with an investment of only $10 billion in end-use efficiency.

Cattle ranching is, in Central America and Brazil at least, a significant cause of tropical forest loss (Leonard 1988). Ranching has been encouraged by tax incentives and low-cost credit structures established by the national governments. There is absolutely no physical need for cattle to be grown on large, extensive areas the way it is done in Central America and Brazil. Cattle can be grown very efficiently in smaller feedlot areas; ideally the feed could be nutritious fodder from trees, which could be grown on the land that is now kept in pasture for the cattle. This will require a change of culture, orientation, law, and political will, which will be slow and difficult. Brazil has made a start by eliminating the tax and credit schemes for ranches in Amazonia.

Another threat to the forests has come from major government agricultural colonization projects, especially in Brazil and Indonesia (Guppy 1984). These governments are moving people into tropical forests to grow rice, coffee, and other crops that are really not appropriate for the soil. In some cases the governments are realizing that these projects have not worked out as planned. An alternative to destroying additional forests to provide more agricultural land would be to take areas that have already been deforested and degraded, and attempt to rehabilitate them for settlement. These lands will need to be fertilized and nurtured as they are no longer productive at present. If these programs concentrated people, resources, and technology on areas that have already been degraded, then the colonizers would not have to move into the remaining stretches of virgin tropical forests.

Strategies for Implementing Change

The strategies to control tropical deforestation have focused on both the external pressures for destructive development and the internal conditions needed to preserve forest ecosystems. For example, the National Wildlife Federation has been using two strategies to promote the conservation of tropical forests. One has been a campaign to pressure the institutions that are providing the funding for forest development projects. The second has been a campaign to provide partial debt relief in exchange for meaningful conservation programs.

Reforming Destructive Development Practices

Environmental organizations in the United States, and increasingly in other countries as well, are coming to a consensus that any outside funding from aid agencies (such as the World Bank and the InterAmerican Development Bank) should only be for the kind of alternatives that have been outlined here. The National Wildlife Federation has coordinated a five-year campaign to reform the environmental policies of the World Bank, and has recently extended it to include the InterAmerican Development Bank and the other major funding organizations. As part of a worldwide coalition of several hundred organizations that are working together to save rain forests, they have had remarkably successful meetings with

World Bank officials, testified at hearings before the United States Congress, and produced evidence of specific "problem projects" in countries such as Brazil, Indonesia, India, and Botswana (Environmental Impact of Multilateral. . . , 1983; Tropical Forest Development Projects, 1984). It is truly heartening that these activities are beginning to have some impact in this short period of time.

In the United States, both conservative and liberal politicians have joined in this campaign. Interestingly, while many conservatives are not very supportive of foreign aid in the first place, they do feel it should be rational, and produce a positive return for the U.S. investment abroad. They agree that the financial help should reach the targeted population. The goal is to increase the actual living standards of the poor. All too often, rain forest development projects that have been funded by development banks and agencies have not shown these positive results. Some of the projects have actually resulted in the opposite effect. Fortunately, the U.S. government position, as it is represented on the Board of the World Bank and the InterAmerican Development Bank, has changed tremendously in the last few years. The United States has almost 20% of the vote in the World Bank, based on its funding contribution, and 35% of the vote in the InterAmerican Development Bank. The United States, therefore, has a major responsibility to ensure that development projects are effective. Congress has helped the National Wildlife Federation persuade the Treasury Department to begin to analyze projects based on environmental effects and sustainable use of natural resources, and to use that information in the voting process (Aufderheide and Bruce 1988). The impact of that change on the staff and leaders of these banks has been significant.

In recent years, we have seen the beginning of a new policy at the World Bank (Environment and Development . . . 1988). Its president, Barber Conable, made a speech in May 1987 in which he announced the formation of a new environmental department. The World Bank hired additional staff people and has totally reorganized itself internally. Bank staff have begun some new studies on the economic and environmental effects of certain kinds of tax strategies, incentives, and subsidies: What really is the effect of a subsidy for certain kinds of pesticides, or for certain kinds of rice seed? What is the effect of a policy, such as Brazil used to have, that allowed industry to invest 25% of its profits in the Amazon and take that amount of money as a tax credit? (Spears 1988; Markandya and Pearce 1989; Repetto 1988). So far, most of these investments have led to destructive development, such as cattle ranching, which produces little meat, and the production of pig iron using irreplaceable tropical forest as charcoal to fuel the blast furnaces. New policies at the World Bank on funding for pesticides with an emphasis on integrated pest management and wild land protection are two examples of positive change.

Up to this time, the banks have been extremely secretive about their analyses of projects and the process of advising countries on development strategies. Having the U.S. Congress and Treasury Department as watchdogs is causing some remarkable changes, particularly now that government representatives and citizen groups from the United Kingdom, Germany, Scandinavia, and the Netherlands are asking the exact same questions. The World Bank has recently issued a draft policy on public involvement in decision-making with regard to development projects (How the World Bank Works 1990). In the long run, this change may be the most important one. It will be the most difficult, however, because the bank feels that its lending discussions with governments are the equivalent of treaty negotiations and thus have always been secret. Nonetheless, the banks' staff are beginning to learn that the local people, who are affected by the impact of these developments, both good and bad, retain valuable knowledge and have opinions that should be heard in the process of putting a development project together.

The external debt crisis is one of the factors that has aggravated the destruction of tropical rain forests during the last 15 years. Ironically, much of the debt owed by impoverished tropical nations to the big banks and lending agencies of the developed world went to pay for "development" projects such as huge hydroelectric dams and highways that were themselves destructive to tropical rain forests environments. The debts incurred for these development loans were aggravated by the oil price increases and inflation, and they became staggering burdens on the economies of the debtor nations. Debtor nations are obligated to make regular payments in hard currency to their creditors. This cash is usually generated by export earnings, but the potential exports are natural resources, such as tropical hardwoods and minerals, plantation crops such as coffee or sugar, or beef from cattle ranches. All too often, the pressure to maximize export earnings leads to excesses, such as unsustainable levels of logging or conversion of primary forests to marginal plantations and ranches. These debt pressures are linked to other problems, including low commodity prices, high inflation rates, and unfavorable trade conditions, creating an economic quagmire that results in much environmental damage but little economic improvement.

Debt-for-Nature Swaps

Given the harshness of this debt situation, one of the most intriguing mechanisms developed in recent years for the conservation of tropical forests is the "debt-for-nature swap." The initial stimulus for the "debt-for-nature" exchange came from a widely publicized *New York Times* editorial written by renowned conservationist Thomas Lovejoy (Lovejoy 1984). The particular arrangement for each "swap" is unique, but the general pattern involves the purchase of a fixed portion of a nation's debt by a donor organization (for a fraction of its face value). This "debt instrument" is then donated to the nation's central bank and is earmarked for the financing of environmentally beneficial activities, usually administered by a local nongovernmental conservation organization.

The first debt-for-nature swap was implemented by Bolivia in 1987, and involved the purchase of $650,000 worth of external debt (for $100,000). This "debt instrument" was used to pay for the establishment and management of the Beni Biological Reserve. This 500-square mile (130,000 hectare) reserve in the headwaters of the Beni River (a tributary of the Amazon River) has been designated to be a unit of the UNESCO International Biosphere Reserve system. The swap that created the Beni Reserve initially generated a bitter controversy within Bolivia about whether its national sovereignty was being compromised by the terms of the agreement, which was perceived by some critics as neo-colonialist. These concerns were aggravated by some misunderstandings that have since been resolved, and the Beni is now functioning as a significant tropical forest reserve. The Beni swap, however, should also serve as a cautionary reminder against overoptimism regarding what "debt-for-nature-swaps" can achieve. Bolivia's total debt reduction from the deal was insignificant: the $650,000 of relief represents barely more than one-thousandth of its $500 million external debt. The arrangement does provide hard cash for establishing and managing the Beni Reserve, but Bolivia also agreed to take on future management costs that will not be covered in perpetuity by the swap (Ayers 1989).

A different approach to "debt-for-nature swaps" was taken by Ecuador, where $10 million in debt was converted into bank bonds, which pay interest to the local Fundacion Natura for the development and protection of state-owned parklands. This arrangement, in which the debt instrument is converted to interest-paying bonds, generates much more money for environmental projects than was initially invested and has become the preferred mechanism for debt swaps around the world (Patterson 1990).

Costa Rica has made the largest commitment to "debt-for-nature swaps"—over $80 million worth of debt relief as of 1991. Costa Rica has negotiated several different kinds of swaps with different donors. "Traditional" parkland preservation arrangements have been made with international nongovernmental organizations such as the Nature Conservancy, and innovative experiments in agroforestry have been financed with debt relief purchased by the government of the Netherlands. The government of Sweden donated a debt instrument to establish a bond fund for the protection of tropical dry forest in and around Guanacaste National Park on the Pacific slope of northwest Costa Rica. Other projects initiated with debt swap funding have included land acquisition and road building for La Amistad National Park and natural resource inventories and buffer zone planning for various units of the Costa Rican park and forestland system. The duration of debt-swap financing is the life of the bonds (7 years); subsequent costs of programs initiated by the swaps will have to be paid for in some other fashion. The most likely outcome will be the eventual integration of these costs into the government environmental agency budgets, unless new agreements can be arranged with the present donors of the debt instruments (Bramble 1987; Page 1989; Patterson 1990).

Although debt-for-nature swaps were initiated in Latin America, similar programs are being undertaken in a variety of nations throughout the tropical world. Among the nations that have implemented debt swaps, or are in the process of implementing them, are Madagascar, Zambia, the Philippines, Jamaica, and the Dominican Republic (Bedarff et al. 1989; Patterson 1990).

The range of environmental activities that can be financed by debt swaps is limited only by the imagination of the participants. As a funding mechanism, debt instruments can be used to finance health care, social services, and infrastructure projects as well as environmental protection, provided that partners can be found to purchase debt for those purposes. They are an ingenious mechanism for paying for environmental protection in countries suffering from severe economic hardship. Even in the best of situations, however, swaps can only relieve a very small portion of the external debt burden carried by tropical nations.

Conclusions

The tropical forests are an irreplaceable resource of vast importance to the future ecological and economic health of the planet and to the particular countries where they exist. Based on current trends, the remaining forests will be mostly destroyed in a relatively few decades. Feasible alternatives to the current destructive practices exist. While further scientific study of tropical forests can help to delineate some specific strategies, the immediate problems threatening tropical forests are social and political problems rather than scientific problems. The real question is how to quickly develop the social and political will to implement solutions in countries where the social system and incentives currently work to further tropical forest destruction. We don't have much time left.

References

Aldhous, P., "Tropical Deforestation: Not Just a Problem in Amazonia," *Science* 259, 1993, p. 1390.

Allegretti, M. H., "Extractive Reserves: An Alternative for Reconciling Development and Environmental Conservation in Amazonia," *In Alternatives to Deforestation: Steps Toward Sustainable Use of the Amazon Rain Forest,* A. B., Anderson, (ed.), New York: Columbia University Press, 1990.

Alper, J., "How to Make the Forests of the World Pay Their Way," *Science* 260, 1993, pp. 1895–1896.

Aufderheide, P. and R. Bruce, "Environmental Reform and the Multilateral Banks," *World Policy Journal,* Spring 1988, p. 301.

Ayers, J. M., "Debt for Equity Swaps and the Conservation of Tropical Rain Forests," *Trends in Ecology and Evolution* 4, 1989, pp. 331–332.

Bedarff, H., B. Holznagel, and C. Jakobeit, "Debt for Nature Swaps: Environmental Colonialism or a Way Out from the Debt Crisis That Makes Sense," *Verfassung und Recht in Ubersee* 22, on-line from C.A.B., 1990, pp. 445–459.

Bradley, D., "Frog Venom Cocktail Yields a One-handed Painkiller," *Science* 261, 1993,. 1117.

Bramble, B., "The Debt Crises: The Opportunities," *The Ecologist* 17, 1987, pp. 192–199.

Brown, L. and E. Wolf, Reversing Africa's Decline, Worldwatch Paper 65, Washington, DC: Worldwatch Institute, June 1985.

"Case Studies from Africa: Towards Food Security," Non-Governmental Liaison Service, United Nations, 1987.

Clay, J., *Indigenous Peoples and Tropical Forests,* Cambridge, MA: Cultural Survival, Inc., 1988.

Cloud, J., "Cloud Forests, Quetzals, and Coffee," *Animal Kingdom,* July–Aug. 1988, pp. 33–41.

Dixon, R. K., S. Brown, R. A. Houghton, A. M. Solomon, M. C. Trexlor, and J. Wisniewski, "Carbon Pools and Flux of Global Forest Ecosystems," *Science* 263, 1994, pp. 185–190.

Dover, M. and L. M. Talbot, *To Feed the Earth: Agro-Ecology for Sustainable Development,* Washington, DC: World Resources Institute, 1987.

"Environment and Development: Implementing the World's New Policies," Joint Ministerial Committee of the Boards of Governors of the World Bank and the International Monetary Fund, June 1988.

"Environmental Impact of Multilateral Development Bank-Funded Projects," *In Hearings before the Subcommittee on Banking, Finance and Urban Affairs of the House of Representatives, 98th Congress, 1st Sess., 1983.*

Eyre, L. A., "The Tropical National Parks of Latin America and the Caribbean: Present Problems and Future Potential," *Yearbook: Conference of Latin American Geographers* 36, 1990, pp. 15–33.

Flavin, C., *Slowing Global Warming: A Worldwide Strategy, Worldwatch Paper 91,* Washington, DC: Worldwatch Institute, October, 1989.

Geller, H., J. Goldemberg, J. Moreira, R. Hukai, C. Scarpinella, and M. Ysohizawa, "Electricity Conservation in Brazil: Potential and Progress," *Energy* 13, 1988, pp. 469–483.

Goldemberg, J., T. B. Johansson, A. K. N. Reddy, and R. H. Williams, *Energy for Development,* Washington, DC: World Resources Institute, 1987.

Goldsmith, E. and N. Hildyard (eds.), *The Social and Environmental Effects of Large Dams,* Wadebridge Ecological Centre, *The Ecologist,* 1986.

Goodman, B., "Drugs and People Threaten Diversity in Andean Forests," *Science* 261, 1993, p. 293.

Gradwohl, J., *Saving the Tropical Rain Forests,* Washington, DC: Island Press, 1988.

Guppy, N., "Tropical Deforestation," *Foreign Affairs* 62, 1984, p. 928.

Hartshorn, G. S., "Natural Forest Management by the Yanesha Forestry Cooperative in Perusian Amazonia," *In* A. B. Anderson (ed.), *Alternatives to Deforestation: Steps Toward Sustainable Use of the Amazon Rain Forest,* New York: Columbia University Press, 1990.

Holloway, M., "Sustaining the Amazon," *Scientific American* 269(1), 1993, pp. 90–99.

"How the World Bank Works with Non-Governmental Organizations," The World Bank, June, 1990.

Houghton, R. A., J. E. Hobbie, J. M. Mellilo, B. Moore, B. J. Peterson, G. R. Shaver, and G. M. Woodwell, "Changes in the Carbon Content of Terrestrial Biota and Soils Between 1860 and 1980: A Net Release of CO_2 to the Atmosphere," *Ecological Monographs* 53(3), 1983, pp. 235–262.

Kricher, J. C., *A Neotropical Companion,* Princeton, NJ: Princeton University Press, 1989.

Koopowitz, H. and H. Kaye, *Plant Extinction: A Global Crisis,* Washington, DC: Stonewall Press, 1983.

Leonard, H. J., *Natural Resources and Economic Development in Central America,* New Brunswick, NJ: International Institute for Environment and Development, Transaction Books, 1988, Chapters 3 and 4.

Lovejoy, T. E., *New York Times,* editorial section, 4 October 1984.

Markandya, A. and D. Pearce, "Environmental Considerations and the Choice of the Discount Rate in Developing Countries," *Environment Department Working Paper No. 13,* The World Bank, January, 1989.

May, R. M., "How Many Species Are There on Earth?," *Science,* 241, 1988, pp. 1441–1449.

Mendes, C., *Fight for the Forest,* New York: Latin American Bureau, 1990.

Miller, K. and L. Tangley, *Trees of Life: Saving Tropical Forests and their Biological Wealth,* Boston: Beacon Press, 1991.

Myers, N., "Tropical Deforestation and a Mega-Extinction Spasm," *In* M. E. Soule (ed.), *Conservation Biology,* Sunderland, MA: Sinauer Associates, 1986.

Myers, N., "Threatened Biotas: 'Hotspots' in Tropical Forests," *The Environmentalist* 8, 1988, pp. 1–200.

Nepstad, D., C. Uhl, and A. E. Serrao, "Surmounting Barriers to Forest Regeneration in Abandoned, Highly Degraded Pastures: A Case Study from Paragominas, Para, Brazil, *In* A. B. Anderson (ed.), *Alternatives to Deforestation: Steps Toward Sustainable Use of the Amazon Rain Forest,* New York: Columbia University Press, 1990.

Nepstad, D. C., C. R. Carvalho, E. A. Davidson, P. H. Jipp, P. A. Lefbevre, G. H. Negreiros, E. D. da Silva, T. A. Stone, S. E. Trumbore and S. Viera, "The Role of Deep Roots in the Hydrological and Carbon Cycles of Amazonian Forests and Pastures," *Nature* 372, 1994 pp. 666–669

Page, D., "Debt-for-Nature-Swaps: Experience Gained, Lessons Learned," *International Environmental Affairs* 1(4), 1989, pp. 275–288.

Patterson, A., "Debt for Nature Swaps and the Need for Alternatives," *Environment* 32, 1990, pp. 4–13, 31–32.

Raven, P. H., "Our Diminishing Tropical Forests," *In Biodiversity,* E. O. Wilson (ed.), Washington, DC: National Academy Press, 1988.

Raven, P. H., L. R. Berg, and G. B. Johnson, *Environment,* Philadelphia: Saunders College Publishing, 1993.

Repetto, R., "Economic Policy Reform for Natural Resource Conservation," *Environment Department Working Paper No. 4,* The World Bank, May, 1988.

Roberts, L., "Chemical Prospecting: Hope for Vanishing Ecosystems?" *Science* 256, 1992, pp. 1142–1143.

Sanwal, M., "Community Forestry: Policy Issues, Institutional Arrangements and Bureaucratic Reorientation," *Ambio* 17, 1988.

Skole, D. and C. Tucker, "Tropical Deforestation and Habitat Fragmentation in the Amazon. Satellite Data from 1978–1988," *Science* 260, 1993, pp. 1905–1910.

Spears, J., "Containing Tropical Deforestation: A Review of Priority Areas for Technological and Policy Research," *Environment Department Working Paper No. 10,* The World Bank, October, 1988.

Sternberg, R., "Hydroelectric Energy: An Agent of Change in Amazonia (Northern Brazil)," *In* F. J. Calzonetti and B. D. Solamon (eds.), *Geographical Dimensions of Energy,* Norwell, MA: Kluwer Academic Publishers, 1985.

Terborgh, J., *Diversity and the Tropical Forest,* New York: Scientific American Library, 1992a.

Terborgh, J., "Why American Songbirds Are Vanishing," *Scientific American,* 266(5), 1992b, pp. 98–104.

Timberlake, L., "Africa in Crisis: the Causes, the Cures of Environmental Bankruptcy," *Earthscan,* 1985.

"Tropical Forest Development Projects—Status of Environmental and Agricultural Research," Hearing before the Subcommittee on Natural Resources, Agricultural Research and Environment of the House Comm. on Science and Technology, 98th Congress, 2nd Sess., 1984.

Whittaker, R. H., *Communities and Ecosystems, 2nd ed.,* New York: Macmillan, 1975.

Wilson, E. O., "The Current State of Biological Diversity," *In Biodiversity,* E. O. Wilson (ed.), Washington, DC: National Academy Press, 1988, pp. 3–18.

Wilson, E. O., "Threats to Biodiversity," *Scientific American* 261(3), 1989, pp. 108–116.

Wilson, E. O., *The Diversity of Life,* Cambridge, MA: Belknap Press, 1992.

WRI, the World Bank, and the UN Development Program, *Tropical Rain Forests: A Call for Action,* Report of an International Task Force convened by WRI, the World Bank, and the UN Development Committee. Parts I, II and III, 1985.

Chapter 10

Desertification

Howard Horowitz, Eric F. Karlin, and William J. Makofske

In the past few decades, episodes of famine associated with drought in sub-Saharan Africa, northeastern Brazil, and elsewhere have begun to raise the world's awareness about the terrible costs of degradation of once-productive lands. The word *desertification* was first popularized by Aubreville in 1949, although the processes involved are not new. The apparent acceleration of the rate of land degradation in recent years, however, has finally drawn the attention of scientists, resource managers, governmental organizations, and the media.

The United Nation's Conference on Desertification (UNCOD), which was convened in Nairobi, Kenya in 1977, established the following definition of desertification:

> Desertification is the reduction or destruction of the land's biological potential, finally resulting in the appearance of desert conditions. It is one aspect of the generalized degradation of ecosystems under the combined pressures of adverse and uncertain climatic conditions and overexploitation. This overuse has reduced or destroyed the biological potential, that is to say the plant and animal production for multipurpose use, at the very moment when increased production was needed to meet the needs of growing populations aspiring to development (Baumer 1990).

Note that this "official" definition acknowledges that there is a linkage between "desertification" and "development", and asserts that "increased production" is needed from the land to support the needs of growing populations. Triangular linkages between "desertification," "development," and "population" are at the heart of the problem around the world.

The popular press has focused on sub-Saharan African areas, where human suffering is great. Indeed, some of the worst-affected areas include sub-Saharan Africa, Andean South America, and Mexico, West Asia, and Nepal (Matheson 1984a). The expanded definition, however, may include as much as 35% of the total land area of the world, equal to the total land area of North America and South America combined. Large areas of every populated continent, including portions of the United States, the Soviet Union, and Australia, have desertification problems that affect an estimated 850 million people (Tolba 1984).

There are different degrees of desertification. *Moderate desertification* describes productivity losses between 10% and 25%, while *severe desertification* can involve losses up to 50% or even higher. There are areas where the productivity has dropped between 50% and 75%. While it is primarily associated with semi-arid areas, desertification can, surprisingly, occur even in wet areas. Every year, about 20 million hectares are added to the land that would be classified as desertified lands (Dregne 1983; Matheson 1984a, b).

The process of land destruction is not new. While the word *desertification* itself is a fairly recent word, the process is really quite ancient. Plato, in *Critias* (350 BC), describes bare hills in Attica that had been forested 500 years before that time as follows:

> What now remains compared with what then existed is like the skeleton of a sick man, all the fat and soft earth having wasted away, and only the bare framework of the land being left.

The classical Greek and Roman civilizations in the Mediterranean, and even before that, the Phoenicians, living in what is today Lebanon, and the Mesopotamians in the Tigris and Euphrates, were all great civilizations of antiquity that depleted their soils in the course of their empire. They built irrigation systems from their rivers but had problems with salinization. There was massive erosion of soils that led to loss of fertility. Shipbuilding was very widespread in the Mediterranean region when these civilizations flourished; however, the vessels were built from timbers cut down from montane forests that never grew back. Some scholars have suggested that many of the ancient civilizations declined largely because of desertification processes (Hyams 1952). Historians tend to focus on political events, such as kings or invaders, but there is also an ecological underpinning to these declines. The Mediterranean basin has never really recovered its former productivity, biologically or politically.

Surprisingly, in some cases it can be difficult even to verify desertification. There are shifting cycles of drought and wet periods, shifting modes of cultivation and pastoral activity, and shifting dune sheets themselves; all of this fluctuation causes considerable disagreement, even among the experts, about how to interpret the evidence. Some observers see a pattern of rapid desert encroachment, while other observers in the same area may not see the same kind of pattern. A striking example of this uncertainty involves the southward expansion of the Sahara Desert in Northern Sudan. A 1975 field survey led by H. F. Lamprey for the UN indicated much specific evidence of desertification, such as previously mapped *Acacia* groves that could no longer be located, active dune sheet obliteration of previously cultivated areas, and severe reductions in sightings of important wildlife indicator species, such as ostrich, onyx, and gazelle. The survey proposed a series of 11 specific research sites, including three biosphere reserves, along a climatic gradient ranging from 400 mm annual rainfall in the south to 30 mm annual rainfall in the north. The survey concluded that the desert's boundary had shifted southward about 90–100 km in the previous 17 years in certain areas of Kordofan Province (Lamprey 1988). A subsequent study of the same area, conducted by Alf Hellden of the University of Lund (Sweden), came to startlingly different conclusions. Using Landsat data and other remote sensing imagery taken between 1962 and 1979, and also various historical records from earlier periods, Hellden was unable to verify either the rapid southward movement of the Sahara, or the dramatic expansion of the specific dune sheets. The severe Sahelian drought of 1964–1974 was identified as having had a severe impact on both crop yields and natural vegetation, but the subsequent wet period was observed to have triggered a fast recovery of the land's productivity (Hellden 1988). These satellite photo images raise questions about the long-term ecosystem effects of farming, grazing, and drought in these long-occupied regions; are they causing permanent impoverishment of soils, natural vegetation, and wildlife? The view from space seems to suggest vegetation resiliency during wet periods; however, this view may not reflect losses, such as wildlife species, that can only be seen from the ground. In many other regions, however, there is little disagreement about the loss of productivity of cropland and rangeland.

This article explores the reasons why desertification today is having great impact on the peoples of the world. This is partly due to expanding populations and the use and misuse of certain technologies and techniques, mostly for food production and wood harvesting. In other cases, technologies for energy production, such as hydropower or biomass production, are partly to blame. In almost all cases, it is human action that is to blame; desertification today is not primarily a natural phenomena caused by drought or climate change. If people can cause desertification, they can also prevent and even reverse its effects. There are many areas in the world where this problem has been successfully fought; indeed, even the United States has its own battles to fight today, mostly in the drier agricultural regions. Desertification today is worldwide and it has been the United Nations that has taken the international role of organizing the fight against it. We will therefore examine the successes and failures of its attempts to provide leadership and resources to combat the problem.

Causes of Desertification Today

Although desertification is not a new process, it has accelerated in the twentieth century. The most significant factors are probably increasing populations and new technologies, which bring about rapid change of the land. The tools for cutting down the forest are more efficient, the cattle herds are larger, and sedentary cultivators are following new roads into sensitive areas. Nightly news programs showing famine pictures from Sahelian Africa focus on starving children, but if you look you will see bare ground. Those two things are connected. Remember, however, that desertification is not just an African problem; every continent has regions suffering from desertification (Figure 10.1). Here in the United States, we have a substantial desertification problem, particularly in the drier western states (Sheridan 1981; Brown and Wolf 1984; Swanson and Heady 1984). Over one-third of the land surface of the Earth is arid or semi-arid. Semi-arid areas tend to be the most prone to desertification problems because they are fragile yet productive enough to attract exploitation (Dudal 1982; Dregne 1983).

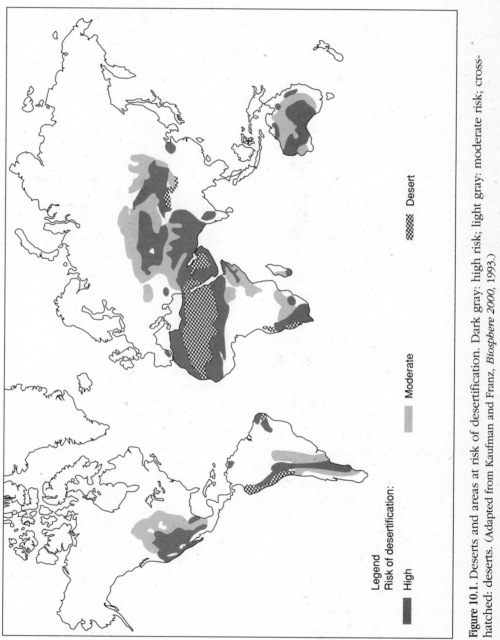

Figure 10.1. Deserts and areas at risk of desertification. Dark gray: high risk; light gray: moderate risk; cross-hatched: deserts. (Adapted from Kaufman and Franz, *Biosphere 2000*, 1993.)

An important key to understanding desertification lies in looking at natural ecosystems in semi-arid regions. Semi-arid areas are parts of the Earth where water is abundant enough to allow somewhat robust ecosystems to exist, but there are seasons when water is in short supply. The plants and animals that naturally exist in these semi-arid areas have the ability to tolerate extended dry periods. They can survive drought for several years and then resume their former productivity. For millions of years, ecosystems have been able to survive in these semi-arid areas because they have the ability to tolerate fluctuations in the availability of water. Today, increasing stresses brought on by human activity have changed these ecosystems, reducing their resilience.

The loss of rangeland productivity, one of the most widespread signs of desertification, may be readily understood by considering changes in a natural ecosystem when domesticated grazing animals are added. If the herds are small, very little impact occurs and the ecosystem is not stressed. If there is a drought, the ecosystem can recover and continue to support plants and animals at normal levels of productivity. If, however, the density of grazing animals exceeds the carrying capacity of the land, the ability of the ecosystem to recover is diminished (Figure 10.2). (See Chapter 2, "Population Growth and the Global Environment: An Ecological Perspective," for an in-depth discussion of carrying capacity.) A drought, which at one time might have been manageable, may trigger long-term destruction. As plant productivity and herd capacity declines, less food is produced. As plant cover decreases, soil erosion is accelerated. If enough soil is lost, ecosystem productivity will remain impaired even when the rains return. A desert is in the process of being formed (Breman and de Wit 1983; Skoupy 1988).

Cultivation can be another culprit. Many of the crops in modern agriculture are not native to the region they are grown in and have special requirements. They are usually annual, unlike the native grasses, which are mostly perennial. Cultivated ground has to be plowed, and often has to be fertilized and watered. People may successfully expand agriculture in semi-arid regions during a period of wet years, but disaster strikes when a dry cycle occurs. It is then too dry to plant, yet the agricultural system has already replaced a natural ecosystem. The ground has already been plowed, leaving bare soil exposed to the wind. Just as with overgrazing, the outcome is massive soil erosion and destruction of the ecosystem. This is what led to the Dust Bowl era which plagued the southwestern United States in the 1930s.

Most of the crop plants used in agriculture are not adapted to drought, and so they are highly susceptible to damage if it gets dry. People often look at areas in terms of the average climate, and assume that the average climate can support these crops. It is the climatic extremes, however, that determine what happens in the long run. While these droughts are well within the normal range of climatic variation, they are now rapidly destroying the ability of these areas to support agriculture.

In other cases, it is the pressure of rising populations causing overcultivation that is largely to blame. Land that was once left idle to recover its fertility is now continuously cultivated, severely depleting the soil. Such conditions are obvious in Northern Sudan, Upper Volta, and increasingly in other places around the world. Generally, increasing population pressure forces the use and misuse of

Figure 10.2. Impact of over-grazing and trampling on a pasture in Iowa, United States. (Grant Heilman, Lilitz, PA.)

marginal lands. Expanding cropland into areas that should be left as pasture or forest ultimately reduces productivity for all uses. On marginal lands, erosion from wind and water have much greater consequences. As the quality of the cropland declines, greater costs are incurred to keep output up. In most cases, poor people cannot afford additional costs to compensate for declining productivity. The quality of life deteriorates; sometimes even life itself is lost (Eckholm 1982; Brown and Wolf 1984).

Another process causing desertification, especially in the agricultural lands in the central valley of California, is salinization of the soil. Precipitation is usually relatively salt and mineral free, and therefore leaches minerals out of soils. In humid regions such as the eastern United States, soils face the problem of having many of their minerals washed out of them by rainfall, requiring fertilization to help replace the minerals that the rain washes away. In semi-arid and arid regions, however, there is not enough rainfall to leach minerals from the soil. Instead, the rain mostly evaporates from the soil surface and the salts become concentrated in the soil. Of course, if there is not enough rainfall to support crops, irrigation is practiced. The water taken from rivers or reservoirs, however, comes from water that has fallen on the soil in the region and then collected in these water basins; this water has picked up salt and minerals as it passed through the soil system. When this water is added to the farmer's ground, it will eventually evaporate and leave all of the salts behind. If this continues year after year, significant amounts of salt are added into the soil, which eventually can no longer support crops. Many areas are facing a major salinization crisis. Although California is one of the major food producing areas in our country, many of its soils are becoming progressively less productive because of the salt problem. While there are potential remedies, they are not easy or cheap (Englebert and Scheuring 1982; Miller 1990).

Another activity that leads to the destruction of ecosystems is deforestation, especially in mountainous areas. An intact forest has several functions in helping to control the water economy of an area. The water falling in a rainstorm gets intercepted by the plants instead of hitting the soil full force, so that it trickles to the soil and percolates into the ground. In many forests, most of the precipitation passes through the soil and enters the groundwater system, which gradually flows to surrounding areas. During a big storm, instead of having a flood, most of the water absorbed by the forest is diverted to groundwater systems and then released to downstream areas over an extended period of time. Downstream from the forest, people have a continuous water supply and are protected from severe flooding.

If that forest is cut down, however, as is happening in many parts of the world, there are no trees to stop the rain; it hits the soil full force. That leads to major soil erosion. In addition, instead of infiltrating into the soil, the water runs off the surface quickly, causing flash floods. After the wet season, water becomes scarce because the groundwater reservoir, which the forest used to feed, is no longer getting sufficient water. All too often the result is earlier and more prolonged periods of soil moisture loss than there used to be. The removal of forests, by accelerating the erosion of soil and unbalancing the water cycle, can often lead to desertification of extensive areas (Jordan 1982). The Indian subcontinent is experiencing this problem today: water supply is feast or famine. When the forests are gone, both floods and dry periods are worsened; there is too much water running off during the monsoon and not enough groundwater later in the year.

Tropical deforestation can also damage the soil by other mechanisms besides simple erosion. Many tropical regions are characterized by oxisols, which are radically different from soils that occur in the temperate regions. These soils are pliable and good for plant growth under natural conditions, although they are relatively infertile. Once exposed to full sun and wind, some oxisols dry into brick-hard material, called *laterite*. If a tropical forest having this soil type is cut down, leaving the soil directly exposed to the sun, the soil hardens into a reddish laterite rock. Thus, instead of providing a desirable agricultural soil, the clearing of the forest results in a seriously degraded soil system (Sanchez and Buol 1975; Jordan 1982). All of these things have little to do with climate directly; they have more to do with human activities.

The Misunderstood Role of Climate in Desertification

The root cause of desertification is not drought or climate change, although these reasons are commonly believed to be primary ones. Drought aggravates the problem and brings attention to immediate food crises. It is the overexploitation of land, however, due to overcultivation, overgrazing, poor irrigation processes, and deforestation, that is the fundamental cause of desertification. These trends are often made worse by overpopulation, bad management, neglect of rural areas, and poverty (Raloff 1985).

The role of climate is widely misunderstood. Climate, by its very nature, is variable. Some years will be too wet; some years will be too dry. Sahelian Africa and many other areas around the world have had these natural fluctuations repeatedly in the past, and humans and ecosystems have for the most part survived them. Why is it that the droughts of previous years caused much less damage than comparable droughts are causing today? The difference today is that many ecosystems have been stressed so much that they are in danger of slipping into irreversible decline. The life-support systems of our civilization—the soil, the vegetation, and the wild animal populations—are deteriorating to the point where the survival of hundreds of millions of people is in jeopardy. Continued overuse of our biological systems will lead to ecological, economical, and ultimately to societal collapse (Brown and Wolf 1984; Brown and Flavin 1988).

There is increasing evidence that the removal of extensive tracts of tropical forests may be linked to a decrease in the amount of rainfall. There is not firm evidence yet, but the tendency seems to be in that direction. Evapotranspiration from plants puts water back into the atmosphere, which falls down as rain somewhere else. If the tropical forests and savanna vegetation are removed on a vast scale, it is certainly conceivable that the overall hydrologic patterns of a whole continent could be changed (Myers 1988).

Desertification could be further exacerbated by any global warming trends brought about by the greenhouse effect. It appears that the global average temperature may already be rising, possibly because of the greenhouse gases, and it is likely to rise more rapidly in the future. There is concern that this global warming will eventually change climate patterns and cause severe ecological and agricultural disruption in many areas (see Chapter 8, "The Loss of Biodiversity" and Chapter 14, "The Challenge of Global Climate Change"). Although these patterns of long-range climate change are likely to have large impacts in the future, the primary source of desertification today is not due to changes in the global climate. Most droughts are not more extreme than those recorded in the immediate historical past. When droughts do occur, however, they highlight the impacts of desertification by making the damage more visible. The increasing desertification is then wrongly attributed to global climate change (Brown and Flavin 1988).

The U.S. Experience

The head of the United Nations Environmental Programme, Mustaf Tolba, aptly described desertification as a harvest of dust (Tolba 1984). In the United States, about 50 years ago, we had a harvest of dust—the Dust Bowl era, when vast areas of the southwest plains became desertified (Figure 10.3). The term was not in use, but that is what happened, with the impact of drought being amplified by poor farming practices that were destructive to the soil. The result was an accelerated desertification episode that had catastrophic impacts on the people that lived in parts of Oklahoma, Kansas, Nebraska, and Texas. Many events that happened in the 1930s are similar to events that have happened in more recent desertification episodes. The mass outmigration to California and elsewhere, as people could no longer earn a living from the land, has parallels in the swelling refugee populations in Africa today. Not only was soil being blown into the air but, with the loss of vegetation cover, the increased runoff caused severe accelerated soil erosion and gullying. The U.S. government, un-

Figure 10.3. Severe desertification in New Mexico, United States. The picture was taken in the mid-1930s during the time of the "dust bowl." Poor farming practices coupled with extensive drought led to significant ecosystem degradation in the Midwestern United States. (The Granger Collection)

der the leadership of President Franklin D. Roosevelt, mounted an aggressive attack to reverse the desertification outbreak. He prioritized the expenditure of federal money and the recruitment of labor, including many Civilian Conservation Corps workers, to stabilize the situation. Millions of trees were planted as shelterbelts in the High Plains, and major efforts were made to reverse soil erosion. Better farming practices, including contour plowing and subsidized conservation set-asides, were designed to reduce soil losses. They were largely successful, and the dust bowl era ended within a decade (Brown and Wolf 1984).

There is a lesson from this episode. Desertification can be halted and degraded land can actually be partially restored if enough effort and money is invested. Even though it was in a depression, the United States in the 1930s was a relatively wealthy country with a lot of resources. The government leadership was strong and could draw on readily available labor and determined will power. That combination of forces succeeded in halting the "dust bowl."

Unfortunately, the United States has seen a renewal of desertification in recent decades. Although it is becoming steadily more serious, we have not yet dealt with it. One reason has been increased erosion and salinization due to the expansion of agriculture into marginal lands. At the same time, large-scale mechanized agriculture has often ignored good soil management practice. Government incentives to implement soil conservation have also not been adequate (Brown and Wolf 1984). In conjunction with 96 nations that participated in the 1977 United Nations Conference on Desertification, however, the United States has prepared a very thorough and candid analysis of its own desertification problems. This analysis was published in final form in 1982 by an interagency team under the auspices of the U.S.

Department of the Interior under the title *Desertification in the United States: Status and Issues* (Sabadell et al. 1982). This report includes detailed maps of critical soil loss areas, overgrazed rangeland, and other problem conditions. It concludes that key pressures leading to worsening desertification in the American West are energy development, mineral extraction, agricultural commodity production, water resource development, and population inmigration.

Desertification in Africa

In the past few years, Africa has tended to get the greatest focus regarding desertification (Timberlake 1985), although this continent certainly is not the only area where productivity loss has reached crisis conditions. The problem is largely confined to a certain zone of Africa, south of the Sahara. The Sahara is already desert, but southward from it is a semi-arid region called the Sahel (stretching from northern Senegal and southern Mauritania on the Atlantic Coast across to Sudan). The Sahel has an annual wet season and dry season, but its savanna ecosystem has undergone tremendous changes in the last century (see Chapter 15, "Predicting Climate Change: The State of the Art"). One of those changes began with the advent of colonialism in the 1880s, when Britain, France, and other European countries divided up the African continent into a number of colonies. The colonies brought about an alteration of agricultural systems, and introduced new systems and tax policies oriented towards growing export crops for markets in Europe. That was very disruptive to the old and established systems of land use, and these changes triggered problems (Franke and Chasin 1980).

The tremendous increase in the population of these countries throughout the twentieth century has gone on without a comparable increase in arable land, so the need to grow more food on the same amount of land meant an increase in the intensity of agriculture. Trying to increase the intensity of agriculture in semi-arid places is very risky, and the increases are difficult to sustain. Even in California, which has inestimable wealth, we are seeing problems that derive from trying to have intense agriculture in semi-arid regions. Sahelian Africa's combination of population increase, cultural and agricultural disruption in the colonial era, and political instability have aggravated the intensity of ecosystem degradation.

Many tribal peoples have had a livelihood that depended on cattle grazing, and that has worked well for thousands of years. Many of these semi-arid lands are well suited for cattle grazing but are easily damaged by overgrazing. There is that fine line between what is good for the land and what is too much for the land. In some of these places that line has been crossed. There has been an actual decline in per-capita food production in parts of Africa over the past several decades. In 1967, for example, the production in Africa was about 180 kg/person, but by 1983/84 (during the severe drought), the food production was down to 120 kg/person. At that time, about 170 million people out of a total population in Africa of 531 million depended upon imported grains for survival. The situation has deteriorated since then (Brown and Wolf 1985).

A combination of factors (overpopulation, wars, poverty, and poor grazing and agricultural practices), in conjunction with very little money available to combat the problem, has resulted in a progressive deterioration (Raloff 1985). We sometimes see in the media the image of the Sahara desert marching southward. That is not a very good metaphor, because deserts do not usually march in a linear fashion. The image of the desert marching 10 miles a year south is false. Instead, there are pockets of degradation, localized places where the land has lost its vegetation and the topsoil has blown away. These pockets are scattered but they tend to grow, and as they grow they begin to coalesce. There is a gradual deterioration in the ability of the land to be productive. The large number of people that live there has caused great environmental stress. During wetter than average years, people are able to produce good harvests. During drier than average years they are not. The present situation in Africa is such that even slight droughts have a severe impact on society, causing starvation (Tolba 1984).

When the food supply is diminished in many places, there are food riots. This occurs particularly in urban areas where there is a large population that is totally dependent on buying their food. In many countries the governments actually have food subsidies, to keep food prices low and prevent political instability. A rise in prices, or just a shortage, often results in tremendous food riots. In 1985, for example, the government of Sudan was actually toppled by food riots. There have been dozens of food riots in many places around the world in the past dozen years. Often government expenditures that might be spent to reduce desertification in rural areas wind up being used to subsidize food in the cities.

One additional factor that is sometimes overlooked in the literature about desertification is warfare. Many of the semi-arid regions in the world have experienced war in recent decades. There are literally dozens of "little wars" that are taking place around the world, particularly in the Horn of Africa. Ethiopia, for instance, is the site of a 17 year war between a military dictatorship and regional subgroups within the country; similar sectarian warfare has raged in the Sudan for over 20 years. Over 2 million people have died in these two countries because of war and famine. Political instability and drought have led to 350,000 deaths in Somalia since 1991. These wars have aggravated the problem of food availability and ecosystem destruction in countries already hard pressed by high rates of population growth, drought, and desertification. Farm fields are burned, planting is disrupted, and the distribution of food is hampered. Furthermore, resources that could have been used to mitigate desertification problems are instead consumed by conflict. Tree planting projects, for example, are not likely to succeed in a war zone. Political instability and desertification are also intensified by the forced migration that occurs because of war. For example, the Sudan now contains over 1.1 million permanent refugees in camps. In Chad, almost 50% of the people were displaced by the combination of drought and war. The relocation of large populations creates a lot of tension, especially if they are refugees from other countries. Often times the host country has few resources to begin with and the additional population adds even more stress to the land. Many of the refugees lost everything they had. Even the seeds that they had saved for planting the next crop were eaten because they were close to starvation. Thus a vicious cycle is started, with drought and warfare leading to more hunger, which in turn leads to increased political and social instability. Clearly, political stability and peace are essential in order to effectively deal with the problems of desertification. Since the strife in Ethiopia and Eritrea ended in 1991, crop yields have increased dramatically.

Stopping and Reversing Desertification

As the causes of desertification are diverse, there is no one simple solution. Rather, a variety of mitigating actions are required. One is better forest management in areas where deforestation is a significant contributing variable. Healthy forests would minimize water loss and soil erosion. Where scarcity of fuel wood has caused deforestation, potential solutions involve community tree planting, the development of more efficient wood stoves, and the use of alternative means of heating and cooking using solar energy.

In semi-arid regions, utilizing plants and animals that can tolerate the natural climatic variations would promote stable agriculture and grazing. Only a small fraction of the total number of plants species on our planet are utilized today. Just six crops (wheat, rice, corn, potatoes, sweet potatoes, and manioc) provide more than 80% of the total calories consumed by humans. There are hundreds of "wild" plant species having great economic potential and they grow in a variety of environments. Instead of growing plants in regions for which they are not adapted (through the use of irrigation, fertilization, etc.), people should utilize plants that grow naturally in that area. The use of native species would require much less intensive resource use and management (Hinman 1984; Vietmeyer 1986).

There are many species of trees and shrubs, native to specific semi-arid regions, that can be managed to provide valuable sources of food, livestock, forage, and natural habitat. The importance of maintaining patches of forest habitat in areas threatened by desertification cannot be overstated. Forest wildlife, including reptiles, birds, insects, rodents, and other mammals, and even snails, make a surprisingly large contribution to the diets of people living nearby; for example, Nigerians living near forested lands consume 84% of their animal protein from such sources of "bushmeat," and similar percentages are found elsewhere in Africa and also in South America (Hoskins 1990). Local villagers in Thailand showed visiting foresters more than 40 species of plants and animals gathered from adjacent forests; together they comprised about 60% of all the food the people living there consumed. Similarly, a large variety of forest products contribute to the regular diet of Sahelian peoples, and this gathered portion of the diet increases in importance during periods of drought and crop failure (Sene 1985).

Actually, despite all of Africa's problems, the agroforestry systems are generally far more sophisticated in Africa today than they are in Latin America. There are a series of exciting, dynamic locally generated publications that illustrate these systems and the efforts being made to implement them (Getahun and Reshidi 1989; Ajan 1990; Baobab #5 1990; Lord 1990). By adopting policies that encourage the planting of appropriate native tree species, both governments and nongovernmental organizations can play a tremendous role in providing relief from desertification. Such plantations can provide multiple benefits to their regions, including accessible firewood that reduces the pressure to destroy existing native forests for fuelwood, and providing a range of forest products for food, medicines, industry, and shelter. Many native species, such as the *Acacias* and *Sesbanias* in Africa and the *Leucaenas* in Southeast Asia, also fix nitrogen and provide other benefits to soil fertility, and can be profitably interplanted with agricultural crops. One obstacle to the more extensive planting of forests in semi-arid areas is the social inequality of land tenure patterns: in regions where most of the land is owned by a few elite, there is little incentive felt by either the landed rich or the landless poor for locally oriented, sustainable forestry activities. The incentives are stronger on communal lands maintained by village custom and tradition, because there the investments in local forest improvement will be seen to provide local returns for the next generation of villagers (Persson 1986; Raintree 1986; Getahun and Reshidi 1989; Baumer 1990). Making farming sustainable will often require diversification so that crop, tree, and animal production are combined to enhance output and preserve soil productivity.

There also has not been enough effort to develop markets for the kinds of crops that potentially could succeed in semi-arid areas (Stiles 1988). The feasibility of this approach has already been demonstrated. For example, the jojoba is a plant native to semi-arid environments that produces seeds that are very rich in oil. The oil represents a large amount of the total biomass of the plant and is of high quality (Hinman 1984).

In many developed countries, investments in dams and irrigation technologies have allowed greater productivity in semi-arid lands but at great ecological cost (see Chapter 7, "Hydro-Development"). In the United States, for example, if you look down out of an airplane onto the old dust bowl area in west Texas, you will see a pattern of green circles; each marks a center pivot irrigation system. Water is pumped up from the Ogallala aquifer for irrigating rice and cotton. While these crops have a high yield and are profitable in the short run to the farmers, in the long run they cannot provide sustainable agriculture for the region. The problem with growing cotton in west Texas is that the annual precipitation averages 15 inches, while cotton requires much greater amounts of water. Therefore, water is being pumped out of the aquifer much more quickly than it is being replaced by precipitation. It is only a matter of time before the groundwater is used up; then the region will be much more degraded than it was before irrigation started. Likewise, in developing countries, many aid programs have been oriented towards growing staples that require irrigation, and tremendous investment has been made to increase water supplies and irrigation. A better investment for these areas is to grow crops that can be sustained by the local rainfall patterns. There is no sense trying to grow water-demanding crops in semi-arid and arid areas without sustainable water supplies (Postel 1989). The solutions must be based on sustainable development. Sound development must focus on primary health care, safe drinking water, sanitation, education, and family planning, as well as agricultural improvements.

World Action on Desertification: Policies and Problems

It is clear that desertification is a significant problem for the world, with great potential for causing misery, famine, political unrest, migrations, movement to cities, and joblessness. It will take a tremendous investment of labor, money, and effort to effectively combat desertification, and that, unfortunately, is not an easy thing to accomplish on a worldwide scale.

The international organization that has taken the lead in this effort is the United Nations. The 96 nations attending the 1977 United Nations Conference on Desertification came out with a series of mandates, one of which was for each participating nation to go home and assess its own desertification problems (Tolba 1984). In addition, a series of programs were set up to attempt to combat desertification on a world scale. The United Nations Environmental Programme estimated that it would cost $90 billion over about a 20-year period to deal adequately with the problem of desertification, or about $4.5 billion dollars a year. That is a lot of money, but the United States annual budget alone is over a trillion dollars, so in the total scheme of things, it is not proportionately an enormous amount. Investment in other areas, such as the military, dwarfs that kind of investment. Moreover, the annual estimated loss from desertification due to crop losses and soil destruction is estimated at $26 billion; the investments in halting desertification would have great economic benefits (Tolba 1984). Unfortunately, whether or not $90 billion over 20 years would have accomplished the task will never be known because in the years since that conference very little money was ever made available. Perhaps one-eighth to one-fourth of the amount of money estimated to be needed to reverse desertification is being spent on the effort.

One of the reasons for this lack of support is that the issues that underlie desertification are of a long-range nature and are not readily solved overnight. Budgets tend to focus on short-range spending that gives an immediate payoff or is politically popular. Support for worldwide efforts in reforestation, erosion control, and improved farming techniques requires longer vision, more persistence, and stronger commitment than is often seen in international aid projects. Too often international aid projects are oriented towards spectacular showpieces: building reservoirs, highways, and the like. Countries faced with famine, starvation, and other crises are more likely to deal with those immediate crises than to focus on longer range efforts to restore degraded ecosystems. The "donor countries," industrialized nations and institutions that tend to have control of most of the world's wealth, must contend with internal political opposition against the idea of foreign aid. Ignorance, hostility, and internal problems create obstacles to generosity from the donor nations. The recipient nations are often unstable, overwhelmed by immediate crises, and not well organized to combat desertification. Technical help, while necessary, cannot solve the problem without taking local social, political, and economic forces into account. The United Nations, severely pressed by both sides, has been trying to focus attention on the problem, and has supported local reforestation and conservation programs in many regions. Yet, tragically, the problem is getting worse (Matheson 1984c; Tolba 1984).

Another politically related problem that indirectly affects desertification is the urbanization crisis in developing countries. The growing urban population, and the problems this growth engenders, receives most of the too-little money available. Meanwhile, desertification is occurring in the rural areas, which are being neglected. The accumulation of problems in rural areas, in turn, is causing the cities to grow even more rapidly. People unable to make a living in rural areas migrate to the cities and aggravate an already unstable situation (Eckholm 1982; Tolba 1984).

Conclusions

Desertification does not occur overnight; it often develops gradually over long time periods. People may not be aware of the desertification process occurring around them until it is far advanced. It is an insidious problem that is becoming more and more visible as the results of overpopulation, overgrazing, overcultivation, and deforestation begin to overwhelm life-support systems in a number of countries. A much greater investment in labor, effort, and international commitment is needed to arrest and reverse this major cause of ecosystem degradation.

References

Ajan, M. A., "Protecting the Bread Basket—An Introduction to Somalia's Coastal Forestry Project," *Baobab #5,* Dakarfann, Senegal: Arid Lands Information Network and Oxfam, 1990.

Aubreville, A., *Climats, Forets, et Desertification de L'Afrique Tropicale,* Paris: Society d'Ed. Geographique, Marrt. et Coloniales, 1949.

Baobab #5, "Project Focus on Organic Farming in Senegal," Dakarfann, Senegal: Arid Lands Information Network and Oxfam, 1990.

Baumer, M., *Agroforestry and Desertification,* Technical Center for Agricultural and Rural Cooperation [CTA], The Netherlands, 1990.

Breman, H., and C. T. de Wit, "Rangeland Productivity and Exploitation in the Sahel," Science 221, 1983, pp. 1341–1347.

Brown, L. R. and C. Flavin, "The Earth's Vital Signs," *State of the World 1988,* New York: W. W. Norton, 1988, pp. 1–21.

Brown, L. R. and E. C. Wolf, *Soil Erosion: Quiet Crisis in the World Economy, Worldwatch Paper 60,* Washington, DC: Worldwatch Institute, September 1984.

Brown, L. R. and E. C. Wolf, *Reversing Africa's Decline, Worldwatch Paper 65,* Washington, DC: Worldwatch Institute, 1985.

Dregne, H. F., *Desertification of Arid Lands,* New York: Academic Press, 1983.

Dudal, R., "Land Degradation in a World Perspective," *Journal of Soil and Water Conservation* 37, 1982, pp. 245–249.

Eckholm, E., *Down to Earth: Environment and Human Needs,* New York: W. W. Norton, 1982.

Englebert, E. and A. Scheuring, *California Water,* Berkeley, CA: University of California Press, 1982.

Franke, R. and B. Chasin, *Seeds of Famine,* Totowa, NJ: Allenheld, Osmun, & Co., 1980.

Getahun, A. and K. Reshidi, *A Field Guide to Agroforestry in Kenya, 2nd ed.,* Rural Afforestation Extension, Ministry of Environment, Nairobi, 1989.

Hellden, U., "Desertification Monitoring," *Desertification Control Bulletin #17,* United Nations Environment Programs, 1988, pp. 8–12.

Hinman, C. W., "New Crops for Arid Lands," *Science* 225, 1984, pp. 1445–1448.

Hoskins, M., "The Contributions of Forestry to Food Security," *Unasylva,* 41, 1990, pp. 3–13.

Hyams, E., *Soils and Civilization,* London: Thames and Hudson, 1952.

Jordan, C. F., "Amazon Rain Forests," *American Scientist* 70, 1982, pp. 394–400.

Lamprey, H., "Report on the Desert Encroachment Reconnaisance in Northern Sudan," *Desertification Control Bulletin #17,* United Nations Environmental Program, 1988, pp. 1–7.

Lord, C., "Pest Control in the Sahel—Lessons from Mali," *Baobab #4,* Dakarfann, Senegal: Arid Lands Information Network and Oxfam, 1990.

Matheson, A., "Latest Assessment of Global Status and Trends in Desertification," United Nations Environment Programme Report Na.84-5176, Assessment of the Status and Trends of Desertification (1978–1984), 1984a.

Matheson, A., "The Institutional and Financial Arrangements to Implement the Plan of Action," United Nations Environment Programme Report Na.84-5175, Assessment of the Status and Trends of Desertification (1978–1984), 1984b.

Matheson, A., "Demographic Evidence from the Arid Lands of the World," United Nations Environment Programme Report Na.84-5171, Assessment of the Status and Trends of Desertification (1978–1984), 1984c.

Miller, G. T., Jr., *Living in the Environment, (6th ed.,)* Belmont, CA: Wadsworth Publishing Company, 1990, pp. 233–237.

Myers, N., "Tropical Forests and their Species: Going, Going . . . ?" In E. O. Wilson (ed.), *Biodiversity,* Washington, DC: National Academy Press, 1988, pp. 28–35.

Persson, J., "Trees, Plants and a Rural Community in the Southern Sudan," *Unasylva* 38, 1986, pp. 32–43.

Plato, *Critias,* 111B, trans. R.G. Bury.

Postel, S., *Water for Agriculture: Facing the Limits, Worldwatch Paper 93,* Washington, DC: Worldwatch Institute, 1989.

Raintree, J. B., "Agroforestry Pathways: Land Tenure, Shifting Cultivation, and Sustainable Agriculture," *Unasylva,* Vol. 38, 1986, pp. 2–19.

Raloff, J., "Africa's Famine: The Human Dimension," *Science News,* 1985, pp. 299–301.

Sabadell, J. E., E. M. Risley, H. T. Jorgenson, and B. S. Thornton, *Desertification in the United States: Status and Issues,* U.S. Bureau of Land Management, Washington, DC: Department of Interior, 1982.

Sanchez, P. A. and S. W. Buol, "Soils of the Tropics and the World Food Crisis," *Science* 188, 1975, pp. 598–603.

Sene, E., "Trees, Food Production and the Struggle Against Desertification," Unasylva, 37, 1985, pp. 19–26.

Sheridan, D., *Desertification of the United States,* Washington, DC: Resources for the Future, 1981.

Skoupy, D., "Developing Rangeland Resources in Africa Drylands," Desertification Control Bulletin #17, United Nations Environment Program, 1988, pp. 30–36.

Stiles, D., "Arid Land Plants for Economic Development and Desertification Control," Desertification Control Bulletin #17, United Nations Environment Program, 1988, pp. 18–21.

Swanson, E. R. and E. O. Heady, "Soil Erosion in the United States," *In* J. L. Simon and H. Kahn (eds.), *The Resourceful Earth,* New York: Basil Blackwell, 1984, pp. 202–222.

Timberlake, L., *Africa in Crisis,* Washington, DC: Earthscan, 1985.

Tolba, M. K., "Harvest of Dust," United Nations Environment Programme Report Na.84-5170, Assessment of the Status and Trends of Desertification (1978–1984), 1984.

Vietmeyer, N. D., "Lesser Known Plants of Potential Use in Agriculture," *Science* 232, 1986, pp. 1379–1384.

Chapter 11

Coastal Ocean Resources and Pollution: A Tragedy of the Commons

John B. Pearce and Angela Cristini

The expression *tragedy of the commons* was first used by Garrett Hardin to describe situations in which people find it in their individual self-interest to abuse the material things or resources that belong to no one, or are shared jointly by all of the citizenry (Hardin 1968). He cited as his first example the abuse of common grazing grounds by shepherds trying to maximize their individual herd sizes, resulting in disaster for all as the range is overgrazed. He gave several other examples of the same phenomenon, including industrial producers who find it in their economic self-interest to dispose of wastes cheaply by polluting a stream and visitors to a national park who arrive in such great numbers that they trample and destroy the qualities that made it worthy of being a national park. For Hardin, the final tragedy of the commons is the freedom to breed and therefore to overpopulate the planet, which he sees as leading to the ultimate disaster as the carrying capacity of the Earth is overwhelmed (see Chapter 2, "Population Growth and the Global Environment: An Ecological Perspective").

The oceans are the largest commons. Covering over 70% of the Earth's surface with salt water averaging 3,000–5,000 meters in depth, they are the dominant physical feature of this planet. Their almost incomprehensible vastness, combined with our very limited knowledge of many fundamental aspects of their dynamics and ecology, has created a condition that favors selfish human actions based on expediency and carelessness, justified by the notion that any damage will be diluted to a negligible level. Although all aspects of use of the oceans by humans has been exploitative—from the all-too-frequent massive oil spills in shipping lanes, to overfishing in international waters, to dumping of radioactive wastes—this paper will focus on one aspect: the alteration of estuarine and nearshore environments by human activities. Our limited understanding of the oceans suggests that their shallow margins covering the shelves of the continents are critical to their health, as well as being vital for many human activities. It is also the coastal areas that suffer most from the tragedy of the commons, from overfishing to dumping of wastes (Weber 1994); the chance that human misuse of coastal zones could cause irreparable harm to the world's oceans is not beyond the realm of possibility.

Pollution of the oceans is in large measure a consequence of the degradation of the coastal environments that form their margins. As fresh water drains to the sea, it brings wastes produced on land along with it; the ocean is the ultimate receptacle. When developed lands lie along the coastal margins, this effect is especially pronounced because the distance between land and sea is short, and because the waters are shallow along the continental shelf. As long as much of the world was still wilderness, or at least relatively undeveloped, the consequences of wastes draining into ocean waters were relatively minor. This is, however, no longer the case. Increases in technology and population during the twentieth century have allowed humans to have dominion over *all* lands, even the Arctic and Antarctic, and have created new kinds of pollutants, such as chlorinated hydrocarbons, that have affected marine life in unexpected ways. Today, humans, as well as the marine ecosystems, have begun to suffer from the consequences. On the eastern seaboard of the United States, the Chesapeake Bay, Jersey shore, Long Island, and Cape Cod are classic examples of overdevelopment. We can find similar examples from all over the world, even in seemingly remote places such as the Arctic and Antarctic.

Marine scientists have ranked *coastal development* as the most significant variable degrading the water quality and health of coastal and estuarine habitats (Weber 1994). In doing this, they have recognized that the urbanization and industrialization associated with development contribute significantly to both toxic contaminant loads and physical degradation of coastal areas. The Club of Rome's report, "The Limits to Growth," noted early that development in itself results in increased use of resources, contamination, physical degradation, and concomitant reduction in carrying capacity (Meadows et al. 1972). In addition, rapid increases in human population have been reported in many coastal regions of the temperate world, for example, Scotland, following industrial growth and improved harbor facilities (Currie and MacLennan 1984). The effect of development must be looked at as a whole, logically including items formerly categorized individually, such as toxic contaminants, nutrient overloading and consequent eutrophication and hypoxia, decline in marine resource diversity, and the dredging of harbors and building of jetties and seawalls. Therefore, this paper discusses these issues in the larger context of development of the coastal zone. By focusing on the New York/New Jersey Harbor Estuary and the New York Bight (the section of the Atlantic Ocean inside a transect between Montauk, Long Island and Cape May, NJ; Figure 11.1) as a case study, it documents the consequences of various types of human activities on oceanic resources. As appropriate, the discussion will be expanded to include other areas around the world.

Development

There have been numerous studies of the effects of harbor and port development. Most major cities have been studied at some time, and reports and popularized articles have been prepared on the resulting data. Increasingly, information neces-

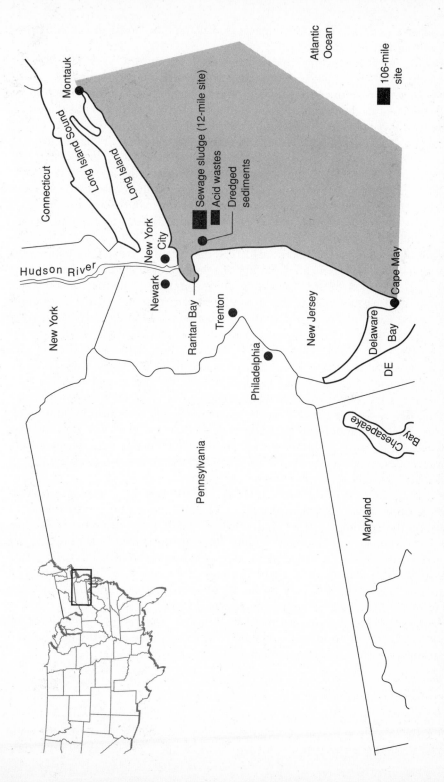

Figure 11.1. The New York Bight and the 106-Mile Site (also known as the Deepwater Municipal Sludge Dump Site). Adapted from Kaufman and Franz, *Biosphere 2000,* 1993.

sary to control or manage negative impacts is available; the problem lies in the implementation of recommendations.

One of the most interesting studies was written about New York City (Koebel and Kruekeberg 1975). It details the urban and suburban development of the region, and suggests relationships between population growth, increased pollution, and environmental decline in the coastal zone. Moreover, the paper considers New York City in the context of "Megalopolis," the extended urbanized tract that links Richmond, VA; Washington, DC; Baltimore, MD; Wilmington, DE; Philadelphia, PA; Newark, NJ; New York, NY; Bridgeport and New Haven, CT; Providence, RI; Boston, MA; and Portland, ME. One of the real hallmarks of the late twentieth century, "Megalopolis" has had very measurable impacts on the coastal zones of the northeastern United States, especially on the principal estuaries in the region.

New York City grew from a village of 270 people in 1628 to a massive complex of 19 million people by 1970 (including suburbs in a three state area). As pointed out by Koebel and Kruekeberg (1975), this was due initially to its fine port and later to its proximity to a canal and rail system that allowed manufactured merchandise to be shipped inland, facilitating the movement of agrarian and forestry products, as well as minerals, to be shipped outward to Europe or the southern colonies of the United States. In later decades growth was fostered by continuing human migration to the city to support elaborate and profitable mercantile and industrial activities. The Ports of New York, Newark, Philadelphia, Baltimore, and Norfolk have continued to be enlarged to provide the basis for strong economic growth and intensified urbanization. Large increases in pollution and continued decreases in estuarine water quality, with deleterious effects on living marine resources, have accompanied such growth. Some of the first recognized pollution effects were due to tainting of shellfish with petroleum products and metallic substances such as copper (Goode 1880). These contaminants were released from early nineteenth century refineries and smelters located on the shores of the Port of New York and the adjunct Raritan Bay in New Jersey.

Until about the time of the first Earth Day (1970), there had been relatively little reported on the total or *cumulative effects* of development on habitat quality or "health," and the conditions of associated living aquatic resources. One of the first studies that evaluated contaminant loading from an urbanized area (Providence, Rhode Island) located on a well-defined estuarine system (Narragansett Bay) was carried out by the Marine Ecosystems Research Laboratory of the University of Rhode Island (Oviatt 1980). Two problems in assessing contaminant effects on coastal waterways are lack of methodologies for determining the net result of many small changes and lack of an agreement on what an *unacceptable* environmental change might be. Two approaches dealing with these problems have been developed: assimilative capacity and cumulative effects of development.

Assimilative Capacity

Assimilative capacity was first discussed extensively as an issue during a National Oceanic and Atmospheric Administration (NOAA) Workshop at Crystal Mountain, Washington, in 1979 (Goldberg 1979). While participants found it difficult to define precisely what "assimilative capacity" is, there was a consensus that ". . . it is the amount of material that could be contained within a body of seawater without producing an unacceptable biological impact." One of the objectives of the workshop was to produce "endpoints," which would show when an unacceptable impact had occurred. The deliberations revolved around problems related to estuaries as well as to coastal and oceanic waters. Four panels studied the (1) site-specific problems of an oceanic disposal site for sewage sludge, (2) Puget Sound, (3) New York Bight, and (4) southern California Bight. The four studies were done in a case study mode and the results led to the conclusion that ". . . the waste capacity of U.S. coastal waters is not now fully used." In two areas, however, Puget Sound and the New York Bight, there was evidence that ". . . the assimilative capacities for some substances in some areas have been exceeded." In these two instances, areas affected were in close proximity to developed urban areas (Seattle and New York City) discharging many wastes in point and nonpoint modes.

Cumulative Effects

In recent years, aquatic habitat managers and scientists have turned their attention to assessing the cumulative effects of land development on coastal resources. Dickert and Tuttle (1985) reviewed the literature and identified critical issues that must be addressed in formulating a planning system that incorporates "cumulative impact assessments." This was done using a case study of a coastal wetland watershed, Elkhorn Slough, California. Methodologies were developed and implemented to be used in the assessment of the cumulative effects of accelerated erosion and sediment deposition in a wetland.

In addition, the U.S. National Research Council Board on Basic Biology and the Canadian Environmental Research Council held a binational workshop on Cumulative Environmental Effects, proceedings from which included sections on effects on marine systems. Dayton (1986) stressed that total cumulative effects must be considered in *any* management scheme or program of assessments; legislation, regulations, and management protocols *must* emphasize the additive effects of the various forms of development. All too often, this is not done; however, it is generally not the individual insults but the cumulative effects that degrade marine ecosystems. Waldichuck (1986) noted that ". . . the environmental impact of expansion of an existing structure (or level of contaminant) can be approximated by examining the impact of the existing facility (or generic information)."

A number of research groups have begun to develop centralized, compatible, historical databases that can be used in analyses of water quality and management of aquatic habitats (Stoddard et al. 1986). This approach is presently being used in the New York metropolitan area. The New York/New Jersey Harbor Estuary Program, funded by the Environmental Protection Agency (EPA), has completed several studies aimed at historically characterizing the harbor and estuary in terms of physical characterization; hydrologic modifications; wildlife habitats; pathogen contamination; toxicants; nutrient loadings; dissolved oxygen; and the distribution of benthic invertebrates, fish, and birds. These characterization reports were completed in 1991 and the Harbor Estuary Program is presently developing a management plan for this entire system. The remainder of this paper will focus on this marine urban habitat in terms of these characterization reports, as well as a discussion of these topics in relation to other similar systems worldwide.

Habitat, Physical Characterization, Hydrologic Modifications

Figures 11.1 and 11.2 are location maps for the NY/NJ Harbor Estuary and the New York Bight. The estuary exhibits a complex bathymetry and geometry, which affects the circulation and transport of materials within the system and out of the Bight. Instead of the usual single connection with the sea, this estuary has two connections with the open ocean: one through the Sandy Hook-Rockaway Point transect and the second through the East River to Long Island Sound. There are five

Figure 11.2. The New York and New Jersey Harbor Estuary.

significant sources of fresh water to the system: the Hudson, with an average of 2.3–46 billion gallons of fresh water per day; the Raritan, Hackensack, and Passaic Rivers, with a joint discharge averaging 2.0–5.0 billion gallons of fresh water per day; and sewage treatment plant effluents, averaging 2.7 billion gallons of fresh water per day. During time of low river flows, discharges from sewage treatment plants are an important source of fresh water. The system contains three tidal straits: the East River connects Upper Bay to Long Island Sound, with a rapid rate of flushing persistently directed towards Upper Bay; the short Kill Van Kull also has a rapid flushing toward Upper Bay; and the longer Arthur Kill displays a very sluggish flow, with almost no discernible direction and poor exchange with the rest of the NY/NJ Harbor Estuary. The circulation is driven by the tidal exchanges from ocean connections, flow of fresh water, and local and regional winds (Hires 1988). The complex interaction of these processes has been clarified to a great extent with the aid of a three-dimensional model used to calculate a mean residence time for most of this system of 1–2 weeks (Oey et al. 1985). This relatively short residence time has important implications for water quality within the system, as well as in the Bight.

The importance of New York's excellent port in the early years of the development of the city is well documented in "The Rise of New York Port" (Albion 1939). Since then there have been several studies of the causal role of port development and harbor maintenance in causing change in habitat quality; one example is "Port Facilities and Commerce" (Hammon 1976). At the time Hammon's paper was written, New York was America's busiest cargo port. The paper detailed the extent of various shipping channels, their lengths and depths, and also documented the amounts of several types of cargo moved through the harbor areas. Commodities carried included petroleum products, "basic" chemicals, liquefied gases, a range of ores and minerals, and manufactured products. As the population of the region grew, so did modifications to the shoreline for commercial and residential purposes, as well as to navigation channels and tributaries. These have resulted in profound changes, which will be described under three subheadings: dredging, shoreline modification, and hydrologic response.

Dredging

The U.S. Army Corps of Engineers maintains 78 different waterways in the urbanized New York Harbor. Rates of siltation vary by waterway, so frequency of maintenance dredging also varies. A principal disposal site for dredged sediments is the "mud grounds" (Figure 11.1); the quantity of spoils it receives varies substantially from year to year, but is in the millions of metric tons annually. An undetermined but substantial amount of these dredged materials are heavily contaminated with organic and inorganic toxicants such as cadmium, copper, mercury, lead, petroleum hydrocarbons, and DDT. The considerable variation in quantity of dredge spoils, combined with imprecise knowledge of the degree of contamination of each barge load, makes precise estimates of the amount of toxic loading difficult to develop; therefore, the *quantitative assessment* of ecological consequences is also uncertain, although estimates can be made.

Recent and still ongoing controversies about the dredging of the Port of Newark terminals illustrates these issues. Port shipping channels must be deepened to maintain its viability as a major terminal for deep-draft vessels; however, the dredged materials that would be removed to the "mud grounds" are contaminated with dioxin, a legacy of manufacturing of the herbicide 2,4,5-T during the Vietnam War era. Although the biological effects of small quantities of dioxin residues on human health are not fully understood, it has become clear that elevated levels of dioxin have accumulated in the tissues of blue crabs (*Callinectes sapidus*) in Newark Bay as well as in Raritan Bay (Cristini et al. 1993). Therefore, what is seen as an economic necessity for harbor maintenance could affect the already hard-pressed local fishing industry.

In addition to concerns about routine dredging in the early 1970s, a review was commissioned on possible effects of an enlarged supertanker terminal in the New York metropolitan area. The authors of the final report stated that the operation and construction of such a terminal would be more harmful to the fisheries and other marine life *if* it were located within an enclosed estuary, Raritan Bay, than if placed offshore in the open Bight (McHugh et al. 1972). It will be increasingly necessary to depend on such assessments to manage our coastal waters. This is especially true where large-scale industrial endeavors are involved in development or where numerous, smaller, projects would have collective effects. The latter situation has rapidly emerged as the more difficult one to resolve.

In many harbors throughout the world the accumulation of eroded silts, sands, and contaminants from upstream, as well as leaking or released cargoes and discharged wastes from vessels, requires regular dredging. As always with dredging, dredged materials must be deposited in other habitats. In most harbors the volume of the dredged material is too great, and often too heavily contaminated with toxic materials, to allow for cost effective land disposal; therefore, dredgings are dumped at sea as they are in the NY/NJ Harbor Estuary. Ocean disposal of sediments containing toxic chemicals remains a serious concern of citizens and governments worldwide. Several nations are considering techniques that could be used to isolate spoils dumped into coastal oceanic habitats; capping of contaminated dredged material is one way to reduce the likelihood of escape and movement of contaminated sediments (Bokuniewicz 1983). In theory, several feet of clay placed as a cap onto contaminated sediments should contain them; in practice, however, powerful storms and other such events may remove the cap, allowing toxicants to leach into the surrounding water.

Federal agencies in the United States are investigating ways to identify, categorize, and trace harmful organic contaminants, such as petroleum hydrocarbons, to better manage and regulate the ocean disposal of dredged materials.

Shoreline Modifications

Squires et al. (1991) reported in a characterization study that of the estimated 100 square miles (25,900 hectares) of wetlands once existing in the NY/NJ Harbor Estuary, about 14 square miles (3600 hectares) exist today. Over 75 square miles (19,400 hectares) have been landfilled for development. In addition, some 471 square miles (122,000 hectares) of nearshore underwater lands in New York and New Jersey have been filled and developed. It is clear that such modifications of natural habitats have resulted in alterations to both the population dynamics of the species present and the species composition of this complex of ecosystems. In addition to the significant reduction in amounts of available habitat, plants and animals living in this system today must contend with pressures of human cohabitation, including the pollution of air and water. Current lists of wildlife compiled from local, state, and federal sources indicate that there are presently 51 species of amphibians, 39 species of mammals, and 326 species of birds in this system. Calculation of the number of species that have been lost was not attempted because of the lack of reliable historical data. Many species that are known to be common in other similar, but less developed systems, however, are rare or were extirpated in the NY/NJ Harbor Estuary, while typical urban "generalists" were common or increasing (Squires et al. 1991).

Perhaps the single most debilitating activity affecting the well-being of coastal zones worldwide is the continued degradation and destruction of wetlands. In temperate biogeographic provinces, coastal wetlands being filled, dredged, or physically removed are often salt marshes. These are characterized by a dominance of grasses and grasslike plants; on the eastern seaboard of the United States dominant grasses include the cordgrasses (*Spartina* spp.). In the tropics, mangrove forests (composed of several species of hardwood trees and shrubs) are the most prominent coastal wetlands. As with salt marshes, mangrove forests are especially important in protecting and stabilizing subtropical and tropical coastal zones. Browder et al. (1986) have detailed changes in semitropical wetlands and associated fauna when channelization disturbs them.

While there are hundreds of papers on the subject of wetlands, few have provided the basis for long-term coastal monitoring and management. Herz (1987) reported on the use of aerial photography to verify rapid change in mangroves along the Brazilian coast. These ecologically crucial ecosystems are being destroyed to provide wood for fuels and new agrarian lands. Their exploitation for such purposes is most extensive along the northern coast of Brazil, where mangrove forests are prominent ecosystems. The less extensive mangroves occurring along the industrialized eastern and southern coasts of Brazil are also extremely vulnerable to domestic and industrial pollution.

The use of coastal waters for ecotourism, bathing, diving, boating, fishing, and aesthetics is increasing. In Florida the activity of 5 million residential beach users results in over 36,000 jobs, with an annual payroll of $240 million and total "sales" exceeding $1 billion. The development of marinas, fishing piers, and general tourism adds even more to Florida's economy; tourist uses of swimming beaches alone adds $3.4 billion and 142,000 jobs (Bell and Leeworthy 1986).

Public support of such a large economy obviously results in increasing pressures to develop additional coastal areas for recreational purposes. Private owners of wetlands have tremendous financial incentive to modify properties into harbors for small vessels, or to create bayfront condominium complexes, walking trails (boardwalks), or recreational beaches. The location of coastal construction setback lines are more likely to be governed by beachfront property values than by any value as habitat for living resources or fisheries purposes. Where local governments establish building regulations to protect coastal resources, loopholes often render these regulations largely ineffective. In New Jersey, for example, the strict building codes established by the Coastal Area Facility Review Act (CAFRA) apply only to construction complexes with 25 or more units; consequently, the coastline is filled with condominium complexes consisting of 24 units apiece.

Beaches are often identified as needing modification, such as sand nourishment for widening, or beach "stabilization" through the building of seawalls or jetties. The effectiveness of such programs is questionable at best; however, effects of such activities on habitat quality for fisheries and the health of coastal waters were rarely taken into account in evaluations leading to a decision to go ahead. Nelson (1985) provides guidelines to determine how beach restoration or rehabilitation programs might be accomplished while meeting certain biological criteria.

Hydrologic Response

Both dredging and shoreline modifications such as filling and damming result in hydrologic change in any harbor/estuary. Thatcher (1991) has divided such change into salinity and circulation modifications for the NY/NJ Harbor Estuary. His report indicates that there are no obvious trends in either category for this system because runoff from the Hudson River Watershed is still a predominant factor. Therefore, the natural variation in the rainfall in the region controls salinity and patterns of water circulation, and masks any of the changes that might be the result of human modifications of the system.

In other parts of the world, however, physical effects of dredging and deepening of channels have been shown to have immediate effects on hydrology, resulting in negative biological consequences. Durand et al. (1985) noted that channelization of the Ivory Coast harbor system in West Africa had major effects on the biota. The harbor at Abidjan was originally characterized by relatively fresh waters. With the opening of the Assinie Pass and the digging of the Vridi Canal, the interior waters, like those of San Francisco Bay (Rozengurt et al. 1987), became increasingly saline, resulting in alterations of the flora and fauna.

Pathogen Contamination

The waters of the NY/NJ Harbor Estuary and the New York Bight have received sanitary wastes from surrounding communities since precolonial times. The system also has a long history of importance as a producer of shellfish and as a place for contact water sports. Pathogen contamination resulting from human waste from the now heavily developed area has resulted in impairment of more beneficial uses of the system. Sanitary wastes enter the water in the form of outfalls from sewage treatment, ocean dumping of sewage sludge, raw sewage discharges, storm drains, and combined sewer overflows (CSOs). Combined sewer overflows are systems in which runoff from streets, parking lots, and roofs is diverted into sewage treatment plants, and then to local receiving waters such as the NY/NJ Harbor Estuary. Although the dumping of sludge into the New York Bight ceased in 1986, the long-term environmental affects resulting from the dumping continue to degrade this ecosystem. Therefore, this section will discuss sludge and other sources of sanitary wastes.

Sewage Sludge

The 12-mile sludge dump site received large quantities of untreated wastes that had well-documented ecological effects on the 65-foot deep Christiansen Basin in the New York Bight (Figure 11.1). A "sludge monster" resulting from a half-century of dumping of over 4 million tons of wet sludge annually captured the public imagination in the late 1980s, with much-ballyhooed incidents of sludge balls and hypodermic needles washing up on New Jersey and Long Island beaches. The physical presence of massive amounts of sludge did alter the characteristics of bottom sediments, as well as levels of dissolved oxygen in the bottom waters, so as to create a biologically "dead zone" in this part of the Bight. In addition, elevated levels of contaminants in tissues of marine life and a significant increase in fish diseases, such as fin rot, threatened the entire regional fishery.

In 1986, New York City and other municipal dumping authorities were mandated to move operations from the 12-Mile Site seaward to the 106-Mile Site (the Deepwater Municipal Sludge Dump Site) located near the edge of the continental shelf (Figure 11.1). Nine municipal sewage authorities, including New York City, continued to dump 9 million wet tons of waste annually into 106-Mile Site for 5 years. After several attempts to eliminate dumping of sludge and raw sewage in all offshore U.S. waters, the U.S. Congress passed the Ocean Dumping Ban Act of 1988, prohibiting disposal of ocean sewage after 31 December 1991.

In Canada, raw sewage sludge is still discharged offshore into coastal waters by Victoria, British Columbia; other major port cities in North America have ceased the direct ocean disposal of sewage sludge, and most cities have upgraded their treatment facilities. Nonetheless, pollution associated with sewage waste waters continues to be a serious problem in coastal waters, both in North America and around the world. Capuzzo et al. (1987) reported high levels of the organic contaminant coprostanol and high counts of the bacterium *Clostridium perfringens* in the urbanized coastal environments of Boston Harbor, Salem Harbor, and Raritan Bay; these are indicative of municipal domestic sewage.

Other Sources of Sanitary Waste

The New Jersey Department of Environmental Protection and Energy has produced a characterization study on pathogen contamination in the NY/NJ Harbor Estuary, which results from sewage treatment outfalls, combined sewer overflows (CSOs), raw sewage leaching, and storm drains and other nonpoint sources. The report states that there have been recent improvements in water quality as a result of upgrading sewage treatment plants and elimination of raw sewage discharges and *some* CSOs. Some of the positive effects, however, are being offset by the remaining CSOs and non-point source pollution associated with a highly urbanized area. The remaining sources of pathogens make it unlikely that the water quality standards required for direct harvest of shellfish, which are stringent, will be realized in the near future. Improvement in water quality resulting from reduced levels of pathogens, however, will likely result in an increase in areas of the system suitable for water contact sports (Downes-Gastrich 1991).

The direct discharge of untreated sewage from the tropical city of Abidjan, West Africa, to its harbor has resulted in anoxia and eutrophication. Changed patterns of circulation that resulted from dredging allowed saline waters to move into fjordlike bays; as salt water is denser than fresh water, salty waters then became isolated at the bottom. The saltier waters contained high levels of "sulphur-bearing" matter from the sewage discharges and were rarely renewed or exchanged; the resulting anaerobic conditions led to the formation of hydrogen sulfide, which is inimical to fish life (Dufour et al. 1985).

Similar events have been reported for tropical lagoon systems in Latin America and other continents (Sierra del Ledo et al. 1985). Sen Gupta and Qasim (1987) noted that in India only 50% of the total population bordering the Indian Ocean have access to sanitary arrangements. Similarly, only 17% of 3.8 million people living in coastal eastern African cities are adequately sewered. The result of such conditions is that there are high coliform counts on beaches and in coastal waters. Continued urban and agricultural development in these Third World nations, without provision for proper sanitation and waste treatment, will aggravate existing problems with regard to coastal water quality.

Human wastes released from shipboard have always represented a major form of pollution in harbor areas. Beginning with the Federal Water Pollution Control Act of 1972, the U.S. government took steps to require the promulgation and use of performance standards for marine sanitation devices (MSDs). These are capable of treating and storing human wastes while vessels operate within harbors having restrictive discharge regulations. Vessels (with MSDs) discharging wastes after 30 January 1980 were to meet Environmental Protection Agency (EPA) standards of 2 fecal coliform bacteria per milliliter and 150 mg suspended solids. It has not been established how much effect such contaminants have on overall water quality, or how such wastes affect living marine resources or amenities of interest to society. Reports suggest, however, that such pollution does have an effect, especially in local ports.

Toxicants

Contaminants enter harbors many ways. These include but are not limited to (1) runoff from streets, roofs, and parking areas; (2) pesticides and fertilizers carried in from farmlands, lawns, and right-of-ways; (3) point discharges of wastes from streams and treatment plants; (4) discharges from the air; and (5) discharges from shipping vessels themselves. Most forms of contaminants are then carried from harbor areas to offshore waters by the ebb and flow of tides, or circulated through the estuary, where they combine eventually with suspended matter and settle to the bottom. Once in place and associated with the sediments, many forms of contaminants may remain sequestered with sediments until resuspended by the energy of waves and currents. This may occur as a consequence of powerful storms, or from dredging done to create or maintain depths needed for shipping channels, turning or berthing areas, or dockside facilities.

Squib (1991) summarized existing data on concentrations of organic and inorganic chemicals in sediments, biota, and waters of the NY/NJ Harbor Estuary and the New York Bight and compared these values to existing "standards." This characterization study indicates that of the 11 inorganic and 35 organic chemicals classified as "toxicants of concern" by the EPA (for this system), 5 metals (cadmium, copper, lead, mercury, and nickel), 5 groups of pesticides (chlordane, DDT, dieldrin, heptachlor, and lindane), and 3 industrial organic chemicals (trichloroethylene, polychlorinated biphenyls [PCBs], and 2,3,7,8-tetrachlorodibenzo-p-dioxin [TCDD]) exceeded marine water quality standards and/or tissue standards designed to protect human health and marine life. Presently fishing bans and advisories are in effect for some of the more contaminated areas of the system because of elevated levels of PCBs and TCDD in edible tissues of fish and invertebrates. Sediment concentrations of such toxicants in the NY/NJ Harbor Estuary rank among the highest in the nation, and this study recommends sediment toxicity and bioavailability studies be performed before management plans are developed.

Distributions of toxicants in sediments, water, and biota reflect, of course, the tidal hydrologic patterns in the estuary. The NY/NJ Harbor Estuary complex receives discharges directly from some 50 municipalities, 700 combined sewer overflow outfalls, countless storm sewers, 80 landfills, and 400 industries. Thousands of septic systems also leach seaward into this system. Fresh water inflows from the rivers also carry residual pollutant loadings into the harbor from point and nonpoint sources in upstream drainage areas (St. John 1991). In addition, pollution is associated with shipping carried out in the active sections of the NY/NJ Harbor Estuary. In a study commissioned by the U.S. Maritime Commission, it was projected that 15.8 million liters of "oily wastes" would be generated in the Port of New York alone each day in 1980 (Harris 1973). While much of this waste is treated at shoreside facilities before being released to the water, an undetermined amount escapes without treatment to harbor waters. Atmospheric fallout is also considered an important source of pollutant loading in this highly industrialized area.

Beyond the NY/NJ Harbor Estuary, other U.S. estuaries were reported to be contaminated with organic compounds such as aromatic hydrocarbons (PAHs), DDT, and PCBs, at levels well above "typical values" as measured by the NOAA National Status and Trends Program (NOAA 1987a). Chlorinated hydrocarbons were abundant in sediments collected from Boston Harbor as well as from waters near Los Angeles, probably reflecting effects of large-scale waste discharges, including entrained DDT, in these regions. Casco Bay, forming the Portland, Maine, harbor, also had high values for PAHs, as did San Diego harbor.

Because of interaction between estuaries and adjacent ocean waters, offshore sediments and target biota were surveyed for the presence of organic and inorganic contaminants over broad geographical areas from Georges Bank and Nova Scotia south to the Delmarva Peninsula. Ocean quahogs (*Arctica islandica*) from Georges Banks and Nova Scotia were minimally contaminated with PCBs (2–5 parts per billion [ppb]), while clams taken from inshore habitats in the New York Bight, Rhode Island Sound, and Buzzards Bay were more contaminated, with values up to 25 ppb (Steimle et al. 1986). The highest values for PAHs, 55 ppb, were reported from the New York Bight. The authors also reported the presence of toxic trace metals: ". . . elevated trace metals were also usually associated with known areas of inputs, e.g., waste dumpsites or adjacent to heavily industrialized coastal areas."

In a review of available data on the distribution of contaminants in tissues from living marine resources taken in New England waters, Capuzzo et al. (1987) concluded that ". . . there are serious contamination problems along the New England coastline." Their review of a large data set involving several species and covering extensive geographic areas and temporal scales led them to state that ". . . the most serious evidence of chemical contamination in fish and shellfish populations is found among samples collected from harbor locations."

Other research and monitoring in areas affected by urbanization, waste disposal, agrarian runoff, and other forms of development show similar loadings of contaminants into estuaries worldwide. A detailed study from Southampton, Hampshire, UK, found levels of nutrient ions up to 104 times higher (averaging $10 \times$ higher) than levels found in undeveloped rural areas. In part, the enrichment of ions in urban settings is attributable to disturbances exposing deeper layers of sediments to atmospheric acidity, which consequently mobilized the ions. Other sources of excess ion input include building materials, industrial and domestic pollution, atmospheric deposition from surrounding industry, highways, railways, airports, and garden and lawn fertilizers (Prowse 1987).

Nutrient Loading, Dissolved Oxygen

Nutrients such as dissolved ammonia, nitrate, nitrite, phosphate, and organic carbon have serious consequences for primary and secondary productivity in aquatic ecosystems. The NY/NJ Harbor Estuary receives sewage waste water from one of the most populated regions of the country; high levels of nutrients are always available for primary productivity. This system also is characterized by sparse rooted aquatic vegetation; most primary production is carried out by phytoplankton. Primary productivity in Raritan Bay exceeds values for other systems where phytoplankton are the dominant producers. Growth and reproduction of phytoplankton are limited, however, because their own numbers, as well as the high density of suspended particles in the water column, increase turbidity of water and thereby decrease the penetration of light necessary for photosynthesis. Such limitation results in phytoplankton utilizing only approximately 10% of the available nutrients. The remaining 90% of the nutrients entering this system flow into the Hudson Harbor Estuary plume, which, in turn, affects water quality along the coasts of New Jersey and Long Island (Studholme 1988).

Levels of nutrient loading directly affect primary and secondary productivity, which, in turn, control amounts of dissolved oxygen in bottom waters. Keller (1991) has analyzed long-term distributions of dissolved oxygen, nutrients, and organic carbon in the NY/NJ Harbor Estuary. His data show that summer bottom oxygen concentrations are highly variable. Hypoxia is often a problem, and some areas of the system frequently experience concentrations of dissolved oxygen below that necessary to support marine life. In several areas of the inner harbor, however, there have been significant long-term improvements in dissolved oxygen after discharges of raw sewage were curtailed and treatment plants upgraded. Factors controlling the oxygen levels were found to be a combination of natural variations in temperature (water of higher temperature holds less dissolved oxygen) and the concentrations of ammonia and organic carbon (Keller 1991). In the NY/NJ Harbor Estuary, major sources of nutrients (dissolved ammonia, nitrate, nitrite, phosphate, and organic carbon) have been shown to be the discharges from municipal sewage treatment plants, tributaries (containing their municipal discharges), and to a lesser extent, CSOs (St. John 1991). This points up again the need to gain a better understanding of how treatment and discharge of wastewater affects the ecosystem directly (discharging into the harbor) and indirectly (discharging into tributaries). In all cases, the municipal dischargers are operating in accordance with permits issued by the states, but the cumulative effect is degraded water quality in river basins and the estuary.

Researchers at the University of Rhode Island studying the input of nutrients, toxic organic compounds, and trace elements into Narragansett Bay report that most excess nutrients were transported by way of the Providence and Pawtuxet Rivers. Copper and nutrients largely were introduced from sewage treatment plants, while most lead had origins from the CSOs. The CSOs become overloaded whenever there is an episode of heavy rainfall and the raw sewage and road runoff wastes are flushed directly into riverine and estuarine waterways. Increased nutrient loadings were found to cause summer hypoxic conditions in the lower Providence River, the upper reaches of the bay, and at the estuarine salt wedge. Shortages of dissolved oxygen change a productive estuary into degraded habitat (Oviatt 1980). This example is from Rhode Island, but unfortunately is typical of most urbanized coastal areas of the eastern seaboard; the practice of combining sewage treatment with street runoff inevitably leads to system failure but remains widespread throughout much of the United States.

Reports from the United Kingdom show similar problems with estuarine contamination associated with agricultural and industrial development. Green (1984) reported a 42% increase in nitrate levels in a small river (the Frome in Dorset) over a period of 7 years. Although the amount of fertilizers the farmers were using did not exceed Ministry of Agriculture recommendations, 25% of the applied nitrogen was carried into reservoir waters. Comparable amounts are undoubtedly exported to coastal waters, where they cause reduced summer oxygen values, which in turn may prove lethal to fishes. Stories of similar contamination and habitat degradation have been reported in the Mersey estuary (Handley 1984) and in Swansea Bay (Collins et al. 1980).

Coral reefs, among the most productive and diverse of marine ecosystems, are quite sensitive to nutrient enrichment. Elevated concentrations of nitrogen and phosphorus lead to phytoplankton population explosions. This in turn results in less sunlight reaching the reef, to the detriment of the reef algae. High concentrations of phosphorus have also recently been found to inhibit calcification in coral, a process that is central to reef development. There is little doubt that nutrient enrichment is a significant contributing variable in the worldwide decline of coral reefs (Burke 1994).

Distribution of Benthic Invertebrates, Fish, and Birds

The NY/NJ Harbor Estuary attracts and supports a diverse assemblage of organisms, most depending on this habitat during critical periods in their life history associated with reproduction and development. Typically these organisms, as well as those species that reside for their entire life cycle in the estuary, are tolerant of fluctuations naturally present in their environment, even some that result from anthropogenic activities. Nevertheless, the effects of living in a system surrounded by heavily populated and industrialized regions have resulted in alterations of species diversity and abundance, and their use (Studholme 1988). This section describes these resources under three headings: benthic invertebrates, fish, and marine birds. In addition, the potential ecological damage from the unloading of ballast water from ships will be discussed.

Benthic Invertebrates

The macrobenthos (invertebrates that are retained on sieve openings of 0.5 mm) are often selected to study effects of chemical contamination because the organisms live on or in the sediments. In addition, they are very important to the trophic structure of the estuary, and many species are commercially important. As has been discussed previously, many chemicals, organic and inorganic, bind to particles and accumulate in sediments of estuaries, which become repositories and potential sources of contamination to the fauna. Since the NY/NJ Harbor Estuary has been subjected to many decades of degraded water quality, changes in the structure of the benthic community have occurred.

Cristini (1991) compiled data on the distribution of benthic invertebrates from 18 studies done in this system in the past 20 years. This characterization report indicated that among the species reported to be present in the different studies, the distribution of amphipods (an order of smaller crustaceans) during summer seasons seemed to be an indicator of degraded conditions in bottom sediments of the system. Certain species of amphipods were present in areas of the system that showed lowest concentrations of toxic chemicals and highest dissolved oxygen values, and became rare, or disappeared, completely in the more degraded sections. These data are preliminary; however, they indicate that monitoring programs could be designed to reflect positive and negative changes in the benthic community, simply by collecting data on "toxics of concern" and relating them to distributions of key "indicator" species as well as concentrations present in the tissues of organisms.

Historical comparisons of the distribution of American oyster (*Crassostrea virginica*) and soft shell clam (*Mya arenaria*) indicate that their abundance is greatly reduced and that existing populations are restricted compared to the late 1800s to the 1930s (Studholme 1988). Although sections of the NY/NJ Harbor Estuary are closed for the harvesting of blue crabs (*Callinectes sapidus*), because of dioxin and PCBs, and of hard clams (*Mercenaria mercenaria*), because of pathogens, these organisms and other large invertebrates are still prevalent. There is a commercial fishery for the blue crab in the open sections of the estuary, and hard clams can be harvested from certain areas of the estuary, to be subsequently relayed to waters of New York and New Jersey with high water quality for 30 days, and then sold for human consumption. This information highlights the state of this natural system that has been the victim of the "tragedy of the commons" for more than a century. There is still some resiliency left; hardy species can and do survive and support a now marginal fishery that is important to the residents of the area; however, a large percentage of the biotic resources are unusable because of pollution.

Similar situations can be found in most other parts of the world. In the Orient there have been numerous reports of the negative effects of development on the nearshore environments. One early report from Japan detailed how paper pulp mills had a "cascading" effect on Suruga Bay: the organic residues discharged from paper mills resulted in reduced oxygen levels and clogged shipping lanes, which in turn required dredging and the placement of contaminated dredge spoils in other areas of coastal waterway (Nakai et al. 1972). Japanese scientists noted that human and industrial wastes have affected both the phytoplankton and the zooplankton, as well as bottom-dwelling animals and finfishes. These effects included uptake of organic and inorganic contaminants, changes in the abundance of plankton and fish species, and the presence of benthic indicators of pollution, including bacteria and the polychaete worm *Capitella capitata*. The latter are now recognized as classic signals of pollution-induced changes and are present in contaminated estuaries around the world.

Fish

A report prepared by Woodhead (1991) summarized 35 fisheries data sets for the NY/NJ Harbor Estuary compiled between 1979 and 1989. It states that more than 100 species of finfish have been recorded from this system, making it host to one of the most diverse communities for estuaries along the eastern seaboard. This high diversity is a result of the mixing of fish species from two zoogeographic provinces, the Boreal (or Arcadian) and the Virginian. The report further states that marine species dominate (70%), with fresh water, estuarine, and migratory species each comprising about the same share (10%) of the community. The occurrence and distribution of these species are seasonal and related to their temporal life history patterns.

The system supports several species of full-time residents, such as the silverside, mummichog, white perch, and the bay anchovy, which serve as forage or bait fishes for coastal species, often attracting them into the estuary. Other species make spring spawning runs into the fresh and brackish waters of the harbors and estuary. Still other fish, such as the blue fish, weakfish, fluke, and winter flounder, depend on both estuarine and marine habitats during different stages of their life cycles (Studholme 1988).

This range of species has supported commercial and recreational fisheries over a 200 year period. The recreational fishery is still very active; however, the commercial fisheries peaked in the 1940s and then steadily declined. There are several reasons for the decline in the commercial fishery: (1) Physical alteration of the habitat, such as dam construction and filling marsh areas; (2) decline in water quality because of sanitary and industrial discharges; (3) failure of reproduction and recruitment; (4) overfishing on reduced brood stocks; (5) improvement of fishery technology (more efficient vessels and gear); and (6) gear restrictions within the estuary are all somewhat responsible for the decline in this industry (Studholme 1988). In addition, the commercial fishery for striped bass has been closed because of elevated levels of PCBs in edible tissues; the taking of bass by recreational fishing has been prohibited in the Newark Bay Complex because of dioxin. Such closures have also created a major perceptual problem extending throughout the whole area and involving *all* species. Many fish and shellfish caught in the NY/NJ Harbor Estuary and the New York Bight have reduced desirability and market value compared to the same species caught in what are perceived to be cleaner waters (e.g., Chesapeake blue crabs, Maine lobsters). Consumers perceive local fish to be of lower quality, even though they do not violate existing standards for chemical contamination in edible portions (Cristini and Reid 1988).

Elevated levels of PCBs in sediments and waters of estuaries along the eastern seaboard prompted the federal government to examine levels of PCBs in the flesh of bluefish along the entire U.S. East Coast. The results of this survey, conducted in 1984–1986, showed that fishes less than 20 inches long did not have levels above the FDA tolerance limit of 2 ppm. At least some larger fish sampled from every East Coast site, however, exceeded limits. Temporal distribution data indicate that the large, contaminated bluefish migrate and could represent an increase in human health risk to consumers taking fishes during times the fish move into various coastal areas (NOAA/FDA/EPA 1987).

Marine Birds

The NY/NJ Harbor Estuary supports nesting sites for 25–50% of wading birds (i.e., herons, egrets, and ibises) in New York, New Jersey, and Connecticut. In addition, several other important species of aquatic birds nest in colonies throughout the system. The feeding ranges of these (some fish eating) birds often cover large sections of the estuary; they tend to be long lived and concentrate available toxicants via trophic interactions. These species therefore have considerable potential for monitoring biological effects and the health of this system (Parsons 1991). Although such potential exists, Parsons (1991) points out that data and assessments pertaining to the levels of pollutants to which the birds are exposed are not complete. In addition, although some data exist on the presence of heavy metals in waterfowl and wading birds, no systematic study of the levels of "toxicants of concern" has been attempted. Furthermore, the effects of pollutants on birds using this system has not been documented. During the coming years it is hoped that the new NY/NJ Harbor Estuary Program will implement a biomonitoring program that will characterize exposures and ecological effects of pollutants on the key avian residents.

Ballast Water

Although not an identified problem in the NY/NJ Harbor Estuary and the New York Bight at present, one of the most troubling sources of ecological damage to critical estuarine habitats has come from the routine unloading of ballast water from transoceanic shipping vessels. Daily, numerous ships transfer ballast waters from regions of origin to their harbor of destination. In the process, hundreds of nonindigenous marine organisms are introduced into local waters. These introduced species range from meroplankton and holoplankton to small fish and larval forms of mussels, worms, and snails. The shipping industry has become a remarkable vector for incidental exchange of exotic species. Although only a very small percentage of these introduced species survive, those that do may be capable of disrupting local ecosystems at the expense of native species. The arrival of zebra mussels (*Dreissena polymorpha*) into eastern North America in the mid-1980s, presumably from ballast water, has led to massive and costly clogging of intake pipes, heat exchangers, and valves at power plants and other facilities. The economic disruption and environmental degradation that zebra mussels are causing has become a severe burden for the Northeast (Ross 1994). Similar disasters are, in the long run, inevitable so long as ballast water is carried from harbor to distant harbor.

Although the individual ship may bear legal responsibility for violations of pollution control ordinances, in the case of ballast water discharge no law is being broken. Since individual ships incur no financial responsibility for consequences of ballast water discharges, and the shipping industry as a whole is not prepared to change voluntarily methods of operation (that are presently cheap and familiar), it is the ecosystem, and the society that depends on it, that bears the brunt of yet another tragedy of the commons.

Conclusions

It is ever clearer that a wide range of ecologically stressful factors are associated with development. To date, however, very few managers in charge of estuaries and scientists have attempted the difficult task of relating changes due to development with field and laboratory measurements of biological effects (such as changes in the distribution and abundance of important living marine resources), as well as using such changes to develop a comprehensive management scheme for the estuarine system. The NY/NJ Harbor Estuary Program has completed the first phase of this process; characterization reports have resulted in the development of centralized, compatible, historical databases for the program. These databases are being used to direct research necessary to provide missing information relating changes due to development with ecological effects, as well as to develop a management plan for the estuary and bight. The application of historical data to resource and habitat management questions is crucial if steps are to be taken to reverse the documented losses that have and are occurring as a result of coastal development.

One of the first attempts at multijurisdictional, multiple-use management in the United States is taking place in Chesapeake Bay. This large estuary is bounded by three states and the District of Columbia. Several major cities, including Baltimore, Washington, and Norfolk, are sited directly on its shores; others, including Philadelphia and Richmond, affect its water quality by virtue of river basin drainage, tides, and currents. Literally scores of smaller cities are sited on the shores of the bay or in its watershed area. Based on the focused study of the principal issues of low dissolved oxygen levels and increasing contaminant discharges, the U.S. Environmental Protection Agency (1983) developed a "Framework for Action" that specified steps to be taken in future decades to rectify the consequences of years of unplanned development. This led to the "Chesapeake Bay Restoration and Protection Plan," prepared by the states of Maryland, Pennsylvania, and Virginia, as well as the U.S. EPA (1985). Based on this plan, interagency studies were initiated that resulted in reports to be used in planning future development. One paper from the Chesapeake Bay Living Resources Task Force, "Habitat Requirements for Chesapeake Bay Living Resources," provides a basis for water quality standards necessary to manage the principal fish species threatened by shoreline development (Maryland Department of Natural Resources 1987). A second paper, "Vegetated Filter Strips for Agriculture Runoff Treatment," provides information on how to manage agrarian habitats so as to reduce erosion, thus preventing excessive nutrient loading (Magette et al. 1987). Continued multiple use management depends on consensus regarding the nature of the conflicts, agreements upon standards or criteria to resolve the conflicts, monitoring to provide evaluation of mitigation and abatement efforts, and feedback to the citizenry to ensure future support by the people and their elected representatives.

A similar program is underway in the Gulf of Maine (GOM), a basin bounded by three states and two Canadian provinces. Beginning with an international conference, the GOM Council has developed monitoring, action, and data management programs based on its original vision statement (Van Dusen and Johnson Hayden 1989).

Many national governments also recognize the need for coastal and estuarine management, and have taken steps to counter the effects of continued unplanned development. For example, in the Thames River estuary, United Kingdom, remedial steps were initiated over two decades ago, and improvements are now highly visible (King and Kendall 1987). Nations should, as emphasized by Waldichuck (1986), develop a conceptual framework wherein cumulative multijurisdictional environmental effects of contaminants and physical degradation can be evaluated. Without such an approach, bolstered by appropriate standards for water quality and the criteria to judge these by, there can be no control of adverse aspects of development. Mitigative steps will range from source controls, recycling, vehicle inspections, and new ways of port and harbor development to absolute prohibitions on new development within especially sensitive areas of the coastal zone. Syntheses and assessments such as the NOAA Land Use Characteristics Data Atlas (NOAA 1987b), as well as the compilation of data from studies using data from remote sensing environmental satellites, must be used as benchmarks against which future change can be compared worldwide. If there is sufficient public awareness and the political will to bring about more vigorous protection policies, these data could provide the basis for enhanced zoning and management of sensitive coastal habitats. As recently suggested by the United Nations Environmental Program (Pearce 1991), it is the *collective effects* of human activities that compromise the sustainable use of the coastal zones and estuaries worldwide. It is only through the education of the general citizenry, and regional and international collaboration and coordination, that real progress can be made. This need for education was recently stressed by the Water Quality 2000 Program of the Water Environment Federation (formerly the Water Pollution Control Federation) (Water Environment Federation 1992). In its chapters "Getting From Problems to Solutions—The Tools of Change" and "A Management Approach for Solving Water Quality Problems," the education of the public, students, politicians, bureaucrats, and the "perpetrators" was seen as essential for progress. Hopefully the present chapter and volume will, in part, fill this need.

References

Albion, R., *The Rise of New York Port (1815–1860),* New York: Charles Scribners Sons, 1939.

Bell, F. and V. Leeworthy, "An Economic Analysis of the Importance of Saltwater Beaches in Florida," Report No. 82, Tallahassee, FL: Florida Sea Grant College, Florida State University, 1986.

Bokuniewicz, H., "Submarine Borrow Pits as Contaminated Sites for Dredged Sediment," *In* D. R. Kester, B. H. Ketchum, E. W. Duedall, and P. K. Park (eds.), *Wastes in the Ocean* (Vol. II), New York: John Wiley and Sons, 1983, pp. 215–227.

Browder, J., A. Dragovich, J. Tashiro, E. Coleman-Duffie, C. Foltz, and J. Zweifel, "A Comparison of Biological Abundances in Three Adjacent Bay Systems Downstream from the Golden Gates Estates Canal System," National Ocean and Atmospheric Administration Technical Memorandum NMFS-SEFC-185, Miami, FL: NMFS, Southeast Fisheries Center, December 1986.

Burke, M., "Phosphorus Fingered as Coral Killer," *Science* 263, 1994, p. 1086.

Capuzzo, J., A. McElroy, and G. Wallace, *Fish and Shellfish Contamination in New England Waters: An Evaluation and Review of Available Data on the Distribution of Chemical Contaminants,* Washington, DC: Coast Alliance, 1987.

Collins, M., F. Banner, P. Tyler, S. Wakefield, and A. James (eds.), *Industrialized Embayments and their Environmental Problems,* London: Pergamon Press, 1980.

Cristini, A., "Effects of Toxicants on the Distribution of Benthic Invertebrates," Characterization Report, New York-New Jersey Harbor Estuary Program, EPA Region II, New York, 1991.

Cristini, A. and R. Reid, "Effects of Contaminants on the Fauna of the Hudson-Raritan Estuary, and Risks to Humans from Consuming Estuary Fauna," *In The Hudson-Raritan: State of the Estuary* (appendix), Water Quality of New Jersey Coastal Waters Vol.1, Part 2, New Jersey Marine Sciences Consortium, 1988, pp XII1–13.

Cristini, A., Z. W. Chi, and M. Gross, "The Occurrence of Dioxins and Dibenzofurans in the Edible Tissues of the Blue Crab, *Callinectes sapidus* from Newark and Raritan Bays," Final report to the NJDEPE, Trenton, NJ, 1993.

Currie, A. and A. MacLennan, "A Prospectus for Nature Conservation within the Moray Firth: In Retrospect," *In* R. Roberts and T. Roberts (eds.), *Planning and Policy,* London: Chapman and Hall, 1984, pp. 238–253.

Dayton, P., "Cumulative Impacts in the Marine Realm," *In Cumulative Environmental Effects: A Binational Perspective,* Ottawa, Canada: The Canadian Environmental Assessment Research Council (CEARC), 1986, pp. 79–84.

Dickert, T. and A. Tuttle, "Cumulative Impact Assessment in Environmental Planning. A Coastal Wetland Watershed Example," *Environmental Impact Assessment Review* 5, l985, pp. 37–64.

Downes-Gastrich, M., "Characterization of Pathogen Contamination," Characterization Report, New York-New Jersey Harbor Estuary Program, EPA Region II, New York, 1991.

Dufour, P., J. Chantraine, and J. Durand, "Impact of Man on the Ebrie Lagoonal Ecosystem," *In* N. Chao and W. Kirby-Smith (eds.), *Proceedings of the International Symposium on Utilization of Coastal Ecosystems: Planning, Pollution, and Productivity,* Beaufort, NC: Duke University Marine Laoratory, 1985, pp 467–484.

Durand, P., J. Chantraine, and J. Durand, "Research and Development: Some Illustrations and Prospects for the Brackish Waters of the Ivory Coast," *In Proceedings of the International Symposium on Utilization of Coastal Ecosystems: Planning, Pollution, and Productivity,* 21–27 November 1982, Rio Grande, Brazil, 1985, pp. 439–466.

Goldberg, E. D. (ed.), "Assimilative Capacity of U.S. Coastal Waters for Pollutants," Proceedings of a Workshop, Washington DC 29 July–4 August, 1979. Boulder, CO: National Oceanic and Atmospheric Administration, December 1979, second printing, revised, June 1980.

Goode, G., *The Fisheries and Fishing Industries of the U.S. Sect II; A Geographical Review of the Fisheries, Industries, and Fishing Communities for the Year 1880,* Washington DC: U.S. Government Printing Office, 1880.

Green, B., "Landscape Evaluation and the Impacts of Changing Land-Use on the Rural Environment: The Problem and an Approach," *In* R. Roberts and T. Roberts (eds.), *Planning and Ecology,* London: Chapman and Hall, 1984, pp.156–164.

Hammon, A., *Port Facilities and Commerce,* MESA New York Bight Atlas Monograph No. 14, Albany, NY: New York Sea Grant Institute, 1976.

Handley, J., "Ecological Requirements for Decision-Making Regarding Medium-Scaled Developments in the Urban Environment," *In* R. Roberts and T. Roberts (eds.), *Planning and Ecology,* London: Chapman and Hall, 1984, pp. 156–164

Hardin, G, "The Tragedy of the Commons," *Science* 162, 1968, pp. 1243–1248.

Harris, F., *Port Collection and Separation Facilities for Oily Wastes,* Maritime Administration, U.S. Department of Commerce, 5 vols, 1973.

Herz, R., "A Regional Program on Coastal Monitoring and Management of Mangrove in Brazil," *In* O. Magoon, H. Converse, D. Miner, L. Tobin, D. Clark, and G. Domurat (eds), *Coastal Zone '87: Proceedings of the Fifth Symposium on Coastal and Ocean Management,* New York: American Society of Civil Engineers, 1987, pp. 2262–2268.

Hires, R., "Circulation and Transport in the Hudson-Raritan Estuary," *In The Hudson-Raritan: State of the Estuary* (appendix), Water Quality of New Jersey Coastal Waters Vol.1, Part 2, New Jersey Marine Sciences Consortium, 1988, pp III1–16.

Keller, A., "Analyses of the Distribution of Dissolved Oxygen, Nutrients and Organic Carbon," Characterization Report, New York-New Jersey Harbor Estuary Program, EPA Region II, New York, 1991.

King, L. and J. Kendall, "State Capacity for Estuarine Management: The Case of Galveston Bay, Texas," *In Proceedings: Oceans '87,* Halifax, Nova Scotia, 1987.

Koebel, C. and D. Kruekeberg, *Demographic Patterns, Vol. 23,* MESA New York Bight Atlas Monograph, Albany, NY: New York Sea Grant Institute, 1975.

Magette, W. L., R. Brinsfield, R. Palmer, J. Woods, T. Dillaha, and R. Renlau, *Vegetated Filter*

Strips for Agriculture Runoff Treatment. Philadelphia: U.S. Environmental Protection Agency, February 1987.

Maryland Department of Natural Resources, *Habitat Requirements for Chesapeake Bay Living Resources—Chesapeake Bay Program,* Annapolis, MD: Maryland Department of Natural Resources, 1987.

McHugh, L. O., B. Knapp, J. Ginter, M. Greenfield, and A. Tsao, "Possible Effects of Construction and Operation of a Supertanker Terminal on the Marine Environment of New York Bight," Special Report to NOAA and the CEQ, Stony Brook, NY: Sea Grant Program, Marine Sciences Research Center, State University of New York at Stony Brook, 1972.

Meadows, D. H., D. L. Meadows, J. Randers, and W. Behrens III, *The Limits to Growth,* New York: Universe Books, 1972.

Nakai, Z., M. Kosaka, I. Okada, S. Kudoh, F. Hayashida, T. Kubota, M. Ogura, T. Mizushima, and I. Uotani, "Change of Biological Environments in the Suruga Bay," *In Interim Report on Change of Marine Environments Caused by Human Society in the Water Around the Suruga Bay for 1971,* Study Group on Environmental Conditions for Human Survival, Shimizu, Japan: College of Marine Science and Technology, Tokyo University, 1972.

National Oceanic and Atmospheric Administration (NOAA), Progress Report and Preliminary Assessment of Findings of the Benthic Surveillance Project—1984. Rockville, MD: NOAA Office of Oceanography and Marine Assessment, National Status and Trends Program, 1987a.

National Oceanic and Atmospheric Administration (NOAA), *National Estuarine Inventory Data Atlas, Volume 2, Land Use Characteristics,* Rockville, MD: Ocean Assessment Division, Strategic Assessment Branch, 1987b.

National Oceanic and Atmospheric Administration (NOAA), Food and Drug Administration (FDA), and the Environmental Protection Agency (EPA), "Report on the 1984–1986 Federal Survey of PCBs in Atlantic Coast Bluefish: An Interpretive Report," National Marine Fisheries Service, NOAA, FDA, EPA 1987.

Nelson, W., "Guidelines for Beach Restoration Projects, Part I," *Biological Report No. 76,* Melbourne, FL: Florida Sea Grant College, 1985.

Oey, L. Y., G. L. Mellor, and R. L. Hires, "A Three Dimensional Simulation of the Hudson-Raritan Estuary. Part III," *Journal of Physical Oceanography* 15, 1985, pp. 1676–1692.

Oviatt, C., "Some Aspects of Water Quality in and Pollution Sources to the Providence River," Report for U.S. EPA, Region I, Contract #68-04-1002. Kingston, RI: University of Rhode Island Graduate School of Oceanography, 1980.

Parsons, K., "Effects of Toxicants on Marine Birds," Characterization Report, New York-New Jersey Harbor Estuary Program, EPA Region II, New York, 1991.

Pearce, J. B., "Collective Effects of Development on the Marine Environment," *In Oceanologica Acta, Proceedings of the International Colloquium on the Environment of Epicontinental Seas,* Lille, France, 20–22 March 1990, Vol. 11, 1991, pp. 287–298.

Prowse, C., "The Impact of Urbanization on Major Ion Flux through Catchments: A Case Study in Southern England," *Water, Air, and Soil Pollution* 32, 1987, pp. 277–292.

Ross, J., "An Aquatic Invader is Running Amok in United States Waterways," *Smithsonian* 24(11), 1994, pp. 40–50.

Rozengurt, M., M. Herz, and M. Josselyn, "The Impact of Water Diversions on the River-Delta-Estuary-Sea Ecosystems of San Francisco Bay and the Sera of Azov," *In* NOAA Estuary-of-the-Month Seminar Series, No. 6. San Francisco Bay: Issues, Resources, Status, and Management. 22 November 1985. NOIAA Estuarine Programs Office, Washington DC, October 1987, pp. 35–62.

Sen Gupta, R. and S. Qasim, "The Indian Ocean—An Environmental Review," *In* R. S. Sharma (ed.), *The Oceans—Realities and Prospects,* New Delhi, India: Rajesh Publishing, 1987.

Sierra de Ledo, B., J. Rocha Gre, and E. Soriano-Sierra, "Fishery Production, Anthropogenic and Natural Stress in Conceicao Lagoon, Santa Catarinas, Brazil," *In* N. Chao and W. Kirby-Smith (eds.), *Proceedings of the International Symposium on Utilization of Coastal Ecosystems: Planning, Pollution, and Productivity,* Beaufort, NC: Duke University Marine Laboratory, 1985, pp. 485–496.

Squib, K., "Characterization of Toxic Contamination," Characterization Report, New York-New Jersey Harbor Estuary Program, EPA Region II, New York, 1991.

Squires, D. F., J. Barclay, R. Craig, and C. Kaiatus, "Inventory of Habitat and Wildlife Populations in the New York/New Jersey Harbor Estuary," Characterization Report, New York-New Jersey Harbor Estuary Program, EPA Region II, New York, 1991.

St. John, J., "Inventory of Pollutant Loadings," Characterization Report, New York-New Jersey Harbor Estuary Program, EPA Region II, New York, 1991.

Steimle, F., P. Boehm, V. Zdanowica, and P. Bruno, "Organic and Trace Metal Levels in Ocean Quahog *(Arctica islandica,* Linn) from the North West Atlantic," *U.S. Fish. Bull.* 84, 1986, pp. 133–140.

Stoddard, A., J. O'Reilly, T. Whitledge, T. Malone, and J. Hebard, "The Application and Development of a Compatible Historical Data Base for the Analysis of Water Quality Management Issues in the New York Bight," IEEE Oceans "86 Conference Proceedings, 1986, pp. 1030–1035.

Studholme, A., "Biological Resources of the Hudson-Raritan Estuary," *In The Hudson-Raritan: State of the Estuary* (appendix), *Water Quality of New Jersey Coastal Waters* Vol. 1, Part 2, New Jersey Marine Sciences Consortium, 1988, pp. X1–22.

Thatcher, M. L., "Inventory of Hydrologic Modifications," Characterization Report, New York-New Jersey Harbor Estuary Program, EPA Region II, New York, 1991.

U.S. Environmental Protection Agency, *Chesapeake Bay: A Framework for Action,* Annapolis, MD: US EPA, Chesapeake Bay Program, 1983.

U.S. Environmental Protection Agency, *Chesapeake Bay Restoration and Protection Plan,* Annapolis, MD: US EPA, Chesapeake Bay Liaison Office, 1985.

Van Dusen, K. and A. Johnson Hayden, *The Gulf of Maine: Sustaining Our Common Heritage,* Augusta, ME, Maine State Planning Office, 1989.

Waldichuk, M., "Management of the Estuarine Ecosystem Against Cumulative Effects of Pollution and Development," *In Proceedings of the Workshop on Cumulative Environmental Effects: A Binational Perspective,* Ottawa, Canada: The Canadian Environmental Assessment Research Council (CEARC), and Washington, DC: The United States National Research Council, 1986, pp. 93–105.

Water Environment Federation, *A National Water Agenda for the 21st Century: Phase III Report ("Solutions to Pollution"),* Alexandria, VA, 1992.

Weber, P., "Safeguarding Oceans," *In* L. R. Brown et al. (eds.), *State of the World 1994,* New York: W. W. Norton, 1994, pp. 41–60.

Woodhead, P., "Inventory of Fish Populations," Characterization Report, New York-New Jersey Harbor Estuary Program, EPA Region II, New York, 1991.

Chapter 12

World Air Pollution: Sources, Impacts, and Solutions

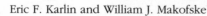

Eric F. Karlin and William J. Makofske

In the past century, the increase of industrialization, population, and living standards in developed nations has been accompanied by ever increasing levels of air pollution. By the midcentury, public recognition of its hazards resulted in the beginning of national programs designed to alleviate the worst conditions. While air pollution was first thought to be mostly a local concern, concentrated in urban or specific industrial areas, the problem is now recognized to be much more complex, involving local, regional, and even global components. Indeed, the very attempt to solve local air pollution by dispersing pollutants over larger regions created a wide range of environmental impacts, often in other nations. Air pollution does not respect political boundaries. While localized air pollution problems still exist, we now have regional photochemical smog and transnational acid deposition. On the global level, stratospheric ozone depletion and global climate change have recently become world public issues. Air pollution generally may be regarded as a severe and growing problem everywhere around the globe.

The primary human activities in industrial countries that release air pollutants are electrical power production, industrial production, and transportation. In the United States, the breakdown among these sources is shown in Figure 12.1. In developing countries, in addition to these activities, the burning of wood, dung, and crop residues in rural areas contribute significantly to air pollution. Acid deposition, an international problem affecting many industrialized countries today, is growing in severity in many developing areas.

Many past attempts at solving air pollution have failed because they focused on one isolated problem. For example, the use of taller smokestacks to dilute and disperse local air pollution created a much wider air pollution problem in the form of acid deposition. The technological solutions that have been attempted have been piecemeal and have often not controlled pollution successfully because population growth, increased industrial activity, and poor planning have overwhelmed the reductions achieved by the technology. The technological fix has often won the battle only to lose the war. Moreover, the international and global air pollution problems that we face today are substantially more complex,

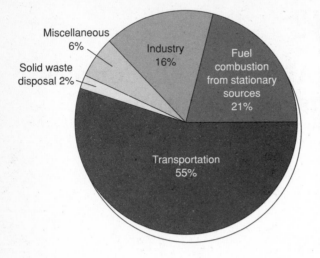

Miscellaneous 6%

Solid waste disposal 2%

Industry 16%

Fuel combustion from stationary sources 21%

Transportation 55%

Figure 12.1. Sources of air pollutants in the United States. Transportation is the largest source, with fuel combustion from stationary sources, such as coal-fired electric generating plants, the second-largest source. The combustion of fossil fuels is the most significant source of anthropogenic air pollution. (From Kaufman, D. G. and C. M. Franz, *Biosphere 2000: Protecting Our Global Environment,* New York: HarperCollins College Publishers, 1993.)

requiring much greater understanding of interactions among pollutants and long-term atmospheric dynamics and chemistry. Implementing solutions to these problems will also require multinational and global agreements, often a difficult task. It is time to discard our tunnel vision about air pollution and to look at the problem from a broader perspective.

The first part of this chapter provides an overview of the current air pollution situation by analyzing individual air pollutants—their formation, impacts, and controls—while noting the often complex interactions among them. Natural sources of air pollution and their magnitudes relative to anthropogenic sources are presented to give a balanced perspective. The second part of the chapter provides a summary of air quality policy efforts in the United States, and some recent international policy initiatives on acid deposition. In addition, a brief perspective on the continuing debate over air quality control in the United States is presented.

A simplified framework for visualizing the contributants to several types of air pollution problems is provided in Figure 12.2, showing how four major air pollutants (sulfur dioxide [SO_2], nitrogen oxides [NO_x], aerosols and particulates[1], and hydrocarbons [HC]) combine to form categories (industrial, photochemical, haze, and acid deposition) of air pollution problems. Figure 12.3 shows, in addition to the four primary air pollutants listed in Figure 12.2, other important air pollutants and their direct and indirect interactions leading to global air pollution problems. Dotted lines show pathways by which certain primary pollutants indirectly contribute to a given problem. Although this chapter focuses on the problems shown in Figure 12.2, connections and discussion relating to the broader set of problems shown in Figure 12.3 are also presented. Global air pollution problems (ozone depletion and global climate change), however, are only briefly considered, since

[1]An aerosol is a suspension of fine solid or liquid particles, generally less than 1 micrometer (μm) in size, which allows it to remain suspended in the air (Koren 1991). Particulates are solid particles of any size in the atmosphere.

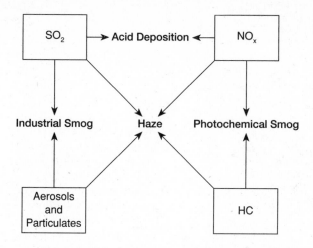

Figure 12.2. Primary air pollutants combine to create air quality problems. Primary air pollutants are in blocks; problems are in bold print. See text for further explanation.

they are covered elsewhere in this book (see Chapter 13, "Stratospheric Ozone Depletion"; Chapter 14, "The Challenge of Global Climate Change"; and Chapter 15, "Predicting Climate Change: The State of the Art" for a more complete treatment of these issues). Other air quality issues (lead, toxic air pollutants, and indoor air pollution) are beyond the scope of this chapter.

Pollutants

Carbon Dioxide (CO_2)

Although it is present in very small quantities (350 parts per million [ppm]), carbon dioxide is the fifth most abundant gas in the atmosphere. The large quantities of carbon dioxide released by respiration and natural wildfires (about 100 billion metric tons of carbon a year on a global basis) are largely balanced by the removal of atmospheric carbon dioxide by photosynthesis. Human activities have displaced this balance, however, and atmospheric concentrations of carbon dioxide have increased by 25% (from 280 to 350 ppm) over the last 140 years. The major source of anthropogenic carbon dioxide is the combustion of fossil fuels, with biomass burning, deforestation, and cement production also contributing significant amounts (for a total of about 7 billion metric tons of carbon annually). Current trends in the use of fossil fuels lead to projections that carbon dioxide concentrations will reach 600 ppm well before the end of the twenty-first century (Mooney et al. 1987; Schneider 1989; Quay et al. 1992).

Carbon Monoxide (CO)

Carbon monoxide results from the incomplete combustion of carbon compounds, with most anthropogenic emissions coming from motor vehicles and biomass burning. Another source of carbon monoxide is the oxidation of atmospheric

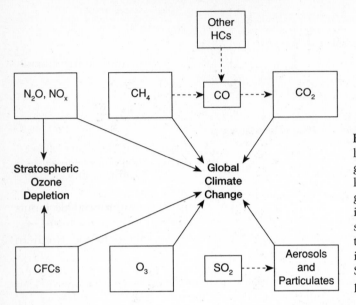

Figure 12.3. Major air pollutants contributing to global air problems. Pollutants are in blocks; global problem areas are in bold type. Solid lines show direct contributions; dashed lines show indirect contributions. See text for further explanation.

hydrocarbons. Having a lifetime in the atmosphere of one to several months, carbon monoxide is ultimately oxidized to carbon dioxide (Fig. 12.3). Unpolluted sites have carbon monoxide concentrations of 20–50 parts per billion (ppb), while polluted areas may exceed 400 ppm. Anthropogenic carbon monoxide emissions are comparable to natural emissions on a global basis, with a total of 1 billion metric tons being released annually. Atmospheric concentrations of carbon monoxide are currently increasing (Mooney et al. 1987; Seinfeld 1989).

The oxidation of carbon monoxide to carbon dioxide involves hydroxyl (OH^-) radicals, as does the oxidation of hydrocarbons. Increased fluxes of carbon monoxide (and hydrocarbons) into the atmosphere results in an increased consumption of hydroxyl radicals, which leads to a decline in its atmospheric concentration. Thus a positive feedback loop develops. As carbon monoxide (and hydrocarbon) atmospheric levels rise, their longevity, and therefore their concentration, increases even more because of the declining concentrations of hydroxyl radicals. Many other atmospheric chemical reactions that involve hydroxyl radicals are also affected (Elsom 1987).

Sulfur Dioxide (SO_2)

Natural emissions of sulfur are for the most part hydrogen sulfide (H_2S) and dimethyl sulfide [$(CH_3)_2S$], both of which are produced by living processes. Hydrogen sulfide is produced in anaerobic environments, such as may occur in wetlands, and dimethyl sulfide is released by marine phytoplankton. Other natural sources of reduced forms of sulfur include methyl mercaptan (CH_3SH), carbonyl sulfide (COS), and carbon disulfide (CS_2). All of these reduced forms of sulfur are oxidized in the atmosphere to sulfur dioxide. Another significant, although sporadic, natural source of atmospheric sulfur is volcanic gas, which contains both hydrogen sulfide and sulfur dioxide (Mooney et al. 1987; Brady 1990; Charlson et al. 1992; Horgan 1992; Hassett and Banwart 1992; Charlson and Wigley 1994).

Anthropogenic emissions of atmospheric sulfur are primarily sulfur dioxide, and are mostly derived from the purification and combustion of fossil fuels and the smelting of sulfide ores. As the amount of sulfur found in fossil fuels varies, (coal: 0.5–6.0%; crude petroleum: <1%; natural gas: trace amounts), sulfur dioxide emissions resulting from the use of fossil fuels also vary (ReVelle and ReVelle 1988). Burning natural gas releases far less sulfur dioxide than the combustion of an equivalent amount (per unit of energy released) of bituminous coal. Likewise, low sulfur coal releases much less sulfur dioxide than high sulfur coal. Over 80% of the sulfur dioxides emitted in the United States are released by power plants (ReVelle and ReVelle 1988; Cunningham and Saigo 1992). Anthropogenic inputs of atmospheric sulfur are comparable to natural sources on a global basis, with about 94% of the anthropogenic sulfur dioxide being emitted in the Northern Hemisphere (Charlson et al. 1992; Hassett and Banwart 1992; Smith 1992; Charlson and Wigley 1994).

Atmospheric sulfur dioxide is oxidized to form sulfates (SO_4) and sulfuric acid (H_2SO_4) and is removed from the atmosphere by precipitation and dry deposition. Having a mean residence time of 1–3 days in the atmosphere, sulfur dioxide and its oxidation products quickly cycle out of the atmosphere (Schwartz 1989).

Nitrogen Oxides (NO_x) and Nitrous Oxide (N_2O)

Two nitrogen gases that are associated with air pollution are nitrogen oxides (nitric oxide [NO], nitrogen dioxide [NO_2]) and nitrous oxide. The major natural sources for these gases are abiotic nitrogen fixation, nitrification, denitrification, and wildfires. Nitrous oxide is largely derived from bacterial denitrification, an important natural process that results in the conversion of mineralized nitrogen (nitrates: NO_3) into nitrogen gas (N_2), nitrous oxides, and a small amount of nitric oxide (Mooney et al. 1987; Brady 1990).

Anthropogenic nitrogen oxides and nitrous oxide are primarily produced by the high temperature combustion of fossil fuels. Some of the nitrogen emitted is derived from the fossil fuels (coal is typically 1% nitrogen and oil/gas average 0.2–0.3% nitrogen), but much is also derived from atmospheric nitrogen gas combining with oxygen during the combustion process (ReVelle and ReVelle 1988). Thus nitrogen oxides would be produced even if the fossil fuel being burned contained no nitrogen. Motor vehicles and stationary fuel combustion are the major sources of anthropogenic nitrogen oxides. These two sources currently generate 48% and 47%, respectively, of the total anthropogenic nitrogen oxide emissions in the United States (Cunningham and Saigo 1992).

Human activities also enhance natural emissions of nitrogen oxides. For instance, humans have doubled the amount of mineralized nitrogen added to soils on a global basis through the use of fertilizers. This enrichment of soil nitrogen automatically leads to greater amounts of nitrates being available for denitrification, with a corresponding increase in nitrous oxide emissions. The extensive destruction of natural ecosystems, particularly tropical forests, is also a significant source of nitrous oxide (Raloff 1986; Mooney et al. 1987; Brady 1990; Matson and Vitousek 1990).

Approximately half of the 100 billion metric tons of nitrogen oxides and 10 billion metric tons of nitrous oxides emitted annually into the atmosphere on a global basis are anthropogenic. Nitric oxide and nitrogen dioxide are rapidly oxidized into nitrates and may form nitric acid (HNO_3). These are quickly removed from the atmosphere by precipitation and dry deposition (usually within 1–4 days). Nitrous oxide, on the other hand, is a relatively stable gas having a lifetime of about 150 years in the atmosphere. Consequently, the amount of nitrous oxide in the atmosphere has been increasing by 0.2% a year. It is eventually converted into nitric oxide by photolysis in the stratosphere (Mooney et al. 1987; McElroy and Salawitch 1989; Schwartz 1989; Cunningham and Saigo 1992).

Aerosols and Particulates

Small solid and liquid particles (<1 micrometer [μm]) that are suspended in the atmosphere are called *aerosols,* while *particulates* refers to solids of any size. Natural sources of aerosols and particulates include biogenic sulfur compounds (mostly produced in anaerobic environments), biogenic carbon compounds, ash and gases from volcanic activity and wildfires, sea salt, and dust. Many aerosols first enter the atmosphere as gases and are subsequently converted into solids. For instance, the gases sulfur dioxide and hydrogen sulfide are oxidized to sulfate. Larger particulates (>10 μm) tend to settle out of the atmosphere relatively quickly, and even aerosols usually do not stay in the atmosphere for very long: dust and sea salt have a mean residence time of <1 day, while sulfates have a mean residence time of 1–3 days. Aerosols that reach the stratosphere, however, could remain there for much longer periods (Mooney et al. 1987; Schwarz 1989; Charlson et al. 1992).

Anthropogenic aerosols and particulates are derived from a diversity of sources, including ash from the burning of coal and wood, dust from soil erosion, asbestos from brake linings, and sulfur dioxide emissions from the combustion of fossil fuels and biomass. Although present in very small quantities, radioactive elements, released from nuclear explosions, accidents at nuclear power plants, and the combustion of coal, have a significant impact on health. The contribution of anthropogenic aerosols and particulates on a global basis (100 million metric tons) is 100 times less than that of natural contributions (10 billion metric tons) (Mooney et al. 1987; Elsom 1987; ReVelle and ReVelle 1988; Cunningham and Saigo 1992).

Hydrocarbons (HC)

Hydrocarbons are released into the atmosphere by a variety of natural and human activities. Most are oxidized, via reactions involving light and hydroxyl radicals, to carbon monoxide within a few days, although some hydrocarbons, such as methane, have an atmospheric residence time of about 10 years (IPCC 1990). Methane is produced by anaerobic decomposition (which commonly occurs in landfills, rice paddies, and wetlands) and the digestive tracts of many animals (i.e., cattle, termites, and humans). Methane emissions are also associated with the use and processing of fossil fuels (methane released from oil wells, coal mines, leaks from gas pipes, and gas appliances). Some 400 million metric tons of methane are released into the atmosphere each year on a global basis, and anthropogenic sources form a significant proportion of this. Atmospheric concentrations of methane have doubled within the past 200 years and are currently increasing at the rate of about 1% a year (Mooney et al. 1987; IPCC 1990).

Plants, especially trees, produce large quantities of hydrocarbons (830 million metric tons annually), with isoprene and terpenes being the most common compounds. Nonmethane anthropogenic hydrocarbons (100 million metric tons annually) are largely derived from the processing and use of oil and natural gas, especially the operation of motor vehicles (exhaust fumes, evaporation of gasoline). They include alkanes, alkenes, carbonyls, alcohols, carboxylic acids, and aromatics (probably the most reactive of the hydrocarbons in the urban atmosphere) (Mooney et al. 1987; Seinfeld 1989; Cunningham and Saigo 1992).

Environmental Impacts

The combustion of fossil fuels and biomass results, directly and indirectly, in a broad spectrum of environmental impacts. These include local and regional impacts (acid deposition, photochemical smog, industrial smog, haze) as well as global impacts (global warming, global cooling, and stratospheric ozone loss) (see Figures 12.2 and 12.3). As most fossil fuel combustion occurs in the Northern Hemisphere, it is not surprising that this is generally where the most severe impacts are observed.

Acid Deposition

Sulfur dioxide and nitrogen oxides and nitrous oxide are the chemical compounds primarily responsible for acid deposition. Although these compounds are not acidic, they can be chemically transformed into strong acids when exposed to water and oxidizing agents; sulfur dioxide is changed to sulfuric acid and nitrogen oxides are converted to nitric acid. When these acids form in the atmosphere, they may be absorbed by the atmospheric water, causing it to become more acid (Mohnen 1988; Schwartz 1989).

Atmospheric water is naturally acidic (unpolluted rain is typically about pH 5.6). This is primarily due to the presence of carbonic acid, which forms via the interaction of water and carbon dioxide. Naturally occurring sulfur dioxide and nitrogen oxides also cause some acidification. Anthropogenic inputs of sulfur dioxide and nitrogen oxides, however, have greatly increased the atmospheric loading of sulfur dioxide and nitrogen oxides. This has led to much higher concentrations of sulfuric and nitric acids, which in turn cause precipitation to become far more acid. The northeastern United States now has precipitation with an acidity that is 25–40 times more acid (pH 4.0–4.2) than it had been before the advent of acid deposition. The pH of precipitation has been observed to be as low as pH 2.1, which is 3,000 times more acid than natural precipitation! The deposition of these acids into ecosystems via precipitation is called *wet deposition* (Elsom 1987; Mohnen 1988; Enger and Smith 1992). Often, spring runoff or drought-ending rainfalls can have an enormous impact on streams and lakes.

Sulfur dioxide and nitrogen oxides can also be added to ecosystems via dry deposition. This occurs by the entry of the gases or particulate matter (such as metal sulfates) directly onto soil, water, or an organism, where they may then be converted into an acid. Dry deposition is difficult to measure, but it is thought to at least equal wet deposition (ReVelle and ReVelle 1988; Schwartz 1989).

Regions having minimal air pollution generally receive 1–2 kg/ha of sulfur and a similar amount of total nitrogen as wet deposition. Human activities, however, have greatly augmented the natural atmospheric concentrations of these chemicals. Regions with moderate to severe air pollution receive 8–50 kg/ha total nitrogen and 5–168 kg/ha sulfur (Brady 1990; Hassett and Banwart 1992; Kauppi et al. 1992).

It has long been recognized that sulfur dioxides and nitrogen oxides and their aerosol products represent a major health risk (see the following discussions on photochemical and industrial smog). The construction of tall smokestacks for coal burning power plants has been one method used to minimize the local impact of sulfur dioxide. The local entrapment and concentration of pollutants by thermal inversions would be minimized by releasing the gaseous byproducts of the combustion of coal above the inversion layer. Release higher in the atmosphere would also allow for a dispersal over large areas, and it was hoped that the pollutants would become so diluted in the process that they would have little, if any, environmental impact. Although exposure to high levels of sulfur dioxide and nitrogen oxides was largely mitigated by this action (but not completely, see below), long-term exposure to acid deposition has been found to be a significant environmental problem. Thus what was once a local problem has been transformed into a large-scale regional problem. In many cases, sulfur dioxides and nitrogen dioxides produced in one country end up being deposited in another country that is several hundred miles away from the source; this has created major political problems (Dovland 1987; Elsom 1987; Schwartz 1989).

The increased acid loading to ecosystems has had significant deleterious impacts, although the scale and nature of the impact varies. Some ecosystems have a high capacity to neutralize (or buffer) the acids added by acid deposition. Such ecosystems are usually associated with areas having limestone bedrock and/or calcium-rich soils. On the other hand, many ecosystems have a limited capacity to neutralize the acids, and thus are sensitive to elevated levels of acids. These are usually associated with regions having granite bedrock and/or highly leached soils. Much of northeastern North America is quite vulnerable to acid deposition, especially the Adirondack region of New York State and much of southeastern Canada. Figure 12.4 shows areas of the world with air pollution emissions that cause acid deposition together with the surrounding regions that are affected by such emissions. Southern Scandinavia is another highly susceptible region that has been severely impacted by acid deposition. Large portions of Brazil, India, and China are also sensitive to acidification, and sulfur emissions from coal burning, particularly in China, are increasing rapidly (Baker et al. 1991; Roberts 1991).

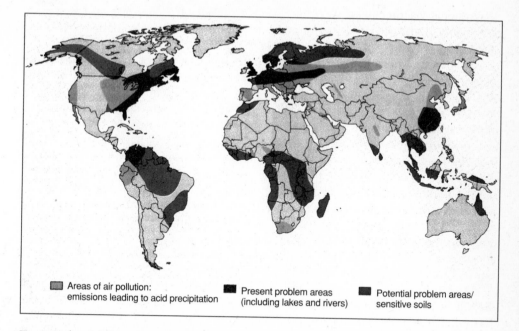

Figure 12.4. Acid deposition and point of origin of its precursors worldwide. Over 60% of anthropogenic emissions come from North America and Europe. Problems caused by acid deposition are particularly severe in the eastern United States, eastern Canada, Scandinavia, northern and central Europe, the United Kingdom, and China. (Modified from Kaufman, D. G. and C. M. Franz, *Biosphere 2000: Protecting Our Global Environment*, New York: HarperCollins College Publishers, 1993.)

In general, fresh water ecosystems are more susceptible to acidification than are terrestrial ecosystems. This is because most soils have a greater capacity to neutralize acids than is the case for aquatic systems. Consequently, the impacts of acid deposition are first evidenced in lakes and ponds, and thousands of these ecosystems have been severely degraded by acidification. These impacts include the loss of much of the original animal community (fish, snails, clams) as well as a change in the plant communities. Among the many impacts are gill damage to fish, death of fish eggs, lack of shell formation in snails and clams, and a reduction in phytoplankton biomass and species diversity (Harvey 1980; Dillon et al. 1984; Elsom 1987; Baker et al. 1991; Roberts 1991).

Terrestrial ecosystems also suffer from exposure to acid deposition, but with their more highly buffered soil systems, it has taken longer for the impacts to become visible. For the most part, these impacts are of a secondary nature. One impact is increased leaching of calcium and other minerals from the soil. This results in soils becoming less fertile over time, and thus less capable of supporting life. Snails, which need calcium to produce their shells, have had major population declines in acidified forests. As many forest birds obtain a significant portion of the calcium needed to produce their eggs from the consumption of snail shells, diminished snail populations directly lead to increases in the number of defective eggs laid by these birds (Graveland et al. 1994). It is thought that stress related to incipient nutrient deficiencies has made many trees more susceptible to other problems (insect pests, plant diseases). A second impact is increased solubility of metals (i.e., aluminum, iron, manganese), which leads to higher concentrations of these metals in the soil solution. High levels of these metals can impair normal plant functioning and even be toxic.

Acid deposition also represents the addition of two plant nutrients (sulfur and nitrogen). The amount of sulfur added is often close to that needed by agricultural systems, and there is actually some concern that farmers will have to add sulfur to their fertilizer regime if acid deposition is significantly reduced. On the other hand, the amount of nitrogen added is usually much less than needed by agricultural crops. Infertile natural ecosystems, however, may have both their nitrogen and sulfur budgets significantly enriched by acid deposition. Some argue that the positive results of this may equal or at least mitigate the negative impacts of acid deposition. But the addition of nitrogen and sulfur to soil systems does more harm than good when associated with soil acidification (Prinz 1987; Brady 1990; Hassett and Banwart 1992; Farmer et al. 1992; Kauppi et al. 1992).

A complicating factor is that terrestrial ecosystems are also exposed to sulfur dioxide, nitrogen oxides, and ozone, all related to the combustion of fossil fuels. The impacts of these are discussed in detail below, but it is often difficult to separate out the impacts of one pollutant from another. This is especially true for high elevation ecosystems. First, these ecosystems have significant contact with damaging levels of sulfur dioxide and nitrogen oxides which have accumulated at higher altitudes because of their release from tall smokestacks. In addition, the interaction of nitrogen oxides, hydrocarbons, and sunlight result in the production of ozone (O_3). Finally, water droplets in clouds are more acid than precipitation and high elevation ecosystems are often directly exposed to clouds and fog for extensive periods of time. Thus ecosystems at high elevations experience all the major detriments of sulfur dioxide and nitrogen oxide pollution, not simply acid deposition. It is not surprising that high elevation forests in the eastern United States and parts of Europe have been so devastated. For these ecosystems, trying to separate out the various impacts of the different pollutants is a mistake; it ignores the reality that one basic source of pollution (sulfur dioxide and nitrogen oxides derived from the combustion of fossil fuels) can result in a myriad of environmental impacts. It is believed that the extensive deterioration of European forests that is currently occurring (called *Neuartige Waldschuden:* novel forest damage) is due in large part to the combined impacts of ozone and acid deposition (Prinz 1987). There is no doubt that greatly reducing the levels of atmospheric sulfur dioxide and nitrogen oxides would greatly enhance the health of forest ecosystems, especially high elevation forests (Elsom 1987; Prinz 1987; Cunningham and Saigo 1992).

When low elevation rural ecosystems are far removed from the site of origin of sulfur dioxide and nitrogen dioxide, they may experience only one major impact—acid deposition. Low elevation ecosystems in or adjacent to urban areas, however, such as the metropolitan areas of northeastern United States, experience both acid deposition, the source of which is power plants several hundred miles away, and photochemical smog derived from locally and regionally produced pollutants.

Acid deposition has numerous health implications. Many atmospheric chemicals (such as sulfuric and nitric acids) are adsorbed onto the surface of particulate aerosols. The inhalation of aerosols and small particulates (<2–3 μm) may cause irritation of the respiratory system, especially if there are harmful chemicals adsorbed to the surfaces of the particulates. Also, small particulates are inhaled more deeply in the lungs and the respiratory system is less able to expel them through natural functions like coughing (Nadakavukaren 1990; French 1992). This is discussed in more detail in the following section. In addition, mercury and heavy metals, like lead and cadmium, are more soluble in acid waters, and may become a health hazard to humans. Fish in many acidified lakes have elevated levels of methyl-mercury, a hazard to human consumers. Copper and lead are leached out of the plumbing of water supply systems by acidified water, making the water delivered by such systems unsafe to drink (Elsom 1987; Kaufman and Franz 1993). These metals have serious health effects. For example, lead can cause anemia, damage to the kidneys, and damage to the nervous system. High levels of lead can cause mental retardation in children; in small amounts lead inhibits the intellectual development of children. Continued exposure to lead, especially in the young, can lead to death (Nadakavukaren 1990; Louria 1992).

Photochemical Smog

Many primary pollutants are chemically transformed in the environment into more harmful forms, often by reacting with other pollutants. Photochemical smog is an example of this. The basic requirements for the creation of photochemical smog are sunlight (wavelengths of 280–430 nanometers [nm]) and high levels of nitrogen oxides and hydrocarbons; these conditions exist in most major urban centers. The major products of photochemical smog are photochemical oxidants such as ozone, peroxyacyl nitrate (PAN), and aldehydes, with ozone usually comprising about 75% of the mixture. Unpolluted tropospheric air has ozone concentrations of 20–50 ppb, while ozone levels in polluted urban areas are as high as 400 ppb (ReVelle and ReVelle 1988; Seinfeld 1989; Kaufman and Franz 1993).

Although naturally produced in significant amounts in the stratosphere, where it plays a major role in filtering out much of the solar ultraviolet radiation, the presence of ozone in the troposphere is largely a result of air pollution (mostly derived from motorized vehicles, electric utilities, and industrial pollution). Because of the presence of nitric oxide (NO), which destroys ozone, only a limited amount of ozone would normally accumulate in the air. As many anthropogenic hydrocarbons serve as a sink for nitric oxide, their presence in high concentrations allows for the accumulation of large amounts of ozone. In addition, one of the byproducts of the oxidation of carbon monoxide and hydrocarbons is nitrogen dioxide, a gas that plays a central role in the chemical reactions that produce ozone (Elsom 1987; Seinfeld 1989).

In the past few decades, tropospheric ozone has been a persistent problem. Despite regulation and control efforts, concentrations have been increasing over many regions of the United States. In 1989, EPA reported that 67 million Americans lived in regions that exceeded the National Ambient Air Quality Standard (NAAQS) for ozone. Although ozone had previously been a concern in and downwind of large urban areas, it is now recognized as a widespread regional problem as well (NRC, 1991). Figure 12.5 shows areas in the United States that are not in compliance with the present NAAQS standard for ozone.

Exposure to tropospheric ozone (and other photochemical oxidants) causes irritation to the respiratory system and eye irritation. People with respiratory or cardiac problems or those who are exercising heavily are at most risk. When ozone is inhaled deeply, it irritates the delicate lining of the lungs, causing a thickening of the terminal air sacs (alveoli). Healing of these injured tissues results in scarring or fibrosis, which inhibits the exchange of gases in the lung and causes respiratory distress (Nadakavukaren 1990; French 1991). Ozone alerts have been developed in many areas in order to warn the population about periods of high ozone concentrations. At the present time, the National Ambient Air Quality Standard for ozone in the United States is 120 ppb (240 micrograms per cubic meter [$\mu g/m^3$]) averaged over 1 hour. Ozone levels exceeding this concentration are considered a human health problem. Ozone levels of only 100 ppb, however, cause eye irritation and at 120 ppb the performance of athletes is impeded. The current standard appears to be too high to provide an adequate safety margin for human health; indeed, the standard was actually significantly lower (80 ppb) in the 1970s (ReVelle and ReVelle 1988). Unfortunately, the Environmental Protection Agency has decided to uphold the current standard (Schneider 1992).

Nitrogen oxides, particularly nitrogen dioxide, also have an impact on human health. Some of the major effects are a reduction in night vision, difficulty in breathing, and inactivation of hemoglobin (like carbon monoxide) (ReVelle and ReVelle 1988).

Tropospheric ozone causes extensive damage to plants, especially trees. It appears to both impede photosynthesis and to enhance leaching of minerals from leaves (Prinz 1987; Mohnen 1988). Plants also appear to be more sensitive to ozone than humans, showing damage at concentrations of 40–80 ppb (80–160 $\mu g/m^3$). Economic loss to ozone is already dramatic, with ozone induced lowered yields for just four United States crops being estimated to be between $1.9 and $4.5 billion dollars annually. Natural ecosystems as well as crops are being affected by exposure to elevated levels of ozone (Sigal and Nash 1983; Prinz 1987; ReVelle and ReVelle 1988; Smith 1992).

On top of the impacts discussed above, photochemical smog forms extensive hazes, which dramatically reduce visibility (see the section on "Haze" below). Many urban areas often experience hazy days on a regular basis. Such conditions are common in many cities around the world. Even relatively isolated areas, such as the Grand Canyon, now suffer from significant reductions in visibility because of the haze from photochemical smog (Crawford 1990).

Recent reassessments of the tropospheric ozone problem have found that anthropogenic hydrocarbon emissions have been seriously underestimated, and that biogenic hydrocarbon emissions may also be an important source in some regions of the country. In addition, the relative effectiveness of hydrocarbon (HC) and NO_x controls for reducing ozone in a particular area depends on the ambient HC/NO_x ratios. Previous reduction strategies have assumed that HC control would be more effective than NO_x control. This assumption now needs to be examined more criti-

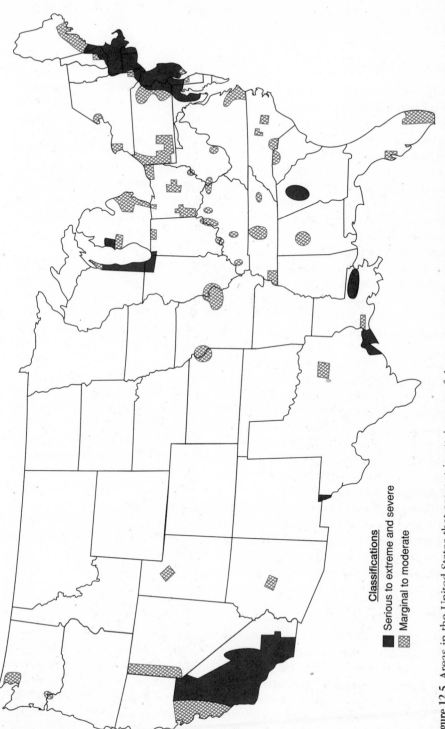

Figure 12.5. Areas in the United States that are not in attainment of the ozone standard. (Modified from National Research Council, *Rethinking the Ozone Problem in Urban and Regional Air Pollution*, Washington, DC: National Academy Press, 1991.)

Classifications
■ Serious to extreme and severe
▨ Marginal to moderate

cally. Reductions in NO_x, either singly or in conjunction with HC, may be needed as well for many regions (NRC 1991).

Industrial Smog

When coal and heavy oil are burned in large quantities without appropriate pollution controls, vast amounts of ash and sulfur dioxides are released into the atmosphere. This mixture forms industrial smog, which consists primarily of sulfur dioxide, sulfuric acid (produced from some of the sulfur dioxide), and particulates. Such smogs have caused many human deaths (see below). As with photochemical smog, there may be a haze associated with industrial smog. Pollution control technology and regulation have eliminated industrial smog from most areas in more developed countries (MDCs). China and many eastern European countries, however, still have severe problems with industrial smog. Unfortunately, one of the technologies used to prevent industrial smog (tall smokestacks) has worsened the acid deposition problem.

The impacts of industrial smog often include the combined effects of sulfur oxides and particulates. Their primary impact is to aggravate existing respiratory and cardiac diseases, especially in the elderly. Greater frequencies of colds and bronchitis are also noted with industrial smog. Very high levels of sulfur dioxide and particulates have even led to premature deaths. More than 4,000 deaths were attributed to the pollution episode that affected London in 1952. Some particulates are suspected to be carcinogenic, especially when they have polycyclic aromatic hydrocarbons adsorbed on their surfaces.

The health impacts of industrial pollution appear to be considerable. Recent epidimeology studies by EPA and the Harvard School of Public Health suggest that small particulates (<10 μm) are causing 50,000 to 60,000 deaths per year in the United States. The health impact is greatest on children with respiratory problems, asthma sufferers, and the elderly with lung-related illnesses such as bronchitis, emphysema, and pneumonia (Hilts 1993).

High concentrations of sulfur dioxide have a major impact on plant health, with entire ecosystems having been destroyed in some cases (i.e., the International Nickel Company smelting operation at Sudbury, Ontario; Copperhill and Ducktown in Tennessee). Bryophytes and lichens are quite sensitive to sulfur dioxide and can be used as biological indicators of sulfur dioxide levels (Gilbert 1970, 1992; Hawksworth and Rose 1970; Farmer et al. 1992; Lee and Studholme 1992).

Haze

Haziness or visibility reduction is mostly associated with aerosols in the atmosphere. While aerosols are composed of six components (sulfates, organics, elemental carbon, ammonium nitrate, soil dust, and aerosol-bound water), it is the sulfates formed from SO_2 emissions that are the major contributors to haze in the eastern half of the United States, while ammonium nitrate also is important in the West. Reducing SO_2 emissions will improve visibility, although the improvement is nonlinear (NAPAP 1991).

Haze may also be produced naturally by hydrocarbons (terpenes) released by vegetation combining with ozone to form a small (0.2 μm) aerosol particle. This results in the blue haze characteristic of places like the Blue Ridge Mountains of Virginia. Larger aerosols, particularly fine dusts and salts that involve hygroscopic water, scatter all wavelengths of light to provide a white haze characteristic of ocean environments (Ahrens 1988). Since aerosols are involved in all forms of air pollution to some degree, you find some loss of visibility associated with photochemical and industrial smog, and with sulfate aerosols associated with acid deposition.

Global Climate Change

The combustion of fossil fuels and biomass for energy plays a major role in global climate change, primarily in the release of carbon dioxide and nitrous oxide, which enhance greenhouse gas warming. These two gases account for about 49% and 6%, respectively, of the annual input of anthropogenic greenhouse gases. Methane is also a significant greenhouse gas (16% of the annual input), although anthropogenic inputs of methane are not principally derived from the use of fossil fuels (Kaufman and Franz 1993).

Tropospheric ozone, generated by photochemical smog and by nitrogen oxides released by airplanes, is another greenhouse gas associated with the combustion of fossil fuels. Its atmospheric concentration is currently increasing at 1% a year. Ozone contributes about 8% of the annual input of greenhouse gases (Beardsley 1992).

CFCs used in aerosol propellants, coolants, and solvents accounts for 20% of the annual input of greenhouse gases. CFCs are also implicated in the destruction of stratospheric ozone (see Chapter 13, "Stratospheric Ozone Depletion").

Increasing atmospheric concentrations of carbon monoxide and hydrocarbons play an indirect role in global warming. First, these gases are eventually oxidized to carbon dioxide and thus add to the global flux of this gas in the atmosphere. Second, higher concentrations of carbon monoxide lead to a decline in hydroxyl radicals (see above). This results in a greater longevity of atmospheric gases, which are oxidized by hydroxyl radicals. Methane is one such gas, and as its residence time in the atmosphere increases so will its impact on global warming (Elsom 1987). For an in-depth discussion of global climate change, see Chapter 14, "The Challenge of Global Climate Change" and Chapter 15, "Predicting Climate Change: The State of the Art."

Global Cooling

Although anthropogenic aerosols and particulates represent only 1% of the total global budget, some exceed amounts emitted by natural sources. For instance, anthropogenic sulfur emissions currently are greater than natural emissions of sulfur-containing gases on a global basis (Charlson et al. 1992; Charlson and Wigley 1994). Such significant increases in sulfates carry major environmental impacts. Given the short residence time of aerosols and particulates in the atmosphere, most of their impacts are of a regional nature, although some have global effects.

For instance, the Mt. Pinatubo volcanic eruption (Philippine Islands) in June 1991 ejected enough SO_2 gases and particulates into the stratosphere to lower the global temperature for a couple of years, perhaps by as much as 0.5°C (Horgan 1992; Kerr 1992a).

Recent studies show that the climate forcing by anthropogenic sulfate aerosols causes a cooling of the atmosphere that is significant, estimated from about one-third to almost equal to the warming brought about by greenhouse gas forcing in the Northern Hemisphere (Charlson et al. 1992; Kerr 1992b; Wigley and Raper 1992). The aerosol-induced cooling is brought about directly and indirectly. Sulfate aerosols have the capacity to strongly reflect sunlight, and increased concentrations of aerosols in the atmosphere would directly lead to less sunlight reaching the Earth's surface. This accounts for about half of the cooling caused by anthropogenic aerosols. The other half is derived indirectly by the impact of sulfate aerosols on clouds. A high density of smaller than normal water droplets develops in clouds when sulfate aerosols serve as cloud condensation nuclei. This results in a higher cloud albedo and also enhances cloud lifetime, both of which lead to less sunlight reaching the Earth's surface. The climate forcing by anthropogenic sulfate aerosols may be one reason why global warming has been less than predicted by global climate computer models (Charlson et al. 1992; Kerr 1992b; Charlson and Wigley 1994).

Although most anthropogenic aerosols and particulates are emitted (or created[2]) on a regular basis, there are potentially significant sources that are of a sporadic nature. The extensive oil fires that were ignited during the 1991 war in Kuwait (more than 700 wells were flamed) are an example. Although there was concern that it might have a short-term impact on the global climate, no appreciable climatic impact was noted (Hoffman 1991). The primary environmental impacts that resulted from these extensive fires were regional and they were health related. Such would probably not be the case, however, with a full-scale nuclear war. It is believed that the nature and amount of aerosols and particulates generated by a full-scale nuclear holocaust would be sufficient to cause a "nuclear winter" (Kaufman and Franz 1993).

Stratospheric Ozone Depletion

Chlorine from CFCs and other compounds are thought to be the major contributor to stratospheric ozone loss. While CFC emissions are decreasing due to their worldwide phase out, their long lifetime in the atmosphere will lead to substantial stratospheric ozone depletion over the next century. Nitrous oxides, increasing in the atmosphere at 0.25% per year, also have the capacity to destroy stratospheric ozone (IPCC 1990). This topic is discussed in detail in Chapter 13, "Stratospheric Ozone Depletion."

---

[2]Most aerosols found in the atmosphere are not directly emitted but are formed by photochemical reactions with other gaseous pollutants (Koren 1991).

Solutions to Air Pollution

Various approaches to air pollution control may be taken. One approach is called *input control;* it tries to prevent the pollution problem from occurring in the first place. Some examples of this approach are controlling population growth, recycling, reducing energy use, using energy more efficiently, switching energy sources to less polluting types, and reducing pollution in the production process itself. A second approach is *output control;* it tries to clean up or reduce the effect of the pollution after it is created. Examples are scrubbers and electrostatic precipitators on smokestacks, the use of tall smokestacks, and liming lakes damaged by acid deposition. While such approaches are often useful and necessary in the short term, output methods seldom work in the long term and often create new problems. For example, pollution control equipment, whether scrubbers on power plants or catalytic converters on cars, are overwhelmed by increases in usage. The residue from pollution control equipment, such as the wet ash from scrubbers, is hazardous material, which must then be disposed of in some fashion. Liming lakes becomes increasingly expensive as more lakes reach the critical stage and must be treated. However successful such techniques are, they are merely expensive technological band-aids that do not deal with the underlying root causes of air pollution (French 1990).

In keeping with this article's focus on primary air pollutants, specific technological output control strategies for each particular pollutant are first considered. Then several broader input control strategies that may offer longer range and more permanent solutions are discussed.

Output Control Strategies for Particular Pollutants

Sulfur Dioxide

The initial approach to reducing local concentrations of SO_2 in the 1960s and 1970s was to increase the height of smokestacks above the stagnant air layer, thereby dispersing the pollutant over wider areas downwind of the plant. While this technique did alleviate nearby SO_2 concentrations, it is generally recognized as having also created the acid deposition problem.

The primary technology that has been used to reduce SO_2 from coal power plants has been scrubbers, or more formally, flue-gas desulfurization (FGD). This approach removes the SO_2 from the smokestack before it is released into the air. The main techniques are the dry regenerative approach, which absorbs SO_2 as it passes through a bed of absorbent material; the wet regenerative approach, which absorbs SO_2 as it passes through a liquid; and a wet nonregenerative approach, in which the gas is passed through a water slurry of lime or limestone. The regenerative approaches allow sulfur or sulfuric acid to be recovered and sold. Because of its retrofit ability and some other advantages, the third technique (nonregenerative wet scrubbing) is most common. Unfortunately it also produces toxic sludge (scrubber ash), a hazardous substance that is currently sent to landfills or deposited in holding ponds, reduces the overall efficiency of the plant, and raises the cost of electricity by 8–18% (Park 1987). The technology removes from 70% up to 95% of the SO_2. FGD has been used in many industrialized countries; Organization for Economic Cooperation and Development (OECD) countries had 140,000 megawatts of plant capacity with FGD either installed or under construction by 1988. By 1990, about 30% of U.S. coal-fired capacity was equipped with scrubbers. The technology is also beginning to be used in other countries such as Czechoslovakia and China (Park 1987; French 1990).

Other means are available to reduce SO_2 emissions. Substituting low-sulfur coal (<1%) for high-sulfur coal (typically 1–5%) is an option if there is low-sulfur coal available. Low-sulfur coal, however, has a lower Btu value per pound compared to high-sulfur coal. Therefore, more coal must be burned for the same power output, thereby reducing the low-sulfur benefit. In addition, both availability and distance have proven to be severe constraints. Furthermore, using low-sulfur coal in conventional power plants reduces the efficiency of electrostatic precipitators, leading to greater particulate emissions (Park 1987).

Methods also exist to remove some of the sulfur (typically only 20–50%) from coal before combustion. Coal washing, often used to remove some of the inorganic sulfur, is relatively inexpensive. Chemical coal cleaning, under development to remove the chemically bound organic sulfur, is considerably more expensive (Park 1987). Research is also underway to use natural or genetically engineered bacteria to accomplish this task (Miller 1992).

A very promising clean-coal technology for reducing SO_2 is fluidized bed combustion (FBC), in which powdered coal mixed with crushed limestone is burned by suspending it over hot air from below. This removes up to 90% of the SO_2, reduces NO_x by 50–75%, reduces CO_2 slightly, and increases combustion efficiency. The technology is commercially available in small to medium size units of up to 250 megawatts (Park 1987; Miller 1992). Other technologies, such as limestone injection multiple burning (LIMB) and integrated gasification combined cycle (IGCC), are in various stages of development (NAPAP 1987; IEA 1988; 1991).

Other approaches include substituting natural gas for coal during the summer when oxidation of SO_2 is greatest and there is an excess of natural gas supply. This leads to considerably lessened total sulfate deposition for the year (NAPAP, 1991).

While SO_2 has undergone significant reductions in the United States and other industrialized countries, many of the developing world's cities are experiencing increases, leading the World Health Organization (WHO) and the United Nations Environmental Program (UNEP) to conclude that 625 million of the world's people are exposed to levels of SO_2 higher than WHO standards (UNEP/WHO 1988; French 1990; Miller 1992).

Nitrogen Oxides

As nitrogen oxides are primarily created during the combustion of fossil fuels, it is necessary to modify the combustion process to prevent its formation or remove it after it is formed. For power plants, a simple and effective technique has been to lower combustion temperatures, reducing emissions by up to 50%. For example, Figure 12.6 shows the temperature dependence of nitrogen oxide emissions from coal combustion. Other forms of combustion modification have been used with varying success (Park 1987). Another method, called *selective catalytic reduc-*

tion (SCR), developed in Japan and currently in extensive use there, offers 80–90% reductions (French 1990). There are a number of promising technologies under development, including reburning after combustion (50% reduction), fluidized bed combustion, integrated gasification combined cycle, reaction with isocyanic acid (99% reduction), and reaction with phosphorous to produce NO_x, which is then removed by flue gas scrubbers (Park 1987; French 1990; UER 1992).

For mobile sources, the use of emission control devices, such as the catalytic converter, can provide significant reduction of NO_x. Unfortunately, 60% of such devices on U.S. cars are either not used or not effective. Coupled with an increasing number of vehicles on the road and the relaxation of auto fuel efficiency standards, NO_x emissions have actually increased since 1975 in spite of control efforts (Kaufman and Franz 1993).

Aerosols and Particulates

Several technologies have been used to reduce particulates from combustion. The most popular of these have been baghouse filters, and for power plants, electrostatic precipitators. With removal efficiencies of over 99% by weight, these devices have served to virtually eliminate the black smoke and soot associated with burning fossil fuels. Very small particulates (<3 μm), however, those most likely to enter the lungs and cause severe health effects, still escape to the atmosphere. Many believe that the NAAQS standard for particulates, based on a weight per unit volume, is inadequate to protect human health. After dropping 20% from the 1975

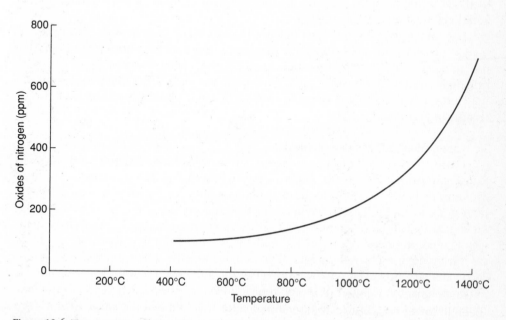

Figure 12.6. Temperature dependence of nitrogen oxide emissions from coal combustion. (Control Techniques for Nitrogen Oxide Emissions from Stationary Sources. Washington, DC: U.S. Dept. of Health, Education & Welfare, 1970, NAPCA Publication No. AP-67.)

value, particulates and aerosols have been increasing in the United States since the early 1980s (Miller 1990, 1992).

Particulates (and aerosols) are not only a threat to developed countries. A UNEP/WHO study of major cities throughout the world found that 37 of 41 monitored for suspended particulate matter had borderline or excessive values. Extrapolating from these results, it is believed that as many as 1 billion people, or one-fifth of the world's population, are exposed to excessive levels of particulates (UNEP/WHO 1988).

Hydrocarbons

Hydrocarbons are a major constituent leading to the formation of photochemical smog. Controlling HCs requires reducing their emission from cars, gas stations, auto painting, petroleum refining, incineration plants, printing shops, and other sources (Kaufman and Franz 1993).

Many methods have been used successfully to control HC emissions. These include destruction using thermal or catalytic incineration, recovery using activated carbon or direct condensation, and concentration using activated carbon or molecular sieves. Incineration techniques can attain better than 95% destruction efficiency, converting the HCs to carbon dioxide and water vapor. Catalytic techniques are more fuel efficient since destruction occurs at lower temperatures. Catalytic incineration, however, may be limited by contaminants that poison the catalyst. Recovery methods absorb or condense HCs from an airstream as it passes through carbon bed containers or over cold heat exchange surfaces. Finally, concentration methods using activated charcoal or zeolite rotors remove HCs from air flows for additional treatment (Droham 1992).

Los Angeles is attempting to drastically reduce HC emissions as part of its air quality management plan. Its major provisions call for (1) controlling automobile use, gradually substituting other fuels to eliminate gasoline engines; (2) strictly controlling industrial HC emissions; and (3) finding less polluting substitutes for household products that release HCs (aerosol cans, paint, and barbecue starter fluids) (Kaufman and Franz 1993).

Carbon Monoxide and Carbon Dioxide

Close to 70% of the anthropogenic carbon monoxide emitted in the United States is from motor vehicles. Technologies to control carbon monoxide from motor vehicles have been developed and implemented. These include increasing the ratio of air to fuel in gasoline engines (which unfortunately increases nitrogen oxide emissions) and the use of catalytic converters. The efficiency of these technologies is quite impressive: cars in the mid-1960s emitted an average of 73 grams of carbon monoxide per mile, while new cars in the early 1980s emitted 3–4 grams. Two variables, however, have largely offset the attempts to control carbon monoxide emissions: the number of motor vehicle miles increases each year and the lack of regulations that require the proper maintenance of emission control systems (see nitrogen oxides above). Compulsory inspections and repair of emission control systems would help, but such regulations are presently implemented in a few localized areas in the United States (ReVelle and ReVelle 1988; Miller 1992).

There are very few control technologies for carbon monoxide release from other anthropogenic sources (mostly from power plants and industry). More complete combustion, as found in fluidized bed combustion, does reduce carbon monoxide emissions compared to typical coal combustion technology (Cassedy and Grossman 1990).

Carbon dioxide, once created by the combustion of fossil fuels, is extremely difficult to remove from the exhaust gases. The major strategy for controlling these gases would be to reduce our dependence on fossil fuels, through efficiency and renewable energy technologies, or to switch from coal and oil to natural gas, which emits less CO_2 per unit of energy produced.

Broad Input Control Strategies

It is the use of fossil fuels in power plants, industry, and transportation that is the primary cause of most air pollution. Rather than focus on technological solutions that reduce only one pollutant or one type of air pollution problem, a more effective approach might be to change our technology so that we depend much less on fossil fuel combustion (French 1990). A long-term energy strategy that provides a transition to energy efficiency and renewable energy resources can provide such a solution (see Chapter 6, "World Energy: Sustainable Strategies"). This approach reduces all the air pollutants at once, even the CO_2, for which there is currently no effective technological band-aid (Flavin and Lenssen 1991).

Such an approach could also provide the short-term control measures that are needed, often in an economical fashion. For example, rather than adding scrubbers to old coal power plants at high cost, end use efficiency could reduce demand so the worst-polluting plants could be closed. A study by the American Council for an Energy Efficient Economy has shown that reducing electrical demand by 15% in the Midwest, where one-third of the SO_2 is emitted, could cost-effectively reduce both SO_2 and NO_x emissions locally as well as acid deposition in the eastern United States (French 1990).

Other researchers envision using natural gas on a much larger scale than today as a bridge to the renewable energy future. While it was thought in the 1970s and 1980s that natural gas was in short supply, recent exploration has found a considerable amount so that reserve estimates today are $3 \times$ higher than in 1984, with more than a 50 year lifetime at current consumption rates (Burnett and Ban 1989). Since the infrastructure for using gas is already in place, its use could be expanded rapidly with significant environmental benefits. Natural gas produces almost no SO_2 and particulate aerosols compared to oil and coal. In addition, the NO_x emissions are much lower, and even CO_2 emissions are 30–60% less per unit of usable energy produced. Technology to burn natural gas for electricity is well established; indeed, half of the plants being built today are gas fired. For vehicles, natural gas is currently used in truck and bus fleets, and is easily used in modified gasoline cars. Not only will natural gas buy time for a transition to cleaner energy sources, but hydrogen, produced from renewable sources such as sun or wind, may even be mixed in the pipeline with natural gas, allowing a gradual transition to a hydrogen-based economy (Flavin 1992).

Other strategies for reducing the pollutants from motor vehicles (NO_x, HC, CO, CO_2, and secondary pollutants such as O_3) are well known. These include a shift to mass transit, bicycles, walking in urban areas, greater car pooling, improved fuel efficiency, and a shift to a variety of less polluting automobile engines and fuels. So far, there has been a noticeable lack of political will in the United States to implement such solutions.

Air Quality Policy Efforts—U.S. and International

Localized air pollution has been associated with human societies for thousands of years, from the wood burning of the Romans to the early use of coal in London in medieval times. The problem worsened greatly with the advent of the industrial revolution in England. Later, the expansion of factories, coal-fired electrical power plants, and the internal combustion engine for transportation created evident air pollution problems throughout the first half of the twentieth century (Regens and Rycroft 1988).

Modern-day recognition of the severity of health effects associated with air pollution came from a number of severe episodes. Air pollution episodes in London in 1952, 1956, 1957, and 1962 are believed to have caused a total of 6,500 deaths. In Donora, Pennsylvania, in 1948, 6,000 people out of its total population of 14,000 became ill and 20 died due to air pollution. During a few day episode in 1963, air pollution was responsible for the deaths of 300 people in New York City. All of these deaths were associated with stagnating weather systems and associated temperature inversions, which trapped pollutants and concentrated them to deadly levels. These events alerted both public and government to the risks of air pollution and created public pressure for government to take a greater role in solving the problem (Regens and Rycroft 1988).

Air Quality Policy Efforts in the United States

Federal efforts to control air pollution began with the Air Pollution Control Act of 1955, which provided funds for research and development (R&D) and aid to set up state government programs. Air pollution was still considered almost entirely a state responsibility, and for the most part, states did not respond (Regens and Rycroft 1988).

In the 1960s, three major pieces of legislation relating to air pollution were passed: the Clean Air Act of 1963, the Motor Vehicle Pollution Act of 1965, and the Air Quality Act of 1967. The first two acts increased the role of the federal government, focusing mostly on motor vehicle emissions. The last act greatly expanded the federal role, setting minimum national standards for air quality and designating air quality control regions. Standard setting and enforcement power, however, still fell to the states, whose seeming inability to improve the nation's air quality led to further public pressure for the federal government to act (Regens and Rycroft 1988).

In the 1970s, the increasing federal role was strengthened through the Clean Air Acts of 1970 and 1977. These acts required the recently formed EPA to set two types of national ambient air quality standards (NAAQS) for seven outdoor air pollutants; primary standards were set to protect human health, while secondary standards were set to protect visibility, crops, buildings, and watersheds. The laws required each state to develop and enforce an implementation plan, subject to review and approval by the EPA. Other policies set by the EPA included prevention of significant deterioration (PSD), national emission standards, and new source performance standards (NSPS). The first prevented a decrease in air quality for SO_2 and particulates in regions where the air is cleaner than required by NAAQS; the second required establishing standards for other less-common toxic air pollutants; the last established uniform emission standards for new sources of pollution (Regens and Rycroft 1988).

In the 1980s, the Acid Precipitation Act of 1980 created an Interagency Task Force and authorized a 10-year national study on acid deposition. The Clean Air Act, scheduled for reauthorization in 1981, was stalemated in Congress for the rest of the decade (see section, *The U.S. Policy Debate Over Air Pollution*). The stalemate was finally broken by the 1990 Clean Air Act. This act required coal power plants to reduce SO_2 and NO_x emissions by essentially half by the year 2000, or 2005 if certain clean-coal technologies are used; industries to reduce emissions of 189 toxic chemicals by 90% by 1995–2003; cities to meet ozone standards by certain deadline dates; new cars to reduce emissions of HC by 35% and NO_x by 60% by 1994; cleaner burning gasoline to be made available in the nine dirtiest cities by 1995; and trucks and buses to reduce particulates by 90% by 1998. In addition, an emissions trading policy was set up that allows companies to buy and sell pollution rights for SO_2 (Kaufman and Franz, 1993). In early 1993, SO_2 emission allowance sales were initiated through the Chicago Board of Trade.

International Air Quality Policy Efforts

Acid Deposition in Europe

The effects of acid precipitation on lakes and fish in Norway and Sweden was noted in the early twentieth century. However, it was not until the late 1960s that observers began to correlate acidification of lakes with sulfur emissions from other countries. Sweden made an unheeded plea for international control efforts at the United Nations Conference on the Human Environment in 1972. The OECD (Organization for Economic Cooperation and Development) did respond by setting up a program to cooperatively collect air pollution data. By 1977, the collected data showed that acid deposition was occurring in many parts of Europe and transboundary air pollution was at least partly to blame. In addition, certain countries, such as Great Britain, West Germany, and Italy, were identified as major net exporters of SO_2 emissions (Park 1987).

The first major international initiative began in 1979 when 35 members of the Economic Commission for Europe (ECE) signed a Convention on the Long-Range Transport of Air Pollutants. The convention merely encouraged countries to work together to limit air pollution; even so, it remained unratified for 4 years. Pressure to ratify the agreement increased at the Stockholm Conference on Acidification in 1982 when West Germany reversed its position and strongly supported control measures. Finally, the convention was ratified in 1983, as reports of continued forest damage across Europe filtered in and as several countries acted unilaterally to reduce emissions. At the Munich International Conference on Environmental Protection in 1984, 18 countries (out of 32) agreed to cut their SO_2 emissions by 30% by 1993, and a binding protocol agreement was signed by 21 countries at Helsinki in 1985 (Park 1987).

In 1984, the Council of Ministers of the European Economic Community (EEC) agreed to adopt a framework to limit air pollution from large plants and vehicles. The approach taken with regard to vehicle emissions, however, has been criticized on the grounds that the standards chosen were somewhat lax and that there was too much variability in how member countries could comply (Park, 1987). Since then, greater emphasis has been placed on a more uniform regulatory approach, although debate on the catalytic converter as a primary solution still continues (Boehmer-Christiansen 1990; Heise 1990; Walsh 1990). While the use of scrubbers to reduce SO_2 emissions increased substantially in many European countries (French 1990), there was concern that the 30% agreement on SO_2 emissions would not be sufficient. In 1988, the Environmental Ministers of the EEC reached an agreement that required a 60% reduction in SO_2 emissions by 2003 (Kemp 1990). In 1992, they ordered cutbacks in the sulfur content of diesel and heating oil (Simons 1992).

Acid Deposition in North America

Canadian concern about acid deposition became a major issue in the late 1970s as noticeable damage to forests and lakes spread in southeastern Canada. It was evident to Canada that a significant portion of their acidification was imported from the United States. During the Carter years, several documents committing the United States and Canada to begin to work towards a solution were signed. The impasse in U.S. air quality legislation (see section below) preventing action on acid deposition during the 1980s, however, soured relations with Canada. During this time, they unilaterally committed themselves to 30–50% reductions in SO_2 and acted as a strong advocate for international treaties that would address transboundary air pollution. Despite all of these efforts, it was not until the Clean Air Act of 1990 passed that tensions between the United States and Canada over acid deposition were lessened (Park 1987; Regens and Rycroft 1988; Miller 1992).

The U.S Debate Over Air Pollution Policy

The 1977 Clean Air Act mandated the use of scrubbers for all new coal-fired power plants, creating a coalition between environmentalists and high-sulfur coal producers that was not to last. By the end of the 1970s, air quality policy in the United States had reached a stalemate. Several competing interests, including industry, government and scientific organizations, and environmental and public interest groups, strongly disagreed about approaches to solving air pollution issues, who should pay, and even about the scientific basis for policy. Regens and Rycroft (1988) have divided these major interests into three different perspectives—economic efficiency, environmental effectiveness, and social equity.

Industry, arguing for economic efficiency, critiqued the Clean Air Act on several grounds: the cost of pollution controls greatly exceed the benefits, regulatory inefficiency is great, and controls negatively impact productivity and innovation. They claimed that air quality improvements might better be correlated with industrial cycles, fuel switching, and economic slowdown rather than controls. Recommendations included relaxing regulations or modifying them with market mechanisms such as marketable rights to pollute, emission offsets and fees, and a bubble concept.

Government and scientific agencies, on the other hand, have made environmental effectiveness or attainment of clean air goals a major focus. They argue that implementation problems found in the current legal/administration approach are just as likely to occur in market incentive approaches. Furthermore, economic efficiency and air quality goals may be contradictory. Recommendations included optimizing the current air quality approach to try to balance both economic and equity issues.

Environmental and public interest groups claimed that the primary purpose of regulation was to achieve social equity values such as fairness and justice. They generally supported regulatory and opposed market-oriented approaches. They viewed cost-benefit analyses as inherently inequitable, improperly valuing the future and undervaluing the environment.

With local air quality legislation stalemated in the early 1980s as environmentalists and industry battled over attempted rollbacks in the Clean Air Act, the debate broadened away from local air pollution and moved to the interregional and international issue of acid deposition. Regens and Rycroft (1988) argue that the major interest groups may have focused on the acid deposition problem as a way of achieving their aims. With certain and large costs, and with uncertain and less tangible benefits, acid deposition offered a better opportunity for industry to win the economic efficiency debate. Furthermore, the international character of acid deposition and the inequitable financial burdens on selected industries and regions provide ready-made arguments for proceeding slowly and cautiously. The "go slow" approach also found support among scientific and technological agencies, easily persuaded of the need for studies, R&D and the development of new technology. On the other hand, environmental groups saw the acid deposition issue as a dramatic way to mobilize public support for action on air quality issues. The net result was that acid deposition became the center of attention, and other forms of air pollution were moved to the back burner. Unfortunately, the very real and much better documented costs of localized air pollution were lost in this conflict.

As the debate began to focus on acid deposition, the Acid Precipitation Act of 1980 set up an Interagency Task Force on Acid Precipitation to establish a comprehensive research program to clarify cause and effect. While initially it was thought that this would not delay regulation of acid deposition, opponents of control used a "good science" argument to block action: environmental and health damage must be proven beyond a doubt; cause and effect must be unambiguous before regulatory action could be taken. The Reagan and later the Bush Administrations' position was clear: study the problem but don't initiate any controls. Meanwhile, proponents of control could not agree how to allocate costs to states and regions. Regional conflicts over these economic equity issues and other issues eliminated the possibility of developing another clean air coalition in Congress throughout the 1980s (Regens and Rycroft 1988).

By 1990, public pressure for Congress to act on air pollution intensified. As powerful evidence concerning stratospheric ozone depletion and fears of global warming mounted, public attention shifted from acid deposition to these even more frightening issues. In addition, the warm weather of the late 1980s had intensified local and regional air pollution problems, and tropospheric ozone became a serious public concern. Finally, the 1 billion dollar NAPAP (1991) study was completed, showing that acid deposition was associated with significant environmental damage, although perhaps less than was feared (Stevens 1989). The Bush Administration's position was favorable to new legislation if costs to industry could be kept down. Congress responded by passing the Clean Air Act of 1990 (see previous section for provisions) by an overwhelming majority, which President Bush then signed into law. The act was truly a compromise, pleasing neither industry, environmentalists, nor other groups completely. Industry warned that implementation would be too costly, affecting economic growth and jobs. Environmentalists felt that chlorofluorocarbon (CFC) reductions were insufficient, fuel efficiency standards inadequate, and standards for incinerators weak. Industry did get a provision for an emissions trading policy for SO_2. Some states have since begun to adopt more stringent air pollution laws. (New York Times 1990; Wald 1991).

Conclusions

While considerable progress has been made on limiting some types of air pollution (mainly industrial smog) in the past 40 years, the increasing severity of photochemical smog in cities and surrounding regions, and a host of newly recognized regional, international, and global problems, such as acid deposition, ozone depletion, and global warming, ensures that air pollution issues will remain at the top of national and international agendas in the next century. Continued research on the atmosphere will undoubtedly reveal even more complexities and feedback loops, and more subtle impacts on health and the environment.

In the years ahead, attempts to improve air quality will be sorely tested by rising populations and industrial activity throughout the world. The industrial countries, presently the world's worst polluters, will need to take a multifaceted approach to air pollution that integrates population, resource, energy, and transportation issues. Output control efforts in the short term will need to be augmented with input control efforts for long-term solutions. The same approach in developing countries, where much of the population growth will take place in the next century, may help to limit their growing air pollution problems. Many new technologies allow the possibility for developing countries to bypass the worst polluting technologies of the industrialized nations. Energy efficiency as an immediate strategy appears to offer an economic way of providing pollution control and easing the transition to less polluting energy sources. The rapid deployment of new efficient technologies and renewable energy sources where possible, together with population control efforts and further R&D in nonpolluting energy technologies, offer perhaps the best hope of ultimately limiting the unacceptable impacts of air pollution.

References

Ahrens, C. D., *Meterology Today, An Introduction to Weather, Climate and the Environment,* 3rd ed., St. Paul, MN: West Publishing, 1988.

Baker, L. A., A. T. Herlihy, P. R. Kaufman, and J. M. Eilers, "Acidic Lakes and Streams in the United States: The Role of Acidic Deposition," *Science* 252, 1991, pp. 1151–1154.

Beardsley, T., "Add Ozone to the Global Warming Equation," *Scientific American* 266(5), 1992, p. 29.

Boehmer-Christiansen, S. A., "Putting on the Brakes: Curbing Auto Emissions in Europe," *Environment* 32, 1990, pp. 16–27.

Brady, N. C., *The Nature and Property of Soils, 10th ed.,* New York: MacMillan, 1990.

Burnett, W. M. and S. D. Ban, "Changing Prospects for Natural Gas in the United States," *Science* 244, 1989, pp. 305–310.

Cassedy, E. S., and P. Z. Grossman, *Introduction to Energy: Resources, Technology and Society,* New York: Cambridge University Press, 1990.

Charlson, R. J., S. E. Schwartz, J. M. Hales, R. D. Cess, J. A. Coakley, J. E. Hansen, and D. J. Hofmann, "Climate Forcing by Anthropogenic Aerosols," *Science* 255, 1992, pp. 423–430.

Charlson, R. J. and T. M. L. Wigley, "Sulfate Aerosol and Climate Change," *Scientific American* 270(3), 1994, pp. 48–57.

Crawford, M., "Scientists Battle over Grand Canyon Pollution," *Science* 247, 1990, pp. 911–912.

Cunningham, W. P. and B. W. Saigo, *Environmental Science, 2nd ed.,* Dubuque, IA: Wm. C. Brown, 1992.

Dillon, P. J., N. D. Yan, and H. H. Harvey, "Acidic Deposition: Effects on Aquatic Ecosystems," *CRC Critical Reviews in Environmental Control* 13, 1984, pp. 167–194.

Dovland, H., "Monitoring European Transboundary Air Pollution," *Environment* 29, 1987, pp. 10–15, 27.

Droham, D., "Different Routes to VOC Control," *Pollution Engineering* 24, 1992, pp. 30–33.

Elsom, D., *Atmospheric Pollution: Cause Effects and Control Policies,* Oxford, UK: Blackwell, 1987.

Enger, E. D. and B. F. Smith, *Environmental Science 4th ed.,* Dubuque, IA: Wm. C. Brown, 1992.

Flavin, C., "The Bridge to Clean Energy," *World Watch* 5, 1992, pp. 10–18.

Flavin, C. and N. Lenssen, "Here Comes the Sun," *World Watch* 4, 1991, pp. 10–18.

Farmer, A. M., J. W. Bates, and J. N. B. Bell, "Ecophysiological Effects of Acid Rain on Bryophytes and Lichens," *In* J. W. Bates and A. M. Farmer (eds.), *Bryophytes and Lichens in a Changing Environment,* Oxford: Claredon Press, 1992, pp. 284–313.

French, H., "Clearing the Air," *State of the World 1990,* New York: W.W. Norton, 1990, pp. 98–119.

French, H. F., "You Are What You Breathe," *In* L.R. Brown (ed.), *The World Watch Reader,* New York: W.W. Norton, 1991, pp. 97–111.

Gilbert, O. L., "A Biological Scale for the Determination of Sulphur Dioxide Air Pollution," *New Phytologist* 69, 1970, pp. 629–634.

Gilbert, O. L., "Lichen Reinvasion with Declining Air Pollution," *In:* J. W. Bates and A. M. Famer (eds.), *Bryophytes and Lichens in a Changing Environment,* Oxford: Claredon Press, 1992, pp. 158–177.

Graveland, J., R. van der Wal, J. H. van Balen, and A. J. van Noorduijk, "Poor Reproduction in Forest Passerines From Decline of Snail Abundance in Acidified Soils," *Nature* 368, 1994, pp. 446–448.

Harvey, H. H. "Widespread and Diverse Changes in the Biota of North American Lakes and Rivers Coincident with Acidification," *In:* D. Drablos and A. Tollan (eds.), *Proceedings, International Conference on the Ecological Impact of Acid Precipitation,* Norway, 1980, pp. 93–98.

Hassett, J. J. and W. L. Banwart, *Soils and Their Environment,* Englewood Cliffs, NJ: Prentice Hall, 1992.

Hawksworth, D. L. and F. Rose, "Qualitative Scale for Estimating Sulphur Dioxide Air Pollution in England and Wales Using Epiphytic Lichens," *Nature* 227, 1970, pp. 145–148.

Heise, S., "The Regulatory Approach of the EEC to Air Quality," *Science of the Total Environment* 93, 1990, pp. 81–95.

Hilts, P. J., "EPA and Harvard Find Particle Pollution...," *New York Times,* A, 2:1, July 19, 1993.

Hoffman, M., "Taking Stock of Saddam's Fiery Legacy in Kuwait," *Science* 253, 1991, p. 971.

Horgan, J., "Volcanic Disruption: A Giant Eruption Frays the Tattered Ozone Layer," *Scientific American* 266, 1992, pp. 28–29.

IEA (International Energy Agency), *Emission Controls in Electricity Generation and Industry,* Paris: Organization for Economic Cooperation and Development, 1988.

IPCC (Intergovernmental Panel on Climate Change), *Climate Change: The IPCC Assessment,* J. T. Houghton, G. J. Jenkins, and J. J. Ephrams (eds). Cambridge: Cambridge University Press, 1990.

Kaufman, D. G. and C. M. Franz, *Biosphere 2000: Protecting Our Global Environment,* New York: HarperCollins College Publishers, 1993.

Kauppi, P. E., K. Mielikainen, and K. Kuusela, "Biomass and Carbon Budget of European Forests, 1971–1990," *Science* 256, 1992, pp. 70–74.

Kemp, D. D., *Global Environmental Issues: A Climatological Approach,* London: Routledge, 1990.

Kerr, R. A. "1991: Warmth, Chill May Follow," *Science* 255, 1992a, p. 281.

Kerr, R. A., "Pollutant Haze Cools the Greenhouse," *Science* 255, 1992b, pp. 682–683.

Koren, H., *Handbook of Env. Health and Safety Principles and Practices, Vol. II, 2nd ed.,* Chelsea, MI: Lewis Publishers, 1991.

Lee, J. A. and C. J. Studholme, "Responses of *Sphagnum* Species to Polluted Environments," *In* J. W. Bates and A. M. Farmer (eds.), *Bryophytes and Lichens in a Changing Environment,* Oxford: Claredon Press, 1992, pp. 314–332.

Louria, D. B., "Trace Metal Poisoning," *In:* J. B. Wyngaaarden, L. H. Smith, and J. C. Bennett, (eds.), *Cecil Textbook of Medicine, Volume II, 19th ed.,* Philadelphia: W.B. Saunders, 1992, pp. 2361–2363.

Matson, P. A. and P. M. Vitousek, "Ecosystem Approach to a Global Nitrous Oxide Budget," *Bioscience* 40, 1990, pp. 667–672.

McElroy, M. B. and R. J. Salawitch, "Changing Composition of the Global Stratosphere," *Science* 243, 1989, pp. 763–770.

Miller, G. T., *Living in the Environment, 6th ed.,* Belmont, CA: Wadsworth, 1990.

Miller, G. T., *Living in the Environment, 7th ed.,* Belmont, CA: Wadsworth, 1992.

Mohnen, V. A., "The Challenge of Acid Rain," *Scientific American,* 259(2), 1988, pp. 30–38.

Mooney, H. A., P. M. Vitousek, and P. A. Matson, "Exchange of Materials Between Terrestrial Ecosystems and the Atmosphere," *Science* 238, 1987, pp. 926–932.

NAPAP, *National Acid Precipitation Assessment Program, 1990 Integrated Assessment Report,* Washington, DC: The NAPAP Office of the Director, November 1991.

NAPAP, *Interim Assessment: The Causes and Effects of Acidic Deposition,* Vol. 2, Washington, DC: U.S. Government Printing Office, 1987.

Nadakavukaren, A., *Man and the Environment: A Health Perspective,* Prospect Heights, IL: Waveland Press, 1990.

New York Times, "Lawmakers Agree on Rules to Reduce Acid Rain Damage," *New York Times,* 22 October 1990, sec. A p. 1:1.

NRC (National Research Council), *Rethinking the Ozone Problem in Urban and Regional Air Pollution,* Washington, DC: National Academy Press, 1991.

Park, C. F., *Acid Rain,* London: Methuen & Co., 1987.

Prinz, B., "Causes of Forest Decline in Europe," *Environment* 29, 1987, pp. 10–15, 32–36.

Quay, P. D., B. Tillbrook, and C. S. Wong, "Oceanic Uptake of Fossil Fuel Carbon Dioxide: Carbon-13 Evidence," *Science* 256, 1992, pp. 74–79.

Raloff, J., "Deforestation: Major Threat to Ozone?" *Science News* 130, 1986, p. 119.

Regens, J. L. and R. W. Rycroft, *The Acid Rain Controversy,* Pittsburgh, PA: The University of Pittsburgh Press, 1988.

ReVelle, P. and C. ReVelle, *The Environment: Issues and Choices for Society, 3rd ed.,* Boston: Jones & Bartlett, 1988.

Roberts, L., "How Bad Is Acid Rain?" *Science* 251, 1991, p. 1303.

Schneider, K., "U.S. Rejects Demands to Tighten Limits on Ozone in Smoggy Cities," *New York Times,* A9, August 4, 1992.

Schneider, S. H., "The Greenhouse Effect: Science and Policy," *Science* 243, 1989, pp. 771–781.

Schwartz, S. E., "Acid Deposition: Unraveling a Regional Phenomenon," *Science* 243, 1989, pp. 753–763.

Seinfeld, J. H., "Urban Air Pollution: State of the Science," *Science* 243, 1989, pp. 745–752.

Sigal, L. L. and T. H. Nash, III, "Lichen Communities in Conifers in Southern California Mountains: An Ecological Survey Relative to Oxidant Air Pollution," *Ecology* 64, 1983, pp. 1343–1354.

Simons, M., "Europe Orders Reduction of Pollution From Fuel," *New York Times,* 24 March 1992, sec. A p. 6:3.

Smith, R. L., *Elements of Ecology,* New York: HarperCollins, 1992.

Stevens, W. K., "Researchers Find Acid Rain Impacts Forests Over Time," *New York Times,* 31 December 1989, sec. I. p. 1:3.

UER (Utility Environment Report), New York: McGraw Hill, Aug. 21, 1992, p. 14.

UNEP & WHO (United Nations Environmental Program and World Health Organization), *Assessment of Urban Air Quality,* Nairobi: Global Environment Monitoring System, 1988.

Wald, M. L., "When the EPA Isn't Mean Enough About Cleaner Air," *New York Times,* 21 July 1991, sec. IV p. 5:1.

Walsh, M. P., "Vehicle Pollution Control in Europe; the Local and Global Significance," *Science of the Total Environment,* 93, 1990, pp. 57–67.

Wigley, T. M. L. and S. C. B. Raper, "Implications for Climate and Sea Level of Revised IPCC Emission Scenarios," *Nature* 357, 1992, pp. 293–300.

Chapter 13

Stratospheric Ozone Depletion

William J. Makofske and Eric F. Karlin

In 1974, two scientists warned that the ozone layer might be in danger of depletion by synthetic chemicals that were being added to the atmosphere. The substances suspected of causing this depletion are generally referred to as CFCs (chlorofluorocarbons) and halons (brominated hydrocarbons) and consist of a number of related chemical compounds that contain chlorine or bromine. Ozone, a molecule consisting of three oxygen atoms (O_3), is found in very small amounts in both the lower and upper atmosphere. The ozone of concern here exists in the stratosphere, a layer of the Earth's atmosphere that stretches from about 8 kilometers (km) at the poles, and 17 km at the equator, to nearly 50 km above the surface of the Earth. Figure 13.1 shows the layers of the atmosphere and the location of the stratospheric ozone layer. This gas acts as the Earth's only protective shield against certain wavelengths of harmful solar ultraviolet radiation that would otherwise penetrate to the surface. The consequences of increased ultraviolet radiation from the sun are potentially severe, affecting human health and the productivity of terrestrial and aquatic ecosystems.

The suggestion of ozone depletion caused scientists to focus more theoretical, laboratory, and environmental studies on the processes that governed the structure and composition of the Earth's atmosphere. In fact, two decades of measurement and study since then have compiled a substantial database about ozone in the atmosphere and the chemical behavior of various compounds with ozone. The hypothesis that ozone depletion was being caused by CFCs was initially taken somewhat seriously. Indeed, a major use of CFCs, as aerosol propellants, was banned by the United States, Canada, Sweden, and Norway in the late 1970s because of public concern (Gribbin 1988; Roan 1989).

Yet the issue then lost momentum for a variety of reasons. The theoretical predictions of ozone modification were extremely complicated, and some assumptions were questionable. All the studies and observations were not conclusive in documenting that ozone depletion was occurring. The existing observations, while often showing significant ozone variability, indicated that ozone levels were on average changing little. There also were other potential natural causes of ozone

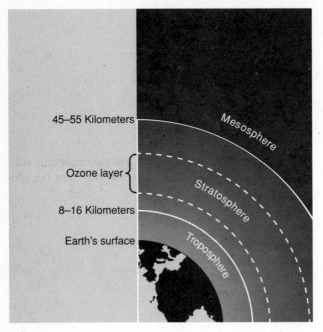

Figure 13.1. The two lowest layers of the Earth's atmosphere, the troposphere (typically 8–16 km thick) and the stratosphere (the upper limit of which typically occurs at 45–55 km above the Earth's surface). The ozone layer occurs in a region of the stratosphere where atmosphere ozone reaches its densest concentration (usually 20–40 km above the Earth's surface). There, ozone acts as a shield to screen out much of the harmful ultraviolet radiation from the sun. Diagram is not drawn to scale.

changes including volcanic eruptions, ocean current changes, and even sunspot activity. In the early to mid-1980s, with reduced budgets for environmental problems, the ozone depletion problem was mostly abandoned by the U.S. government. Worldwide production of CFCs actually increased over this period (Roan 1989). It would unfortunately take some shocking discovery to convince the nations of the world that this was a serious problem.

The shock came in 1985 when data taken in Halley Bay in Antarctica by a British research team indicated that during the spring period (late August to early November) a 40% decrease in ozone occurred (Farman et al. 1985). These results were verified by Nimbus 7 satellite measurements, which showed the area of the resulting ozone hole to be millions of square miles and that it extended halfway to the equator (Shea 1989a). Figure 13.2 indicates the dramatic loss of ozone as measured by the Nimbus 7 satellite. To the general public, it appeared that indeed the sky was falling or at least disappearing.

Since then, improved modeling and data gathering techniques verified that there is significant ozone depletion taking place in both hemispheres because of CFC and halon emissions. While some uncertainties remain, the ozone depletion problem[1]

---

[1]The depletion of stratospheric ozone should not be confused with the problem of increasing levels of tropospheric ozone. The latter is a serious air pollution problem of the lower atmosphere. Exposure to ozone is a health hazard. Thus although ozone in the stratosphere benefits life by absorbing harmful ultraviolet radiation, ozone in the troposphere is damaging to life. Interestingly, humans are faced with the dilemma that their actions are causing both a depletion of stratospheric ozone and the accumulation of tropospheric ozone.

rapidly entered the realm of international politics and economics as the nations of the world debated and chose a course of action.

The quick resolution of the ozone depletion problem is very significant for several reasons. First, the effects are severe and potentially catastrophic. The ozone layer needs to be protected quickly to avoid even greater depletion. Second, it is truly a global problem since, regardless of who produces CFCs, the entire world is affected adversely. Cooperative global efforts to solve the problem are precedent setting. Third, the ozone problem is a solvable one; there are CFC substitutes. The costs, while large, are not overwhelming. Indeed, as we describe in subsequent sections, the world has acted quickly to phase out many ozone-depleting substances. The irony is that even this unprecedented world action may not be adequate to prevent significant detrimental impacts.

Understanding the Nature of Ozone Depletion

Ultraviolet Light and Ozone Balance

The sun emits vast amounts of electromagnetic radiation, which not only heat the Earth but also provide the energy for life. Yet, solar radiation comes in many forms, some of which are not beneficial for life. Visible light (roughly 390–780 nanometers [nm] of wavelength) is the radiation most useful for life. It provides the energy and/or stimulus for many biochemical reactions (such as photosynthesis and vision). Infrared radiation (wavelengths >780 nm) is important for its heat producing capacity. In contrast, ultraviolet radiation (wavelengths <390 nm) carries enough energy[2] to fragment the organic molecules found in cells, and is thus a hazard to life.

Just as visible light comes in many forms (red to blue), there are different forms of ultraviolet radiation.[3] The most powerful is UV-C (<286 nm), which is biocidal. Although less powerful, UV-B (286–320 nm) still has the potential for causing significant cellular damage and can kill cells. The least powerful is UV-A (320–390 nm). Large doses of UV-A are required to cause significant cellular damage (Giese 1976).

Although ultraviolet radiation is a significant component of solar radiation, it represents less than 1% of the solar radiation reaching the Earth's surface. This is because a large percentage of the solar ultraviolet radiation (including all the UV-C and much of the UV-B) is filtered out in the upper stratosphere by oxygen molecules and a thin layer of ozone. Most of the UV-A and some of the UV-B reach the surface of the Earth (Giese 1976; Lloyd 1993).

The UV-C is largely absorbed by oxygen molecules (O_2). Through a process called *photodisintegration,* the energy gained from the absorption of the ultraviolet

---

[2]The energy carried by electromagnetic radiation varies with wavelength. Generally, shorter wavelength radiation carries greater energy and is said to be more penetrating, or powerful, than longer wavelength radiation.

[3]The division of ultraviolet radiation into UV-A, UV-B, and UV-C is somewhat arbitrary, and different authors list slightly different wavelength ranges for each.

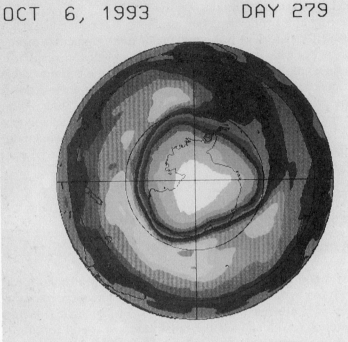

OCT 6, 1993   DAY 279

SOUTH POLAR PLOT

Figure 13.2. NASA Nimbus-7 satellite photo (South Polar Plot) shows the extent of the ozone depletion over Antarctica in the early spring of 1987. The lightest area in the center is the ozone hole. Source: NASA, 6 October 1993.

radiation causes an oxygen molecule to split into two atoms of oxygen (O). If each of these atoms then combines with another oxygen molecule, two ozone molecules are formed. Thus stratospheric ozone is created by the interaction of ultraviolet radiation and oxygen molecules.

Ozone is also capable of absorbing ultraviolet radiation, particularly UV-B and the UV-C not absorbed by the oxygen molecules. The absorption of ultraviolet radiation causes the ozone molecule to split apart into O and O_2. This photo-disintegration does not necessarily result in a net loss of ozone, as another ozone molecule would be formed if the oxygen atom (O) recombined with an oxygen molecule (O_2). A loss of ozone, however, would occur if two oxygen atoms were to combine to form an oxygen molecule (Giese 1976). The natural formation and destruction of ozone is illustrated by the equations in Figure 13.3.

Other natural agents that cause the breakdown of ozone also exist. Nitrogen oxides are one example. Natural processes are involved in both the formation and destruction of ozone. The thickness of the ozone layer thus depends upon the relative balance between these processes. Environmental conditions have allowed for the development and maintenance of an ozone layer sufficient to filter out a large percentage of the UV-B and also the UV-C not absorbed by oxygen molecules for the past several hundred million years (Giese 1976).

FORMATION

$$O_2 + [UV\text{-}C] \rightarrow O + O$$

$$O_2 + O + x \rightarrow O_3 + x$$

DESTRUCTION

$$O_3 + [UV\text{-}B] \rightarrow O + O_2$$

Figure 13.3. Ozone is naturally generated and destroyed by the following reactions. [UV] is ultraviolet radiation while x is any nearby atom or molecule that allows conservation of energy and momentum. See text for further discussion.

Ozone Depletion

Recent human activities have apparently upset the balance between ozone creation and destruction. There is now evidence that the destruction of ozone is occurring at a much faster pace than its formation. This is leading to a thinning of the ozone layer, a process that will ultimately allow for more ultraviolet radiation to reach the Earth's surface. A slight depletion would allow for increased levels of UV-B to reach the Earth's surface. A large depletion of ozone would not only allow for a correspondingly greater dosage of UV-B to reach the Earth's surface, but some UV-C as well (Giese 1976). Increased levels of UV-B reaching the Earth's surface at middle latitudes have recently been documented, and this increase has been strongly linked to stratospheric ozone depletion. Kerr and McElroy (1993) found that erythemally active (sunburn causing) UV-B levels increased 5.3% a year in the winter and 1.9% a year in the summer over a 4 year period (1989–1993) in the Toronto, Canada region.

Like all chemical reactions, the destruction of ozone is accelerated in some environments and hindered in others. Thus ozone depletion is not occurring uniformly across the globe, and it also varies with time. Ozone depletion is generally highest in the early spring months, with ozone accumulation occurring in the summer. There are therefore annual fluctuations in ozone abundance, as well as long-term changes in the global concentrations of this ecologically significant gas. The region having the greatest rates of ozone depletion is the Antarctic, where ozone concentrations have been annually reduced on the order of 50% or more (1987, 1989–1993) (Roberts 1989; Stolarksi et al. 1992; Gleason et al. 1993; Kerr 1993b). In 1993, ozone depletion in the Antarctic reached a record low, with depletion occurring at altitudes both higher and lower than normal (Kerr 1993b). There has been considerable concern about an Arctic ozone hole developing as well. Scientists predicted a possible 50% loss of Arctic ozone in spring of 1992 (Kerr 1992a). Early warming in January and a weakened polar vortex, however, led to only a 10–25% loss, nonetheless a record low for the Northern Hemisphere (Kerr 1992b; Rodriguez 1993). While ozone depletion is potentially the greatest at the poles, there has also been considerable worry about midlatitude depletion, where large human populations are located. Ground-based data for 26–64° N latitude show a statistically significant decrease of 1.8% per decade since 1970, while no significant decrease has been found in the tropics (25°S–25°N) (Stolarski et al. 1992). While it was initially thought that this midlatitude reduction occurred primarily in early spring, new results show that significant depletion is occurring at other times of the year. An international panel of scientists convened by the United Nations found that summer stratospheric ozone decreased by 3% and 5%, respectively, in the temperate zones of the Northern and Southern hemispheres (Stevens 1991). In the winter of 1992–1993, ozone in the midlatitudes was down 9%, while in June 1993 the decrease was 10%, leading to a 13% increase in ultraviolet radiation (Stone 1993).

Human impact on the ozone layer arises primarily from the creation of new chemicals, many of which are far more effective at destroying ozone than naturally occurring chemicals. In addition, human activities have also increased the concentrations of natural agents of ozone destruction, particularly nitrogen oxides. Certain activities, such as nuclear war, could have catastrophic results (NAS 1975; Hastings 1987).[4]

The Role of CFCs

The principal new chemical culprits are CFCs, a class of organic compounds that are used as aerosol propellants, refrigerants, cleaning solvents for electronic components, and foaming agents for plastics. Their use has increased steadily in industrial societies. These compounds are very popular because of desirable chemical characteristics, particularly their lack of chemical reactivity and volatility. This is also the basis for their destructive power. CFCs remain essentially unchanged during use and eventually escape to the Earth's atmosphere where they remain for many decades. The usual methods of atmospheric cleansing, where compounds are broken down by sunlight, oxidized, or dissolved in rain, does not happen readily with CFCs (Rowland 1989).

Eventually, CFCs can extend to greater heights, into the stratosphere, and above most of the ozone. Here they are relatively unprotected. UV-C, not present at lower altitudes, can now strike the CFCs and break the molecule apart, releasing atomic chlorine. Atomic chlorine reacts with ozone to form chlorine oxide (ClO) and O_2. The chlorine oxide then combines with atomic oxygen, formed by the photodisintegration of another ozone molecule, to form atomic chlorine and O_2. These two reactions are repeated again and again with the chlorine acting as a catalyst, as shown in Figure 13.4. Chlorine is thus not destroyed or changed in the reactions; it reappears to cause the process to occur again. After about 3 years (the mean stratospheric residence time of chlorine), it is deposited back to the Earth, but not before destroying tens of thousands of ozone molecules. This large amplification factor means that it does not take huge amounts of CFCs to affect significant numbers of ozone molecules (Rowland 1989; Ko et al. 1994).

In addition to CFCs, carbon tetrachloride and chloroform may also contribute chlorine for ozone depletion. Bromine atoms, found in compounds such as

[4]The massive amounts of nitrogen oxides that would be generated by a major nuclear war would significantly deplete the ozone layer. It is projected that the ozone layer would be depleted by 50–75% if approximately half of the nuclear weapons of the United States and the former USSR were to be detonated. The impacts of this depletion could be more disastrous on a global scale than the impacts of ionizing radiations released by the nuclear explosions.

Figure 13.4. Ozone is destroyed by chlorine in the following set of reactions. In (a), ultraviolet light dissociates a chlorofluorocarbon compound. In (b), the chlorine chemically combines with ozone, thereby destroying the ozone molecule. In (c), chlorine is regenerated through the reaction so that it may again undergo the reaction in (b). Thus, chlorine acts as a catalyst by allowing the reaction (b) to proceed without itself being destroyed. A single chlorine atom is capable of destroying many thousands of ozone molecules.

(a)
$$CCl_3F + [UV\text{-}C] \rightarrow CFCl_2 + Cl$$

(b)
$$Cl + O_3 \rightarrow ClO + O_2$$

(c)
$$ClO + O \rightarrow Cl + O_2$$

halons and methyl bromide (widely used as a biocide and also produced by biomass burning), behave similarly to chlorine (Cicerone 1994; Ko et al. 1994; Mano and Andreae 1994). Other chemicals that destroy ozone via catalytic action include hydrogen oxides and nitrogen oxides. The catalytic action of the hydrogen oxides and nitrogen oxides, however, is not as great because they are removed from the reaction chain more rapidly than chlorine or bromine. It should be recognized that other gases in the atmosphere, such as carbon dioxide, methane, and nitrous oxide, also have effects on ozone depletion and need to be accounted for in any realistic modeling calculations. Table 13.1 shows data listing the emissions, growth rate, atmospheric lifetime, and contribution of the major chemicals to ozone depletion (Shea 1989a).

Health and Environmental Effects of Ozone Depletion

Ozone depletion has serious impacts, not only on the human population, but on plant and animal life as well. While the human impacts alone make a compelling case for ozone protection, it is the ecosystem effects that may be the most significant. This section outlines some of the major impacts that can be expected if only UV-B is increased.

Human Health

Ultraviolet radiation has a number of negative impacts upon human health. As it does not penetrate more than a few layers of cells, the damage wrought by UV-B is primarily limited to the skin and eyes. The most common hazard is sunburn, with maximum damage associated from wavelengths around 280 nm. In addition, the genetic material of cells, DNA, absorbs ultraviolet radiation. The energy gained by the absorption of ultraviolet radiation causes the DNA in skin cells to be fragmented (Taylor 1990; Lloyd 1993).

Table 13.1
Use and Emissions Profiles of Commonly Used Chemicals, 1985

Chemical	Emissions (10^3 tons)	Atmospheric Lifetime[a] (in years)	Applications	Annual Growth Rate (percent)	Share of Contribution to Depletion[b] (percent)
CFC-12	454	139	Air conditioning, refrigeration, aerosols, foams	5	45
CFC-11	262	76	Foams, aerosols, refrigeration	5	26
CFC-13	152	92	Solvents	10	12
Carbon tetrachloride	73	67	Solvents	1	8
Methyl chloroform	522	8	Solvents	7	5
Halon 1301	3	101	Fire extinguishers	n.a.	4
Halon 1211	3	12	Fire extinguishers	23	1
HCFC-22	72	22	Refrigeration, foams	11	0^c

[a]Time it takes for 63% of the chemical to be removed from the atmosphere.
[b]Total does not add to 100 due to rounding.
[c]Contribution of HCFC rounds to zero.

Source: Shea, C.P., "Protecting the Ozone Layer," *State of the World 1989,* Washington, DC: Worldwatch Institute, 1989.

The repair of DNA damaged by ultraviolet radiation can proceed by normal DNA repair mechanisms possessed by all cells.[5] Although cells have the capacity to repair such damage to their DNA, the repair is not always perfect. In those cases where the DNA is not properly reconstructed, the cell could die or possibly become cancerous (Giese 1976; van der Leun 1988; Kripke 1989; Taylor 1990). Skin cancer is strongly linked to exposure to UV-B (Lloyd 1993) and one of the immediate effects of increased ultraviolet radiation at the Earth's surface would be an increased incidence of skin cancer. Although the relationship between skin cancer incidence and UV dose is not linear (Madronich and de Gruijl 1993), for small ozone depletions (less than 5%) it can be assumed that a 1% reduction in stratospheric ozone would lead to a 2% increase in UV-B and a 5 to 6% increase in skin cancer (Shea 1989a). This increase in skin cancer would include a 2 to 5% increase in nonmalignant forms of skin cancer and a 1% increase in malignant melanoma, an often fatal form of skin cancer (EPA 1987; Stevens 1991; Concar 1992). Deaths due to melanoma in the United States (6,800 in 1993) have doubled since 1980 and increasing levels of UV-B are believed to be a major variable behind this phenomenon (Adler 1994).

[5]Many organisms also possess a second system of DNA repair called *photoreactivation* or *photorepair.* This additional system involves photoenzymes that are activated by exposure to UV-A or blue-violet radiation. These photoenzymes repair DNA damaged by ultraviolet radiation. Unfortunately, photoreactivation is not well developed in placental mammals (see Giese 1976).

Based on known ozone depletion rates, in April 1991 the EPA announced that over the next 50 years an additional 12 million Americans would develop skin cancer, and that there would be more than 200,000 additional fatalities (Stevens 1991). For comparison purposes, about 600,000 cases of nonmalignant squamous and basal cell carcinoma and 26,000 cases of malignant melanoma led to over 12,000 skin cancer deaths per year in the United States in the late 1980s (Shea 1988). The 1991 EPA projections indicate that ozone depletion will therefore result in a 38% increase in skin cancer, and a 33% increase in related deaths. This is likely to be a conservative estimate. Subsequent data on higher rates of ozone depletion and the discovery that ozone depletion also occurs in the summer (Stevens 1991), when people are more likely to be exposed to sunlight, will substantially increase these numbers. Any substantial reduction in ozone may require that people wear protective clothing in order to go outside during the periods of more intense sunlight. Another mode of protection would be the use of sunscreen lotions, which reduce exposure to UV-B and help to prevent sunburn. Unfortunately, however, the most commonly used lotions do not provide adequate protection against UV-A (Pennisi 1993) and the immunosuppressive effects of UV-B (Concar 1992; Adler 1994)

Ultraviolet radiation also affects the eye. Exposure to excessive amounts of UV-B, as may occur with sunlight being reflected from snow on sunny days, causes an inflammation of the cornea. This condition is called snow blindness. Long-term exposure to UV-B can also lead to the development of cataracts (Lloyd 1993). Any increase in the amount of UV-B reaching the Earth's surface would cause these eye problems to become more prevalent (van der Leun 1988). The conservative EPA projections indicate that 555,000 to 2.8 million additional Americans born before 2075 will get cataracts due to ozone depletion (EPA 1987; Shea 1988).

UV-B may also damage the body's immune system, which would make the body less able to fight disease. Humans with UV-B damaged immune systems could be more susceptible to skin cancer, parasitic infections of the skin, herpes, hepatitis, and other ailments (Alderson and Kent 1986; Miller and Mintzer 1986; EPA 1987; Kripke 1993).

Ecosystem Impacts

The various species present on Earth have mechanisms that allow them to tolerate the current levels of UV-B.[6] Increases in UV-B could overwhelm the ability of many species to deal with it and result in increased incidences of injury, mutation, and death. Organisms currently having minimal UV-B defenses would be at most

---

[6]Current levels of ultraviolet radiation do have some positive influence on life. Many flowers have intricate patterns on them which reflect UV-A. Invisible to humans, these patterns serve as landing and/or nectar guides for insects (many of which can see UV-A) which visit the flower (Meeuse 1961). Some organisms require exposure to UV-B for the synthesis of Vitamin D. UV-B has also been effectively used in the treatment of some skin diseases, especially prior to the advent of antibiotics. Psoriasis is one example. As viruses and bacteria are more sensitive to ultraviolet radiation than are human cells, it is believed that the UV-B kills the pathogens on the skin surface before much damage is done to the host cells. As noted above, UV-A also plays a role in photoreactivation (photorepair), a process that repairs UV-B damaged DNA (Giese 1976; Roberts 1989).

risk to elevated levels. For instance, the recent widespread dieoffs of many amphibians (frogs and salamanders) is in part due to increased levels of UV-B (Blaustein and Wake 1990; Blaustein 1994). On the other hand, species having a better defense capacity against UV-B may only be slightly affected, although even they would eventually succumb if exposure to UV-B and UV-C were to continue to increase because of a progressively more depleted ozone layer. Many species (e.g., nocturnal animals, organisms dwelling at the bottoms of lakes and oceans) would not be directly affected by elevated UV-B levels.

Single-celled organisms are highly affected by exposure to UV-B. In fact, exposure to UV-B is lethal to most bacteria. This fact has long been appreciated in medicine, where UV lamps are used to help provide a more sterile environment in operating rooms. This ability to kill bacteria also makes ultraviolet radiation useful in disinfecting water supplies (Carlson 1985). If UV-B levels were to increase significantly, there would most likely be a major negative impact on many microorganisms, both pathogenic and beneficial (Giese 1976).

Multicellular organisms are not as impacted by UV-B, although there is still an effect. As noted in the discussion on human health impacts above, ultraviolet radiation primarily affects the outer cell layers of multicellular organisms. At most risk would be the gamete and young larval stages. Studies have found that UV-B damages crab and shrimp larvae, and also fish larvae and juveniles (Worrest 1989). Most mammals have similar responses to UV-B exposure as that found in humans. Plants have reduced photosynthetic levels and damaged DNA when exposed to damaging levels of UV-B (Virhanski 1989). Many crops, including commercially important species such as soybeans, appear to be sensitive to UV-B radiation (Shea 1989a).

Most organisms have evolved mechanisms to minimize damage by ultraviolet radiation, especially those that are exposed to full sunlight on a regular basis. Mechanisms such as materials that reflect or safely absorb ultraviolet radiation (hair, feathers, scales, waxy layers on leaves), pigments (melanin), and behavioral responses all serve as effective means to minimize exposure to UV-B. Organisms vary, however, in their ability to handle ultraviolet radiation. Plant species typically found in shaded environments are much less able to tolerate exposure to UV-B than are plant species normally found growing in full sunlight. Plant species growing at high altitudes (which are naturally exposed to higher levels of UV-B) have much more resistance to damage from ultraviolet radiation than do species found in the lowlands (Virhanski 1989). Humans, with their naked skin, are far more susceptible to UV-B damage than are those mammals that have a thick fur coat.

Because water absorbs ultraviolet radiation, aquatic ecosystems are more protected from ultraviolet radiation than are terrestrial ecosystems. Even so, exposure to elevated levels of UV-B would still affect aquatic organisms. One group of microorganisms that may currently be at risk are the planktonic algae of the Antarctic Ocean. This region is already experiencing elevated levels of UV-B because of the extreme depletion of ozone that occurs there in the winter. Exposure to elevated levels of UV-B is known to lower the photosynthetic rates of these algae and to also reduce their reproductive capacity. The resulting drop in the algal population would have far-reaching effects, as they are the basis of the oceanic food web. Krill feed upon the algae, and the krill then serve as food for fish and whales. Any significant reduction in numbers of algae could thus lead to a major disruption of the Antarctic marine ecosystem. Some scientists point out that although the more UV-B sensitive algal species might be decimated, other, more UV-B resistant, species of algae may become more abundant and replace them. Although we do not yet know what the final outcome will be, it is clear that the existing populations of planktonic algae in the Antarctic oceans are likely to change dramatically (Roberts 1989; Worrest and Hader 1989; Smith et al. 1992). As the above discussion indicates, even if only a few species in an ecosystem were to be directly affected by elevated levels of UV-B, the change in their ecological status could have repercussions on the entire ecosystem (Bothwell et al. 1994).

Ultraviolet radiation has the potential to have even more significant impacts upon terrestrial ecosystems. A 25% depletion of stratospheric ozone could cause a 20 to 25% drop in soybean harvests and reduce the growth rate of loblolly pine such that this species would no longer serve as a viable source of timber (Virhanski 1989). If the ozone layer were to be largely depleted or destroyed a large percentage of the species on Earth would probably become extinct (Ellis and Schramm 1995).

Climate Change

Reducing the amount of ozone in the stratosphere could have potentially dramatic effects on the climate. One effect could be increased warming of the lower atmosphere and the Earth's surface brought about by increased amounts of ultraviolet and infrared radiation (both absorbed by ozone) being transmitted. This would exacerbate the global warming problem. In addition, as the absorption of solar radiation by ozone produces heat, reducing ozone would cause a significant cooling of the upper stratosphere. The overall consequences of this for the climate at the Earth's surface are not well understood (Giese 1976; Watson 1988). (See also subsequent section, *CFCs as Greenhouse Gases.*)

Scientific Studies on Ozone Depletion

The SST and Space Shuttle Controversies

In the late 1960s and early 1970s, the development of a supersonic transport (SST) aircraft by Britain and France, named the Concorde, and the possibility of an American counterpart, caused scientists to initiate studies of what the impact of stratospheric emissions would be on the Earth's ozone layer. A 3-year study, called the Climatic Impact Assessment Program, implicated nitrogen oxides as potentially causing 10% ozone depletion in the Northern Hemisphere. Later, in 1977, recalculations showed that the depletion effect was less than thought (Gribbin 1988).

A related but less public controversy occurred over the effects of the space shuttle in the mid-1970s. Each launch of a space shuttle releases 200 metric tons of chlorine molecules, one-third of which are emitted into the stratosphere (Ko et al. 1994). While the studies indicated an ozone reduction of less than 1% based on 60 flights per year, perhaps the most significant finding was that chlorine was implicated in ozone depletion (Gribbin 1988).

CFC Controversy Initiates Further Scientific Studies

After Molina and Rowland's (1974) theory that CFCs were reaching the stratosphere, releasing chlorine and depleting the ozone layer was made public, a long and bitter debate began. The principal players in this debate were scientists, the CFC industry, government policymakers, and environmental groups. After House Subcommittee hearings in 1974, the federal government organized the Committee on the Inadvertent Modification of the Stratosphere (IMOS 1975), which described the issue as a "legitimate cause for concern" and promptly handed it over to the National Academy of Sciences in 1975. The NAS study consisted of two separate tasks and two committees: one was to determine the validity of the Rowland–Molina theory; the other was to determine what policies ought to be implemented. The NAS (1976a,b) report, released in September 1976, concluded that ozone depletion by CFCs would occur, with an eventual loss of somewhere between 2% and 20%. The ozone depletion would also increase skin cancer due to increased ultraviolet intensity. The other committee advised that the government should study the issue for 2 more years before any policy decisions were made.

Later in the 1970s, chlorine oxide was detected in the atmosphere, confirming the Rowland–Molina theory, and a second NAS report estimated ozone depletion ultimately reaching 16% (NAS 1979a,b). There were, however, still no actual measurements showing ozone depletion. Satellite data reported in 1981 showing about 1% depletion of ozone over a decade (Norman 1981) was discounted because of potential instrument problems. A third NAS (1982) report in 1982 reduced the projected ultimate level of ozone depletion to 5–9%, while a fourth NAS (1984) report in 1984 lowered it to 2–4%. On the other hand, an international UNEP-sponsored WMO-NASA (1986) report in 1985 had concluded that average ozone depletion would range from 4.9% to 9.4% by 2050.

The Ozone Hole Discovery

In mid-1985, the British Antarctic Survey led by Joseph Farman published a paper in *Nature* documenting their ground findings that the ozone at Halley Bay, Antarctica, and at Argentine Island had been showing a serious depletion (a remarkable 40% in 1984) at the end of each austral winter over the past several years (Farman et al. 1985). While the large ozone depletion was initially greeted with skepticism, NASA's Goddard Space Flight Center quickly confirmed that the Nimbus 7 satellite data showed a precipitous drop in ozone the size of the continental United States and extending from the South Pole to almost 45°S latitude. It turned out that NASA's computer analysis programs had discounted very low ozone data, which was thought to be in error. Because of this and a backlog of data that had not been analyzed, NASA failed to make one of the most important scientific discoveries about ozone depletion—that it was indeed happening, and on a global level (Roan 1989; Shea 1989a). While this shocked scientists and the public around the world, the critical question was why this was happening.

Several groups provided explanations that implicated chlorine together with the harsh Antarctic winter conditions for the depletion; other scientists suggested that downward air movement in the polar vortex was weakened so that ozone at lower levels was not being replenished; still other scientists blamed the depletion on the sunspot cycle (Crutzen and Arnold 1986; Schoeberl and Krueger 1986; Solomon et al. 1986). Regardless of who, if anyone, was right, the ozone hole discovery had suddenly brought the ozone issue back to life (Kerr 1986).

The First NOZE Study

In 1986, Robert Watson, director of NASA's stratospheric research programs, organized an emergency research team of 13 American scientists to take measurements on the ozone hole. Called NOZE, the National Ozone Expedition, the team had only 5 months to prepare for their trip. Under grueling conditions, they were able to obtain some excellent data. Although it was preliminary, they concluded that their data suggested that CFCs were to blame for the ozone hole. The debate between various theories for the ozone hole continued unabated, however, and a second much larger trip to Antarctica was set for 1987 (Watson 1988; Roan 1989).

The Second NOZE Study

Meanwhile in 1987, both satellite data and an analysis of the Dobson stations,[7] mostly in the Northern Hemisphere, showed about a 4% loss over a 6 to 7 year period, a result not predicted by any models (Roan 1989). With the possibility that ozone depletion could be occurring more rapidly than anyone had anticipated, the second trip to Antarctica took on great importance. NOZE II, a 10 million dollar venture to Antarctica involving 150 people, was to collect data by flying aircraft through the hole in addition to taking more ground and satellite measurements (Watson 1988; Roan 1989). In spite of risky flight conditions, the project succeeded in measuring the critical chlorine compounds in the ozone hole, which confirmed beyond any doubt that it was human-made chlorine chemicals that were causing the problem. The mechanism that explained the detailed chemical reactions that caused ozone depletion was reproduced in the laboratory (Molina et al. 1987). On ice particles similar to those found in the polar stratospheric clouds, hydrochloric acid and chlorine nitrate were found to react rapidly to release chlorine radicals, which attack ozone. The ozone hole was due to a combination of ozone-depleting chemicals and the extreme weather conditions.

The Ozone Trends Panel Report

The NASA (1988) Ozone Trends Panel report made headlines throughout the United States. The panel found significant ozone losses in the Northern Hemisphere. The loss, between 1.7% to 3%, was already greater than the 2% the EPA

---

[7]Dobson stations are ground-based spectrophotaic instruments measuring ultraviolet light, and thus indirectly, ozone. It was the Dobson station run by the British in Antarctica that first discovered the ozone hole.

had predicted would occur by 2075, assuming that the provisions of the 1987 Montreal Protocol were implemented (see section on *Post-1985 Worldwide Efforts*).

An Arctic Ozone Hole

In 1988, after finally understanding the Antarctic ozone loss, researchers began to focus on the Arctic region (Harris et al. 1988). While temperature conditions in this region were not as severe, a scaled down version of the Antarctic phenomenon might be possible. Early measurement found excess chlorine compounds, suggesting that a similar mechanism might be present. A major airborne expedition to the Arctic was carried out in January 1989. The scientists confirmed that chlorine chemicals were indeed causing an observed 5% depletion over the Northern Hemisphere and that, depending on the weather conditions, considerably greater ozone depletion could occur (Roan 1989). In 1991, a 5-month mission to the Arctic to determine the possibility of greater Arctic ozone losses was begun (Planes to Search . . . 1991). The results were disquieting. Scientists found relatively high levels of ClO and there was concern that sulfuric acid aerosols (from anthropogenic sources and from the Mt. Pinatubo eruption) might lead to severe depletion. A possible 50% drop in Arctic ozone levels was predicted; a significant 10 to 25% drop in ozone actually occurred (Kerr 1992a, b; Rodriguez 1993).

Midlatitude Ozone Depletion

Data released in 1991 by the EPA (Kerr 1991) and a United Nations panel of scientists (Stevens 1991) confirmed that total ozone loss over the midlatitude regions in both the Northern and Southern hemispheres is occurring much more rapidly than scientists had previously thought. Moreover, the depletion is occurring over the late spring and summer months, leading to potentially much greater health and ecological effects. In the winter/spring of 1992, in the latitude range from 50 to 60°, ozone was 12% below normal—the lowest yet seen (GCCD 1993; Gleason et al. 1993). In late 1992, the region between 30 and 60° in the Northern Hemisphere was 9% below normal (Kerr 1993a). The mechanism for ozone depletion outside of the polar region is still uncertain. Such loss could be due to transport of ozone-depleted air from the poles, or perhaps caused by in situ ClO-related chemical reactions (Stolarski et al. 1992). It is possible that volcanic debris could catalyze the destruction of NO_x that normally prevents the release of chlorine into its active form (Kerr 1993a).

A Brief History of Political Efforts to Protect the Ozone Layer

Early U.S. Efforts—The 1970s to 1985

The Aerosol Ban

Even while NAS was debating the issue in 1975, bans on CFC-containing aerosols were introduced in several states; Oregon became the first state to pass such legislation. All across the nation, consumers were reducing their use of aerosol sprays. In June 1975, Johnson Wax announced that it would terminate use of CFCs in its products.

In spite of the apparent lack of urgency suggested by the NAS (1976b) report, both the Food and Drug Administration and the Environmental Protection Agency moved to ban nonessential uses of CFCs, primarily the use of CFCs in aerosol spray cans, by the end of 1978. The U.S. ban on aerosols succeeded in reducing the total worldwide use of CFCs by 25% (Roan 1989). Only three countries, Canada, Norway and Sweden, followed suit.

After the Ban

The nonaerosol uses of CFCs increased dramatically over the next few years while the issue retired to the back burner. There were a variety of reasons for this. CFC use in refrigerators, air conditioning, foam packaging, and the semiconductor and microchip industries was considered essential. There was still a great deal of scientific uncertainty about ozone depletion as more atmospheric chemicals were considered. Perhaps more importantly, the regulatory climate in Washington was rapidly eroding in the early 1980s (Roan 1989).

Moreover, the CFC industry had regrouped and continued their adamant opposition to any controls. They lobbied extensively for more research time for developing alternatives, although in fact they were discontinuing such efforts. They also argued that any action must be worldwide; it would be unfair to penalize U.S. industries and give other countries an economic advantage. Their arguments were viewed sympathetically within the administration (Roan 1989).

Early Worldwide Efforts—The 1970s to 1985

The groundwork for international efforts on ozone was laid at the United Nations Conference on the Human Environment in 1972. These meetings led to the formation of the United Nations Environmental Programme (UNEP) and the adoption of a principle of national responsibility for environmental damage beyond its own borders. Early meetings sponsored by UNEP in 1977 and 1978 led to a commitment to cap CFC production, but at amounts well above current production. At an international meeting in Oslo in 1980, a number of countries, including the United States, called for a reduction of CFCs from all sources. The EPA even announced its intent to freeze CFC production at 1979 levels, a move that rallied the CFC industry to lobby against their proposals to further restrict CFC usage. Prospects for international controls improved in 1983 when Norway, Sweden, and Finland proposed a draft protocol, known as the Nordic Annex, to ban all aerosols and to control other uses of CFCs. Political changes within the EPA allowed that agency to strongly support this protocol in spite of reluctance at the State Department. The European Economic Community (EEC), however, strongly opposed the Nordic Annex. By 1985 a very weakened protocol, the Vienna Convention, was signed that only specified that negotiations would resume within a specified time period, although it did include provisions for research, monitoring, and exchange of information that was later to prove valuable (Roan 1989).

Post-1985 Worldwide Efforts

International negotiations were resumed in December 1986 in Geneva, but the idea of a phase-out was not received favorably by the European Economic Community or Japan, and ultimately negotiations broke down. By the middle of 1987, however, it appeared that agreement might be reached on a 50% reduction (Roan 1989).

In September of 1987, the UNEP-sponsored international meeting reconvened in Montreal. Despite some initial serious disagreements, a historic agreement, known as the Montreal Protocol, was reached. CFC production and consumption would be frozen at 1986 levels by 1990, there would be a 20% reduction by the beginning of 1994, and there would be an additional cutback of 30% by the beginning of 1999. While there were exemption clauses and loopholes that weakened the agreement, it did specify that the nations would meet again and ultimately eliminate ozone-depleting chemicals (Roan 1989).

By 1989, international pressure mounted for a new protocol. While the Montreal Protocol was to take effect in 1989, the new evidence on the cause of the ozone hole and the newly found depletion in the Northern Hemisphere required that the negotiators reconvene quickly and strengthen the agreement. At a meeting in Brussels in March of 1989, the EEC announced they were willing to eliminate CFCs by the year 2000. President Bush then called for a ban by the end of the century but carefully based his statement on the availability of safe alternatives (Roan 1989). Other countries, including the Soviet Union, China, India, and Japan, balked at this faster timetable. It seemed that the world community could not reach an accord (Roan 1989).

In June 1990 at the London Saving the Ozone Layer Conference, agreement was reached on the following major provisions: a complete phase-out of CFCs by the industrialized nations by the year 2000, a fund of 240 million dollars to help developing nations convert to other technologies, an accelerated timetable for CFC elimination in 13 industrialized countries, methyl chloroform phase-out by 2005, and a meeting to reconvene in 1992 to assess the problem (Browne 1990).

In early 1992, the United States accelerated the timetable for phasing out CFCs to 1995 when measurements made by the Airborne Arctic Stratospheric Expedition and NASA's Upper Atmosphere Research Satellite found large levels of chlorine and aerosols in the Northern Hemisphere's stratosphere that potentially could lead to a large loss of ozone (Taubes 1993). Globally, ozone reached record lows in 1992 (Gleason et al. 1993). Finally, in late 1992, at an international ozone conference held in Copenhagen, agreement was reached that would terminate production of CFCs, carbon tetrachloride, and methyl chloroform at the end of 1995, halons at the end of 1993, freeze methyl bromide levels at 1991 levels by 1995, and place controls on HCFCs (IER 1992; Ko et al. 1994).

An Uncertain Future

In spite of the unprecedented international success on an agreement to phase out CFCs and other ozone-depleting compounds, there is a great deal of uncertainty about whether the world acted soon enough to prevent severe ramifications. The CFCs in the atmosphere today, together with those entering the atmosphere in the future, commit the Earth to continued depletion for almost a century before the ozone shield stabilizes. Scientists simply do not know whether we may be in for more surprises. The same chemistry that occurs in Antarctica occurs in Arctic regions, although the effects are not as pronounced. And the severe ozone hole over Antarctica could become worse, reaching further into a number of Southern Hemisphere countries. The increasing amount of methane in the atmosphere and the sunspot cycle could affect the rate of ozone depletion, possibly in unpredictable ways. It has even been speculated that dust particles from volcanoes might also enhance ozone depletion by providing surfaces for ozone-depleting chemical reactions. Indeed, satellite measurements in 1992–1993 indicate an ozone drop of 4% below the average of the past dozen years, and this appears linked to the eruption of Mt. Pinatubo (Kerr 1993a). Increasing chlorine dioxide levels in Antarctica may also be a result of the volcano eruption (Solomon et al. 1993). On the other hand, it is thought that chlorine from volcano eruptions will have only a limited impact on ozone depletion (Tabazadeh and Turco 1993). Nonetheless, the depletion that we are now seeing was not predicted by the scientific models so it would be foolish to think that we understand and can predict exactly what may happen in the future.

Ironically, while the scientific understanding of stratospheric ozone depletion has grown, there has also emerged a popular critique that has minimized the importance of CFCs for ozone depletion, and has instead emphasized the role of natural sources of chlorine from volcanoes, biomass burning, and sea salt. Indeed, some popular books and talk shows have argued that the ozone depletion crisis is a "scam" by scientists and NASA to ensure research funding (Taubes 1993). Atmospheric scientists, on the other hand, are quite adamant that these attacks are based on a lack of understanding of how CFCs and ozone behave in the atmosphere. They argue, based on a number of careful scientific measurements, that most of the chlorine (more than 99%) from natural sources does not make it into the stratosphere (Tabazadeh and Turco 1993). Despite initial fears, the Mt. Pinatubo eruption has not had drastic effects on ozone depletion (Kerr 1992b). While there are some uncertainties and ambiguities surrounding stratospheric ozone depletion, it would appear that the present criticisms of CFCs as major causal agents of stratospheric ozone depletion are not scientifically substantiated (Taubes 1993).

Stopping the Flow of CFCs and Halons to the Atmosphere

A wide range of strategies are needed to eliminate the damaging chlorine and bromine compounds used in refrigeration, air conditioning, aerosols, foams, solvents, and fire extinguishers. These approaches generally can be classified as substitution of non- or less-damaging chemicals; improved maintenance, recycling, and disposal strategies; and substitute technology and products.

Chemical Substitutes

Chemical substitutes should have a number of desirable properties, including effectiveness, reasonable cost, nontoxicity, and environmental safety. Although substitutes must undergo extensive testing, increasing the time before they are available for use, many substitutes are available now.

CFCs are readily removed from aerosol products. Although several countries, including the United States, banned CFC propellants in most aerosols in the late 1970s, present international agreements will eliminate them globally in 1996. This action has a large impact since one-third of all the CFC-11 and CFC-12 used worldwide in the late 1980s came from aerosol use (Shea 1989b).

Various substitutes for CFC-12 in air conditioning and refrigeration are either in production (HCFC-134a) or under study (HFC-134a). HCFCs (hydrochlorofluorocarbons) have about 5% of the ozone depleting ability of the chemicals they replace (Shea 1989b; Stone 1992). They react with OH in the troposphere, which reduces their atmospheric lifetime and thus limits their effect on the stratosphere. Present international agreements limit the production rates of HCFCs and provide for their phaseout by 2030 (Ko et al. 1994). HFCs (Hydrofluorocarbons) do not appear to harm the ozone layer (Ravishankara et al. 1994). HFC-134a is a greenhouse gas, however, with about one-fourth the greenhouse gas contribution of the CFCs it replaces (Stone 1992; Ravishankara et al. 1994).

For CFC-113, used as a solvent, there appear to be a number of substitutions. AT&T has found the compound, Bio-Act EC-7, manufactured from terpenes in citrus fruit rinds, to be an effective substitute in its electronics manufacturing processes (Shea 1989b). Finding replacements for CFCs in the United States electronics industry has been relatively easy, and it is expected that this industrial use of CFCs will be mostly eliminated by early 1994 (Young 1994).

Maintenance, Recycling, and Disposal Strategies

Although substitutes are being developed and deployed, a number of actions can keep CFCs and halons already in use from entering the atmosphere. These include sealing leaks in refrigeration and air conditioning systems to prevent losses; recovering CFCs when refrigeration and air conditioning systems are serviced, repaired, and discarded; and better design of these systems to reduce losses and minimize system maintenance. The automobile air conditioning market is currently the largest user of CFCs in the United States. CFC recovery technology now exists to remove, purify, and reinject coolant back in the auto air conditioning system (Shea 1989b).

In the manufacture of flexible foams, newly designed ventilation systems can recover as much as 85–90% of the CFCs. In the electronics industry, it is possible to recover up to 95% of CFC-113 used to clean computer chips. Incineration technology may be used for destroying CFCs in rigid foams, which breaks down their ozone-depleting potential (Shea 1989a).

Current emissions of halons in fire fighting may be sharply curtailed by changing testing procedures, which often release the gas, and by better design of systems to avoid accidental discharge. Another major use, the release of halons in fire fighting training, can be eliminated by using simulators. This has recently been incorporated into U.S. military training (Shea 1989a).

Substituting Technology

In the longer run, new technology and product designs may eliminate the need for CFCs totally. For example, helium-cooled refrigerators, extensively used in space and military applications, are now produced for consumer use. They have an additional benefit: their high energy efficiency results in substantial monetary savings and a reduction in energy consumption. Vacuum insulation is a promising technology that could eliminate the need for rigid foam insulation in refrigerators and freezers (Shea 1989a).

CFCs as Greenhouse Gases

In addition to depleting the ozone layer, CFCs also act to warm the Earth through the greenhouse effect. It is well known that the outgoing infrared radiation emitted by the Earth and its atmosphere is absorbed and reemitted in all directions, keeping the Earth and lower atmosphere warmer. This is caused by certain atmospheric gases such as carbon dioxide, ozone, and water vapor, which have internal vibrational frequencies in the infrared region. The normal amounts of these gases in the atmosphere provide about an additional 60°F of temperature compared to an Earth without an atmosphere. The change in temperature caused by adding additional trace gases to the atmosphere is commonly termed the *enhanced greenhouse effect* (see Chapter 14, "The Challenge of Global Climate Change"). While carbon dioxide is the major greenhouse gas, a number of other gases, including methane, nitrous oxide, and CFCs, account for half of the projected warming.

Given the small quantities of CFCs (CFC-11: 0.28 ppb; CFC-12: 0.48 ppb) in relation to carbon dioxide (353 ppm) in the atmosphere (Kiehl and Brieglab 1993), it is somewhat surprising to find that they are significant greenhouse gases. Sufficient carbon dioxide, however, is already present in the atmosphere to capture much of the infrared radiation that lies within its absorption range. Consequently, additional carbon dioxide has a limited impact in adding to atmospheric heating. On the other hand, CFCs absorb strongly in optically transparent infrared regions of the atmosphere and are about 10,000 times more effective, per molecule, in absorbing infrared radiation compared to carbon dioxide. As current concentrations of CFCs are absorbing only a small percentage of the infrared radiation that lies within their absorption range, small increases in atmospheric CFCs lead to significant inputs into atmospheric heating. They therefore are of major concern as a greenhouse gas in addition to the ozone depletion that they cause. The role of CFCs in the greenhouse effect provides a further independent argument for their rapid phase-out (Rowland 1989).

There are other couplings that are less appreciated. Ozone depletion in the stratosphere allows more UV light to reach the troposphere, which enhances the formation of tropospheric ozone by enhancing reactions between other gases in the atmosphere. In this region, ozone acts as an infrared absorber and emitter, contributing to the greenhouse effect. It has also been hypothesized that reductions in stratospheric ozone lead to cooling of the lower stratosphere, which would in turn cause a negative radiative forcing of the surface-troposphere system. This may act to compensate for the positive forcing resulting from the greenhouse gas effect of the CFC gases (Ramaswamy et al. 1992; Wigley and Raper 1992; Schwarzkopf and Ramaswamy 1993; Isaksen 1994; Toumi et al. 1994). Lower stratospheric cooling could also enhance polar ozone holes (Zhong et al. 1993). Other greenhouse gases, such as carbon dioxide, methane, and nitrous oxide, also affect ozone depletion in the stratosphere. Changes in ozone will affect the Earth's climate, while changes in climate will affect ozone. The net result is that we have a complex strongly coupled system in which changes in one component have consequences for the behavior of the system. Humans are currently conducting a massive global experiment on the Earth's atmosphere without understanding its ramifications (Postel and Brown 1987; Watson 1988).

Conclusions

Ozone depletion is a significant threat to the global environment and warns humans that they cannot continue their present course. But it is only one of many signs that the fabric of life on the planet is at risk. As human numbers and activities continue to grow rapidly, the risk of exceeding the life-support systems of the planet becomes greater. Also, because we do not fully understand these natural systems, we may find long-term irreversible effects that have no easily detectable warnings before it is too late (Graedel and Crutzen 1990). We thus have a high risk of doing irreparable damage.

One cannot but be struck by the similarities between greenhouse warming and ozone depletion. Both are atmospheric phenomenon related to minor constituent gases that have been increased by human activity. Both have long lag times between the release of pollution to the beginning of a measurable result. Both have the potential of severely affecting life on the planet. Both require early action to minimize the damage; both depend heavily on theory and modeling to predict the future effects. Both have been controversial and fought heavily by industry and others because of economic dislocations that would occur. Both are clearly global in scope requiring global action for their solution.

At the same time, there are differences that are significant. There is currently more scientific controversy over both the rate of global warming and its ultimate magnitude (Fernau et al. 1993). There is less agreement about what ought to be done. The greenhouse effect has much greater implications for dislocations in the economies of nations, since we are so fossil fuel dependent. At least in the short run, a minor warming might provide benefits to some nations so there is a perception that there could be winners and losers in any agreement (Morrisette 1990). So far there has not been any early warning like the ozone hole to emphasize the potential seriousness of the situation.

To their credit, humans have responded globally to the ozone disaster. Precedents have been set on international agreements, and the United Nations has played a crucial role in the process. The ozone hole is a lesson to us about the fragility of natural systems and that they do not always change slowly or predictably. Let us hope we take the lesson to heart; it is much more prudent to avoid or prevent a crisis than to deal with one after it has been unleashed.

References

Adler, T., "Sunscreen Can't Give Blanket Protection," *Science News* 145, 1994, pp. 54–55.

Alderson, M. R. and D. Kent, "The Effect of Ultraviolet Radiation on Humoral Immune Response to T-Dependent Antigens," *Environmental Research* 40, 1986, pp. 321–331.

Blaustein, A. R., and D. B. Wake, "Declining Amphibian Populations: A Global Phenomenon?," *Trends in Ecology and Evolution* 5, 1990, pp. 203–204.

Blaustein, A. R., "Amphibians in a Bad Light," *Natural History* 103(10), 1994 pp. 32–37.

Bothwell, M. L., D. M. J. Sherbot, and C. M. Pollock, "Ecosystem Response to Solar Ultraviolet-B Radiation: Influence of Trophic-level Interactions," *Science* 265, 1994, pp. 97–100.

Browne, M. W., "93 Nations Move to Ban Chemicals That Harm Ozone," *New York Times,* June 30, 1990, p. 1.

Carlson, D. A., *Ultraviolet Disinfection of Water for Small Water Supplies,* EPA Report 600/S2-85/092, 1985.

Cicerone, R. J., "Fires, Atmospheric Chemistry and the Ozone Layer," *Science* 263, 1994, pp. 1243–1244.

Concar, D., "The Resistable Rise of Skin Cancer," *New Scientist* 134, 1992, pp. 23–28.

Crutzen, P. and F. Arnold, "Nitric Acid Cloud Formation in the Cold Antarctic Stratosphere: A Major Cause for the Springtime Ozone Hole," *Nature* 324, 1986, pp. 651–55.

Ellis, J. and D. N. Schramm, "Could a Nearby Supernova Explosion Have Caused a Mass Extinction?" Proceedings of the National Acadamy of Sciences 92(1), 1995, pp. 235–238.

EPA, *Regulatory Impact Analysis: Protection of Stratospheric Ozone,* Vol. I, Washington, DC, 1987.

Farman, J. C., B. G. Gardiner, and J. D. Shanklin, "Large Losses of Total Ozone in Antarctica Reveal Seasonal ClOx/NOx Interaction," *Nature* 315, 1985, pp. 207–210.

Fernau, M. E., W. J. Makofske, and D. W. South, "Review and Impacts of Climate Change Uncertainties," *Futures,* Oct. 1993, pp. 850–863.

Giese, A. C., *Living With Our Sun's Ultraviolet Rays,* New York: Plenum Press, 1976.

Gleason, J. F., P. K. Bhartia, J. R. Herman, R. McPeters, P. Newman, R. S. Stolarski, L. Flynn, G. Gabon, D. Larko, C. Seftor, C. Wellemeyer, W. D. Kohmyr, A. T. Miller, and W. Planet, "Record Low Global Ozone in 1992," *Science* 260, 1993, pp. 523–526.

Graedel, T. E. and P. J. Crutzen, "The Changing Atmosphere," *Scientific American* 261(3), 1990, pp. 58–68.

Gribbin, J., *The Hole in the Sky,* New York: Bantam Books, 1988.

Harmon, D. L. and W. J. Rhodes, "Overview of Controls for Chlorofluorocarbons," *In* T. Schneider et al. eds., *Atmospheric Ozone Research and its Policy Implications,* Amsterdam, Netherlands: Elsevier Science, 1989, pp. 765–774.

Harris, N., F. S. Rowland, R. Bojkov, and P. Bloomfield, "Winter-Time Losses of Ozone in High Northern Latitudes," Aspen, CO: NASA Polar Ozone Workshop, 1988.

Hastings, A. H., "The Ecological Dimension of Nuclear War," *Environmental Conservation* 14, 1987.

IER (International Environmental Reporter), Current Reports, 15, Dec. 16, 1992.

IMOS (Federal Task Force on Inadvertent Modification of the Stratosphere), *Fluorocarbons and the Environment,* Council on Environmental Quality, Federal Council for Science and Technology, 1975.

Isaksen, I. S. A., "Dual Effects of Ozone Reduction," *Nature* 372, 1994, pp. 322–323.

Kerr, J. B. and C. T. McElroy, "Evidence for Large Upward Trends of Ultraviolet-B Radiation Linked to Ozone Depletion," *Science* 262, 1993, pp. 1032–1034.

Kerr, R. A., "Taking Shots at Ozone Hole Theories," *Science* 234, 1986, pp. 817–818.

Kerr, R. A., "Ozone Destruction Worsens," *Science* 252, 1991, p. 204.

Kerr, R. A., "New Assaults Seen on Earth's Ozone Shield," *Science* 255, 1992a, pp. 797–798.

Kerr, R. A., "Ozone Hole: Not Over the Arctic—For Now," *Science* 256, 1992b, p. 734.

Kerr, R. A., "Ozone Takes a Nose Dive After the Eruption of Mt. Pinatubo," *Science* 260, 1993a, pp. 490–491.

Kerr, R. A., "The Ozone Hole Reaches a New Low," *Science* 262, 1993b, p. 501.

Kiehl, J. T. and B. P. Brieglab, "The Relative Roles of Sulfate Aerosols and Greenhouse Gases in Climate Forcing," *Science* 260, 1993, pp. 311–314.

Ko, M. K. W., N.-D. Sze and M. J. Prather, "Better Protection of the Ozone Layer, *Nature* 367, 1994, pp. 505–508.

Kripke, M. L., "Health Effects of Ozone Depletion," *American College Toxicological Journal* 8, 1989, pp. 1083–1089.

Kripke, M. L. and A. Jeevan, "Ozone Depletion and the Immune System," *Lancet* 342, 1993, pp. 1159–1160.

Lloyd, S. A., "Stratospheric Ozone Depletion," *Lancet* 342, 1993, pp. 1156–1158.

Madronich, S. and F. R. de Gruijl, "Skin Cancer and UV Radiation," *Nature* 366, 1993, p. 23.

Mano, S. and M. D. Andreae, "Emission of Methyl Bromide from Biomass Burning," *Science* 263, 1994, pp. 1255–1257.

Meeuse, B. J. D., *The Story of Pollination,* New York: Ronald Press, 1961.

Miller, A. S. and I. M. Mintzer, *The Sky Is the Limit: Strategies for Protecting the Ozone Layer,* Washington, DC: World Resources Institute, November 1986.

Molina, M. J., and F. S. Rowland, "Stratospheric Sink for Chlorofluoromethanes: Chlorine Atom-Catalyzed Destruction of Ozone," *Nature* 249, 1974, pp. 810–812.

Molina, M. J., T. L. Tso, L. T. Molina, and F. C. Y. Wang, "Antarctic Stratospheric Chemistry of Chlorine Nitrate, Hydrogen Chloride and Ice: Release of Active Chlorine," *Science* 238, 1987, pp. 1253–57.

Morrisette, P., "Negotiating Agreements on Global Change," *Resources* RFF, 99, Spring 1990, pp. 8–11.

National Academy of Sciences (NAS), *Longterm Worldwide Effects of Multiple Nuclear-Weapons Detonations,* P. Handler (ed.), Washington, DC, 1975.

NAS (Committee on Impacts of Stratospheric Change), *Halocarbons: Environmental Effects of Chlorofluoromethane Release,* 1976a.

NAS (Panel on Atmospheric Chemistry), *Halocarbons: Effects on Stratospheric Ozone,* 1976b.

NAS (Committee on Impacts of Stratospheric Change), *Protection against Depletion of Stratospheric Ozone by Chlorofluorocarbons,* 1979a.

NAS (Panel on Stratospheric Chemistry and Transport), *Stratospheric Ozone Depletion by Halocarbons: Chemistry and Transport,* 1979b.

NAS (Committee on Chemistry and Physics of Ozone Depletion), *Causes and Effects of Stratospheric Ozone Reduction: An Update,* 1982.

NAS (Committee on Causes and Effects of Changes in Stratospheric Ozone), *Causes and Effects of Changes in Stratospheric Ozone: Update 1983,* 1984.

NASA, *Present State of Knowledge of the Upper Atmosphere 1988: An Assessment Report,* R. T. Watson and Ozone Trends Panel, M. J. Prather and Ad Hoc Theory Panel, and M. M. Kurylo and NASA Panel for Data Evaluation, NASA Publication 1208, 1988.

Norman, C., "Satellite Data Indicate Ozone Depletion," *Science* 213, 1981, pp. 1088–1089.

"Planes to Search for Ozone Hole in the Arctic," *New York Times,* November 10, 1991, p. 34.

Pennisi, E., "Visible, UV-A Light Tied to Skin Cancer," *Science News* 144, 1993, p. 53.

Postel, S. and L. Brown, "State of the Earth 1987: Life, the Great Chemistry Experiment," *Natural History,* April 1987.

Ramaswamy, V., M. D. Schwarzkopf, and K. P. Shine, "Radiative Forcing of Climate from Halocarbon-Induced Global Stratospheric Ozone Loss," *Nature* 355, 1992, pp. 810–812.

Ravishankara, A. R., A. A. Turnipseed, N. R. Jensen, S. Barone, M. Mills, C. J. Howard, and S. Solomon, "Do Hydrofluorocarbons Destroy the Stratospheric Ozone?," *Science* 263, 1994, pp. 71–75.

Roan, S., *Ozone Crisis: The 15-Year Evolution of a Sudden Global Emergency,* New York: John Wiley and Sons, 1989.

Roberts, L., "Does the Ozone Hole Threaten Antarctic Life?," *Science* 244, 1989, pp. 288–289.

Rodriguez, J. M., "Probing Stratospheric Ozone," *Science* 261, 1993, pp. 1128–1129.

Rowland, F. S., "Chlorofluorocarbons and the Depletion of Stratospheric Ozone," *American Scientist* 77, 1989, pp. 36–45.

Schoeberl, M. and A. Krueger, "Overview of the Antarctic Ozone Depletion Issue," *Geophysical Research Letters,* November 1986.

Schwarzkopf, M. D. and V. Ramaswamy, "Radiative Forcing Due to Ozone in the 1980s: Dependence on Altitude of Ozone Change," *Geophysical Research Letters* 20, 1993, pp. 205–208.

Shea, C. P., *Protecting Life on Earth: Steps to Save the Ozone Layer,* Worldwatch Paper 87, Washington, DC: Worldwatch Institute, December 1988.

Shea, C. P., "Protecting the Ozone Layer," *State of the World 1989.* Washington, DC: Worldwatch Institute, 1989a.

Shea, C. P., "Mending the Earth's Ozone Shield," *Worldwatch* 21, Jan/Feb 1989b.

Smith, R. C., B. B. Prézelin, K. S. Baker, R. R. Bridigare, N. P. Boucher, T. Coley, D. Karentz, S. MacIntyre, H. A. Matlick, D. Menzies, M. Ondrusek, Z. Wan, and K. J. Waters, "Ozone Depletion: Ultraviolet Radiation and Phytoplankton Biology in Antarctic Waters," *Science* 255, 1992, pp. 952–959.

Solomon, S., R. R. Garcia, F. S. Rowland, and D. J. Wuebbles, "On the Depletion of Antarctic Ozone," *Nature* 321, 1986, pp. 755–758.

Solomon, S., R. W. Sanders, R. R. Garcia, and J. G. Keys, "Increased Chlorine Dioxide over Antarctica Caused by Volcanic Aerosols from Mount Pinatubo," *Nature* 363, 1993, pp. 245–248.

Stevens, W. K., "Ozone Losses in Arctic Larger Than Expected," *New York Times,* September 6, 1990, Section A, p. 25.

Stevens, W. K., "Summertime Harm to Ozone Detected Over Broader Area," *New York Times,* October 22, 1991, Section A, p. 1.

Stolarski, R., R. Bojkov, L. Bishop, C. Zerefos, J. Stachelin, and J. Zawodny, "Measured Trends in Stratospheric Ozone," *Science* 256, 1992, pp. 342–349.

Stone, R., "Warm Reception for Substitute Coolant," *Science* 260, 1992, p. 22.

Stone, R., "Ozone Prediction Hits it Right on the Nose," *Science* 261, 1993, p. 290.

Tabazadeh, A. and R. P. Turco, "Stratospheric Chlorine Injection by Volcanic Eruptions: HCl Scavenging and Implications for Ozone," *Science* 260, 1993, pp. 1082–1085.

Taubes, G., "The Ozone Backlash," *Science* 260, 1993, pp. 1580–1583.

Taylor, J. S., DNA, Sunlight and Skin Cancer," *Journal of Chemical Education* 67, 1990, pp. 835–841.

Toumi, R., S. Bekki, and K. S. Law, "Indirect Influence of Ozone Depletion on Climate Forcing by Clouds," *Nature* 372, 1994, pp. 348–351.

van der Leun, J. C., "Effects of Increased UV-B on Human Health." *In EPA U.S.-Dutch Third International Symposium (Atmospheric Ozone: Its Policy Implications)*, Nijmegen, Netherlands, 1988, pp. 803–812.

Virhanski, L., "Ozone Park," *Discover* 10, 1989, p. 32.

Watson, R. T., "Atmospheric Ozone," Paper presented at a seminar at Ramapo College, March 1988.

Wigley, T. M. L. and S. C. B. Raper, "Implications for Climate and Sea Level of Revised IPCC Emission Scenarios," *Nature* 357, 1992, pp. 293–300.

WMO-NASA, *Atmospheric Ozone 1985,* 3 vols, WMO Global Ozone Research and Monitoring Project, Report No. 16, 1986.

Worrest, R. C., "What are the Effects of UV-B Radiation on Marine Organisms?" *In* T. Schneider, et. al. (eds.), *Atmospheric Ozone Research and Its Policy Implications,* Amsterdam, Netherlands: Elsevier Science, 1989, pp. 269–277.

Worrest, R. C. and D. P. Hader, "Effects of Ozone Depletion on Marine Organisms," *Environmental Conservation* 16, 1989, pp. 261–263.

Young, J. E., "Using Computers for the Environment," In L. R. Brown (ed.), *State of the World 1994,* New York: W.W. Norton, 1994, pp. 99–116.

Zhong, W., Haigh, J. D. and J. A. Pyle, "Greenhouse Gases in the Stratosphere," *Journal of Geophysical Research* 98, 1993, pp. 2995–3004.

Chapter 14

The Challenge of Global Climate Change

William J. Makofske

In recent years, there has been great concern that increasing concentrations of greenhouse gases (GHGs) in the atmosphere will lead to global climate change, with global temperature increases, changes in precipitation, and changes in sea level being some of the major consequences. Recognition of vulnerability to climate change has heightened in recent years because of the 1988 midwestern drought in the United States and the decade-long drought in the Sahel region in Africa. The phenomenon of climate change has been extensively studied by scientists and policymakers, including the Environmental Protection Agency (EPA 1989), the U.S. Congress (OTA 1991), the National Academy of Sciences (NAS 1992), and the United Nations (IPCC 1990, 1992). The results of these studies pose dilemmas for scientists and policymakers, and indeed, for societies around the world. The reason is that while such studies agree that potentially severe climate impacts are possible or even likely over the next century, there are large uncertainties in the magnitude, rate of change, and regional distribution of climate change. Since increasing GHGs result from current human activities, reducing such gases enough to stabilize climate will likely require considerable restructuring of societies and the way people live. But it is not only the change in climate and resulting impacts that are uncertain. The economic consequences of reducing GHGs and stabilizing the climate are thought to be relatively high; yet there are studies that indicate that considerable reductions in present GHG emissions may be made at relatively low cost, and that some reductions would even pay for themselves. While climate change potentially poses great risks to human civilization and to ecosystems, it is future generations (but perhaps not very far in the future) who will suffer most of the burden. Given the potential high risks and impacts of severe climate change, the uncertain but potentially large costs, and present climate uncertainties, what should the appropriate response be now? This article provides a broad perspective on climate change, summarizing the scientific problem (greenhouse [GH] effect; sources and sinks of GHGs; climate modeling, global warming verification, climate response, and associated uncertainties), the potential impacts of global warming, possible responses to climate change, current policy initiatives and policy controversies.

Chapter 15, "Predicting Climate Change: The State of the Art," examines the key role of climate models in assessing climate change.

Greenhouse Warming

The Greenhouse Effect

The Earth's average temperature and climate is maintained by an energy balance between incoming solar radiation in the short-wave visible, ultraviolet, and infrared spectrum and outgoing radiation in the long-wave infrared spectrum. As incoming solar radiation travels through the atmosphere, it is absorbed, reflected, and scattered. Solar radiation reaching the Earth's surface may be absorbed, thereby heating the surface and reradiating as infrared radiation, or it may be reflected. The fraction of incoming radiation to the Earth/atmosphere system that is reflected (about 30%) is called the *planetary albedo*. This reflected radiation escapes into space and plays no role in heating the Earth. Surprisingly, much of the warming of the Earth's surface and lower atmosphere occurs, not directly from incoming solar radiation, but indirectly by the absorption and reradiation of long-wave infrared radiation in the atmosphere (Figure 14.1). The atmospheric gases that are the primary absorbers of this outgoing infrared radiation are water vapor (H_2O), carbon dioxide (CO_2), methane (CH_4), nitrous oxide (N_2O), ozone (O_3), and chlorofluorocarbons (CFCs). These gases are often referred to as GHGs, in an (scientifically incorrect) analogy with greenhouses, which let visible light in but retain heat because of the glass window. The behavior of these gases in heating the Earth is commonly referred to as the GH effect. GHGs keep the Earth's average surface temperature about 60°F (33°C) higher than it would be otherwise. Increasing these gases in the atmosphere could lead to what is termed an *enhanced GH effect,* that is, an increase in the average temperature of the Earth, often referred to as global warming. Human activities are increasing the levels of GHGs in the atmosphere at a substantial rate. Determining the ultimate impact of increasing GHGs on climate, however, is not a simple matter. The reason is that the existing climate system (consisting of atmosphere, hydrosphere, lithosphere, cryosphere, and biosphere) reacts to the radiative forcing[1] through a number of complicated changes involving negative and positive feedback loops (see sections on climate uncertainties in this and the following chapter). While the exact response of the climate system is difficult to predict, the net result of increasing GHGs, according to current climate models, appears to be a relatively rapid increase in global temperature and associated climate changes.

Greenhouse Gases: Sources and Sinks

The major GHGs, their pre-industrial and current concentrations, rates of change, atmospheric lifetimes, global warming potentials, and relative contribu-

---

[1]*Radiative forcing* refers to the increase in trapping of outgoing terrestrial infrared radiation resulting from increasing absorption by GHGs.

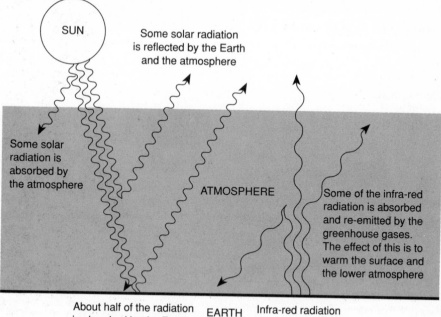

Figure 14.1. A simplified diagram illustrating the greenhouse effect. See text for a more detailed description. (From Intergovernmental Panel on Climate Change, 1990. *Climate Change: The IPCC Scientific Assessment,* Houghton, J. T., G. J. Jenkins, and J. J. Ephraums, (eds.), New York: Cambridge University Press.)

tions to GH warming are summarized in Table 14.1. Global warming potentials take into account the different radiative absorption properties of the gases, together with their lifetime in the atmosphere. While much emphasis is placed on CO_2, the other gases combined account for almost half of the radiative warming impact (Figure 14.2). While CO_2 levels alone may have increased about 25% from pre-industrial levels, the sum total of all GHG additions may be equivalent (in terms of radiative forcing) to doubling CO_2 by the year 2030. Increases in concentrations result from the difference between source and sink processes. Because of the long lifetimes of most of these gases (Table 14.1), even present concentrations of GHGs commit the Earth to warming for centuries. Substantial and immediate reductions in existing GHG emission levels, on the order of 60%, would be needed to stabilize the radiative forcing effect of these gases and thereby stabilize global climate (IPCC 1990). Increased water vapor, resulting from an increase in evaporation associated with global temperature rise, and increased tropospheric ozone, resulting from enhanced photochemical air pollution, depend on the temperature rise and therefore are not shown in Table 14.1.

The primary sources of the increase in atmospheric CO_2 are the burning of fossil fuels (coal, oil, and natural gas), cement manufacturing, and the loss of tropical

Table 14.1

Summary of Key Greenhouse Gases Affected by Human Activities

	Carbon Dioxide	*Methane*	*CFC-11*	*CFC-12*	*Nitrous Oxide*
Units of atmospheric Concentration[a]	ppmv	ppmv	pptv	pptv	ppbv
Pre-industrial (1750–1800)	280	0.8	0	0	288
Present day (1990)	353	1.72	280	484	310
Current rate of change per year in conc. and (percent)	1.8 (0.5%)	0.015 (0.9%)	9.5 (4%)	17 (4%)	0.8 (0.25%)
Atmospheric lifetime in years	(50–200)[b]	10	65	130	150
Direct global warming potential[c]	1	11	3400	7100	270

[a]ppmv = parts per million by volume; ppbv = parts per billion by volume; pptv = parts per trillion by volume.
[b]The way in which CO_2 is absorbed by the oceans and biosphere is not simple and a single value cannot be given.
[c]Based on a hundred-year time horizon.

Source: IPCC, Intergovernmental Panel on Climate Change, *Climate Change: The IPCC Scientific Assessment*, J. T. Houghton, G. J. Jenkins, and J. J. Ephraums (eds.), London: Cambridge University Press, 1990. IPCC, Intergovernmental Panel on Climate Change, *Climate Change 1992. The Supplementary Report to the IPCC Scientific Assessment*, J. T. Houghton, B. A. Callander, and S. K. Varney (eds.), London: Cambridge University Press, 1992.

forests (burning, decay, and reduced photosynthetic removal). CO_2 is removed from the atmosphere by a number of biologically mediated processes, primarily carbonate precipitation in the ocean and photosynthesis (IPCC 1990).

Methane (CH_4) sources include anaerobic decay processes in swamps, rice paddies, and landfills; the burning of forests and grasslands; coal mining; natural gas venting into the environment; and digestive processes in ruminants (livestock), and termites. Methane is removed by a number of atmospheric chemical reactions, primarily reaction with hydroxyl radicals (OH) in the troposphere. Oxidation of methane by OH results in the production of CO_2 and also produces stratospheric water vapor, which contributes to the GH effect. The rate of growth of methane appears to have declined in the last decade, although the reason is still unclear. Human activities account for almost two-thirds of methane in the atmosphere (IPCC 1990, 1992).

Nitrous oxide (N_2O) is produced from the breakdown of nitrates in the soil, livestock wastes, sewage disposal, nylon manufacture, biomass burning, and automobiles with three-way catalysts. The only known sink is stratospheric photodissociation. Human activities are believed to account for one-third of total emissions (IPCC 1992; Khalil and Rasmussen 1992).

CFCs are used in air conditioners and refrigerators, aerosol propellants, industrial solvents, and in the production of plastic foams. CFCs are currently being phased out of production because of their role in stratospheric ozone depletion (see Chapter 13, "Stratospheric Ozone Depletion"). Nonetheless, because of

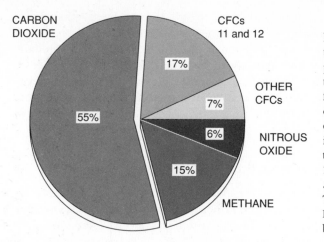

Figure 14.2. Pie chart illustrating the contribution of anthropogenic greenhouse gases to the change in radiative forcing from 1980 to 1990. Ozone's contribution is not easily quantified and therefore is not included. (From Intergovernmental Panel on Climate Change, 1990, *Climate Change: The IPCC Scientific Assessment,* Houghton J. T., G. J. Jenkins, and J. J. Ephraums, (eds.), New York: Cambridge University Press, 1990.)

their long lifetime, they will be significant contributors to GH warming for the next century. As for N_2O, the only known sink for CFCs is stratospheric photodissociation (IPCC 1990).

Climate Change and Uncertainties

Projections of Climate Change Using Modeling

Climate models (called *general circulation models* or GCMs) have been used by scientists to project the effect of increasing GHG concentrations on the Earth's climate (see Chapter 15, "Predicting Climate Change: The State of the Art," for a discussion of climate models, their assumptions, and limitations). These models use computers incorporating the laws of physics for heat, momentum, and moisture, together with simplified descriptions of other climate processes (for example, cloud formation and the ice-albedo feedback) to predict climate variables such as temperature and precipitation. Essentially, they calculate the result of radiative forcing by GHGs on the climate system by taking into account the many feedback loops in the climate system. Both simple atmospheric/ocean and fully coupled atmospheric/ocean models have been investigated. Most models project an equilibrium[2] global warming based on the doubling of CO_2 levels (or its equivalent from all GHGs), which ranges about 1.5 to 4.5°C (about 3 to 8°F). Because of ocean caused delays in warming, the time for realizing the full temperature change may take from decades to a century. Such models generally project that warming will be greater at high latitudes than in the tropics, that the Northern Hemisphere will warm to higher temperatures and more quickly than the Southern Hemisphere, and that temperature change at middle and high latitudes, particularly for the Northern Hemisphere, will be 2 to 3 times the

---

[2]*Equilibrium* refers to the total change predicted by the model with doubled CO_2 compared to a base case run in which CO_2 is not increased. Time lags in climate response may delay the changes.

global average. Both southern Europe and central North America have reduced precipitation and soil moisture in summer, while the Asian monsoon is strengthened (IPCC 1990). There will be enhanced probability of extreme heat waves, drought, and enhanced storm frequency and intensity (Schneider 1990). It is also expected that sea level will rise, both due to thermal expansion of the warmed ocean and from melting of glacial ice and snow. While wide variations in estimates exist, the current best estimate gives an average increase of 18 cm by 2030 and 48 cm by the end of the next century (IPCC 1992).

Uncertainties in Climate Modeling

Most GCMs predict a wide range in calculated global temperature increase for doubled CO_2, a range that varies from significant (1.5°C) to potentially catastrophic (4.5°C) (IPCC 1990, 1992). This wide range in predicted global temperature increase is due to many factors, including uncertainties in GHG emissions, concentrations, and sinks; uncertainties in feedback loops (particularly involving water vapor, clouds, oceans, sea ice and snow, and biosphere vegetation); and limitations in computer technology that affect grid resolution, thereby restricting the use of realistic topography and geography. Regional predictions, in particular, have the greatest uncertainties (Fernau et al. 1993). Some of the sources of uncertainty involving climate system components and feedbacks are depicted in Figure 14.3 (see Chapter 15, "Predicting Climate Change: The State of the Art," for a discussion of major feedback processes). There are some scientists who believe that when improved water vapor feedbacks are incorporated, model temperature predictions will be reduced (Lindzen 1990). The observational studies, however, have so far supported positive feedback: increased water vapor leads to increased temperatures (Raval and Ramanathan 1989; Rind et al. 1991). While there are a number of research programs to gather better information about feedback loops and natural processes and to improve grid resolution with faster and more powerful computers, it is expected that significant improvements in GCM predictions will come slowly, perhaps over several decades (Fernau et al. 1993).

Uncertainties in the Global Warming Signal

While projections of future climate change using climate models show potentially large increases in temperature (called the *global warming signal*) in the next 50–100 years, there are considerable uncertainties in these projections, both in timing and in magnitude. Our confidence in model predictions would be enhanced if the past global temperature record agreed with model temperature predictions. We do have reconstructed global temperature records of the past 100 years that perhaps can offer some information about the relationship between increasing GHGs and global temperature. Indeed, measurements (Figure 14.4) suggest that the global temperature has warmed about 0.5°C (about 1°F) over the past century, at the lower end of what might be expected based on climate modeling. Eight of the 12 years preceding the Mt. Pinatubo eruption have been the warmest in the global record. The trend, however, has not been the smooth one predicted by current climate models, and there

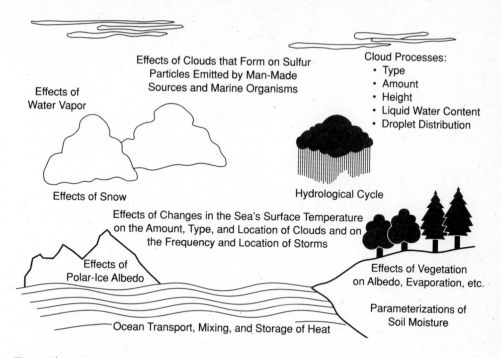

Figure 14.3. Schematic of various imperfectly known variables that complicate global climate change understanding and predictions. Simplification, parameterization, and lack of knowledge of the processes, effects, and feedbacks depicted here lead to uncertainty in the results of general circulation models. (From Fernau, M. E., W. J. Makofske, and D. W. South, "Review and Impacts of Climate Change Uncertainties," *Futures,* October 1993.)

are difficulties in directly attributing this observed warming to increases in GHGs (IPCC 1990).

The basic problem is that there are many factors that could be responsible for the observed temperature trends. There is natural variability in climate due to interactions between components of the climate system (El Niño is the result of such an interaction), cooling from volcanic emissions (due to ash and sulfates), and human-caused pollution (sulfates from fossil fuel combustion and biomass burning), as well as possible changes in incoming solar radiation. Unfortunately, it may be impossible to explain the temperature record of the last century, since reliable data for many of these factors does not exist (Fernau et al. 1993). Indeed, accounting for sulfate aerosol forcing from pollution (Charlson et al. 1992) and ozone-depletion feedback[3] (Ramaswamy et al. 1992) does help explain why the global temperature record might be lower than model predictions (Wigley and Raper 1992). Coupled ocean–atmosphere models suggest that the recent observed warming (0.5°C per

---

[3]Stratospheric ozone loss results in the lower stratosphere cooling, which causes negative radiative forcing of the surface-troposphere system, offsetting the positive radiative forcing of the CFCs.

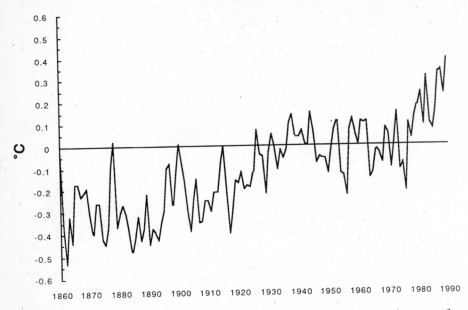

Figure 14.4. Annual average global surface temperature expressed as a deviation from 1950 to 1979 mean in degrees C. Data analyzed by Jones, P. D., T. M. L. Wigley, and P. B. Wright and reported in Boden, T. A., P. Kancirnk, and M. P. Farrell, *Trends '90, A Compendium of Data on Global Change,* Carbon Dioxide Information Analysis Center, Environmental Sciences Division, Oak Ridge National Laboratory Report ORNL/CDIAC-36, 1990. (From Fernau, M. E., W. J. Makofske, and D. W. South, "Review and Impacts of Climate Change Uncertainties," *Futures,* October 1993.)

century) is not a natural feature of the ocean–atmosphere interaction, but likely due to human influences (Stouffer et al. 1994). Nonetheless, the uncertainties in the relative contribution of the various factors to the global temperature record make it impossible to attribute the observed global warming trend to the enhanced GH effect (IPCC 1992).

It is known that volcanic eruptions can have significant short-term consequences. For example, the Mt. Pinatubo eruption in 1991 appears to have had a significant, although temporary, cooling effect on global temperature. With increased monitoring of the various contributors to global temperature changes, scientists hope to be able to explain future temperature trends (Ardanuy et al. 1992).

Potential Climate Surprises

While present climate models project smooth incremental increases in global temperature as a result of increasing GHGs, not all changes in climate will necessarily occur so smoothly. It is possible that if temperature changes too much or too fast, the response of the climate system will be rapid and discontinuous. This could occur because of positive feedback loops that act when the system has changed beyond a certain point. In essence, there could be a fundamental restructuring of the planetary climate system and its interactions. An analogy is the threshold effect in ecology, where even small increases in pollutants beyond a certain point (the threshold) can have large and often devastating effects on ecosystems. As the climate system continues to be investigated, there is increasing evidence that these so-called climate surprises are possible.

There is evidence from paleoclimatology for rapid changes in climate over relatively short time intervals for which there is no readily accepted explanation. While the Milankovitch theory (see Chapter 15, "Predicting Climate Change: The State of the Art"), explains much previous climate change as due to changes in the distribution of incoming solar radiation from periodic and slowly varying Earth orbital variations, it appears that additional mechanisms may be needed (Winograd et al. 1992). Warming could lead to a relatively quick collapse of the West Antarctic ice sheet, which would cause a rapid and large change in sea level (Barrett et al. 1992; MacAyeal 1992). There is also evidence that the North Atlantic thermohaline ocean circulation changed suddenly during the last deglaciation (Lehman and Keigwin 1992). If such a change were to be triggered by global warming, Europe's climate would be drastically affected. It is evident that our knowlege of the mechanisms for climate change is far from complete.

The possibility of a serious impact on North Atlantic ocean circulation also has been found in climate model studies. Using a coupled ocean-atmosphere model, CO_2 was doubled in 70 years in one run and quadrupled in 140 years in another. In both cases, CO_2 was then held constant and the model run until 500 years elapsed. Global temperatures increased 3.5°C in the first case and 7°C in the latter. With doubling, the thermohaline circulation weakens considerably but starts to reintensify at 150 years. With CO_2 quadrupled, the circulation ceases entirely and never reintensifies over the period of study. The latter case represents an adjustment to a new stable state that prevents deep ocean ventilation with serious consequences for the Earth. Unless actions are taken, quadrupled CO_2 or its equivalent would occur by around 2100 (Manabe and Stouffer 1993).

Potential climatic surprises are many and include: rapid release of methane from permafrost and gas hydrates, collapse of the West Antarctic ice sheet with rapid ocean rise, rapid shifts in the North Atlantic ocean circulation, rapid increase in tropospheric ozone buildup from increased temperatures, collapse of local hydrological cycles in tropical rain forests, and an enhanced potential for an Arctic ozone hole (Clark 1991; Austin et al. 1992; Brooks 1992). There may be some that we have not yet recognized. In evaluating the consequences of climate change, we must be aware that current understanding does not rule out the possibility of discontinuous and relatively sudden shifts in climate with potentially very serious impacts.

Impacts of Climate Change

The climate change projected by computer models over the next century, even at the lower range of temperature increase, is of great concern. This is not only because of the magnitude of change (1.5–4.5°C), which exceeds the change that has occurred over the past 10,000 years (about 1°C), but also because of the rapid rate of change (about 0.3°C per decade), which would put great stress on the ability of humans and ecosystems to respond.

A National Academy of Sciences panel (NAS 1992) has studied impacts of climate change on human and natural systems on the United States and similar industrialized countries (see Table 14.2 for a listing of areas affected). The dislocations due to climate change are many and include such possibilities as increases in drought conditions in the midsection of the United States and other areas of the world; changes in insect populations and occurrence of plant diseases; redistribution of water availability due to precipitation shifts and salinity intrusions into groundwater; loss of forests and coastal areas for recreation; damage to human structures due to coastal flooding; costs associated with human migration resulting from loss of work, flooding, and other damage; increased air pollution (tropospheric ozone) and possible enhanced ozone holes (Austin et al. 1992); and political and social stress from migration and other losses (NAS 1992).

The NAS report is optimistic in that it assumes gradual changes in climate and the availability of resources to adapt to such changes. Even so, it finds the impacts on natural ecosystems (forests, grasslands, etc.) and marine ecosystems from climate change to be problematic. It is likely that natural ecosystems will suffer severe consequences and that there is little that humans will be able to do to amelio-

Table 14.2
The Sensitivity[a] and Adaptability of Human Activities and Nature

	Low Sensitivity	Sensitive, but Adaptation at Some Cost	Sensitive, Adaptation Problematic
Industry and energy	X		
Health	X		
Farming		X	
Managed forests and grasslands		X	
Water resources		X	
Tourism and recreation		X	
Settlements and coastal structures		X	
Human migration		X	
Political tranquility		X	
Natural landscapes			X
Marine ecosystems			X

[a] *Sensitivity* is defined as the degree of change in the subject for each "unit" change in climate. The impact (sensitivity times climate change) will thus be positive or negative depending on the direction of climate change. Many things can change sensitivity, including intentional adaptations and natural and social surprises, and so classifications may shift over time. For gradual changes assumed in the study, the Adaptation Panel believes these classifications are justified for the United States and similar nations.

Source: NAS, National Academy of Sciences, National Academy of Engineering, Institute of Medicine, *Policy Implications of Greenhouse Warming: Mitigation, Adaptation and the Science Base,* Washington, DC: National Academy Press, 1992.

rate the damage (see Chapter 8, "The Loss of Biodiversity"). Also, the effects of climate change on developing countries that do not have the resources to respond are likely to be considerably greater than the impacts on developed countries. Even though temperature changes are expected to be smaller in tropical regions where many less developed countries (LDCs) are located, sea level change; changes in precipitation patterns, storm frequency, and intensity; and other climate effects may still have great impact. Finally, it is extremely difficult to determine actual impacts because we do not know the magnitude and, especially, the rate of climate change; regional predictions for climate change are poor; and rapid climate change is beyond the experience of present human societies and scientific studies.

Policy Responses to Greenhouse Warming

The policy responses to GH warming may be classified into two categories: mitigation, that is, reducing or offsetting the emissions of GHGs to avoid climate change; and adaptation,[4] that is, adjusting and adapting to new climatic conditions. Since many scientists believe that increased levels of GHGs have already committed the Earth to future warming (IPCC 1990), it is likely that both approaches will be needed. If climate changes are rapid, however, adaptation will be difficult and expensive. The National Academy of Sciences study (1992) has examined both mitigation and adaptation responses.

Mitigation

A mitigation approach to solving global warming involves reducing or offsetting the emission levels of all the major GHGs. A major criterion for mitigation actions is cost effectiveness, achieving the largest GHG reduction at the least cost. The feasibility of implementation and the impact on other issues are also important. Among the major mitigation strategies are: phase out CFCs, reduce global deforestation, increase reforestation, and change energy policy to reduce CO_2 emissions (Table 14.3) (NAS 1992). CFCs are already in the process of being phased out worldwide (see Chapter 13, "Stratospheric Ozone Depletion"). Each country needs to assess its own GHG emissions and the most appropriate strategy for reducing them; some of the most cost-effective options for reducing GHG emissions may lie in developing countries (NAS 1992). Some of the most attractive options for mitigation in the United States involve energy efficiency improvements in the buildings, transportation, and industrial sectors where estimates indicate that 10 to 40% of current U.S. GHG emissions (which includes CFCs) may be reduced at a net savings.[5] In addition to conservation and efficiency, energy technologies that release less (substituting natural gas for coal), no net (biomass) or no (most renewable sources, including solar, wind, and

---

[4]*Adaptation* used in this context has no connection to the evolutionary concept of adaptation. Rapid climate change would not allow evolutionary adaptation.
[5]Net savings implies that the total discounted cost over the analysis period is less than the discounted net benefit.

Table 14.3
Comparison of Selected Mitigation Options in the United States

Mitigation Option	Net Implementation Cost[a]	Potential Emission Reduction[b]
Building energy efficiency	Net benefit	900 million[c]
Vehicle efficiency (no fleet change)	Net benefit	300 million
Industrial energy management	Net benefit to low cost	500 million
Transportation system management	Net benefit to low cost	50 million
Power plant heat rate improvements	Net benefit to low cost	50 million
Landfill gas collection	Low cost	200 million
Halocarbon-CFC usage reduction	Low cost	1,400 million
Agriculture	Low cost	200 million
Reforestation	Low to moderate cost[d]	200 million
Electricity supply	Low to moderate cost[d]	1,000 million[e]

[a]Net benefit = cost less than or equal to zero; low cost = cost between $1 and $9 per ton of CO_2 equivalent; moderate cost = cost between $10 and $99 per ton of CO_2 equivalent; high cost = cost of $100 or more per ton of CO_2 equivalent. Tons are metric.
[b]In metric tons of CO_2 equivalent per year. This "maximum feasible" potential emission reduction assumes 100% implementation of each option in reasonable applications and is an optimistic "upper bound" on emission reductions.
[c]This depends on the actual implementation level and is controversial. This represents a middle value of possible rates.
[d]Some portions do fall in low cost, but it is not possible to determine the amount of reductions obtainable at that cost.
[e]The potential emission reduction for electricity supply options is actually 1700 Mt (million metric tons) CO_2 equivalent per year, but 1000 Mt is shown here to remove the double-counting effect.

Source: NAS, National Academy of Sciences, National Academy of Engineering, Institute of Medicine, *Policy Implications of Greenhouse Warming: Mitigation, Adaptation and the Science Base,* Washington, DC: National Academy Press, 1992.

hydro) CO_2 can be phased in but are likely to have higher costs than most efficiency options. In addition, there are attractive options for reducing methane, from reducing flaring from oil wells and eliminating distribution losses to capturing coalbed and landfill methane, both nationally and internationally (Claussen 1992; Rubin et al. 1992).

Also included in mitigation are geoengineering responses (large-scale human intervention to affect the climate). These include screening incoming solar radiation by such means as space mirrors or stratospheric dust and enhancing oceanic CO_2 uptake by phytoplankton by iron fertilization. The consequences of these types of actions are not well understood, and they should be looked at only as a last resort should climate change occur too quickly, or global political solutions fail (NAS 1992).

Adaptation

An adaptation approach to global warming involves preparations that would lessen the impacts of climate change. These include agricultural research to develop varieties of plants capable of growing well in the changed climatic condi-

tions, design and construction of robust water supply systems, investment in structures such as bridges and dams that would be able to withstand the extremes of climate change, and measures to slow the loss of biodiversity. While adaptation may
be possible in some areas, unmanaged ecosystems and marine and coastal environments are particularly vulnerable (NAS 1992). Unfortunately, the effectiveness
of adaptation strategies depends on a gradual change in climate with reasonably
accurate regional prediction and no climate surprises. It also assumes that the
money and the technology and knowledge base are adequate to the task. These
conditions may be lacking.

The Policy Dilemma

As we have seen, scientific understanding of the complex climate system is far
from complete so that model projections have large uncertainties, and it is currently difficult to know to what extent the global temperature trends are reflec-
tive of the impact of increasing GHGs. Yet, it is evident that the magnitude and
especially the rate of global climate change, as currently predicted by models,
poses significant risks to human society and to ecosystems. There is also the
possibility of climate surprises, sudden changes in global or regional climate
with significant adverse impacts. How much money should be spent today in order to reduce (mitigate) or to live with (adapt to) climate change? Do we even
know enough about regional impacts from climate change to be able to make
wise choices on adaptation strategies? These and related questions face policymakers at all levels—national energy planners, utility executives deciding on the
energy source for a new power plant, and the road engineer who must decide
on the construction of coastal bridges and roadways. Since climate change affects everyone, how is the global community to decide what actions to take? Despite uncertainties, there seems to be a general agreement among many policy
studies, both nationally and internationally, that a prompt response to prevent
severe climate change is warranted (EPA 1989; OTA 1991; NAS 1992; IPCC 1990,
1992). Difficulty arises, however, when specific agreements, timetables, and actions are considered. Nonetheless, there has been some movement to respond
to the threat of climate change.

Beginning Policy Responses to Global Warming

The Global Community

The United Nations Environmental Programme (UNEP) has taken the lead in
international efforts to deal with climate change. They have sponsored an international group of scientists, the International Panel on Climate Change (IPCC), to
study and prepare reports on the climate change issue (IPCC 1990, 1992). In 1992,
they sponsored the Earth Summit or UNCED, the United Nations Conference on
the Environment and Development, in Brazil. For several years prior to this meet-

ing, the Intergovernmental Negotiating Committee (INC) for a Framework Convention on Climate Change met to develop provisions to control climate change acceptable to all nations. At UNCED, the resulting agreement (referred to as the Global Climate Convention Treaty) proposed to cut back levels of CO_2 emissions to 1990 levels by the year 2000. Although 150 countries pledged support, there were no enforcement provisions in the agreement. Countries only committed to formulating a national plan to mitigate climate change, and in meeting periodically to review progress and share scientific knowledge. More developed countries (MDCs) did pledge some support for less developed countries (LDCs) to meet their commitments, although there was disagreement between industrialized and developing countries on details for financial assistance to LDCs and terms for technology transfer (Brooke 1992; Reinstein 1992).

Since so much of current GHG emissions come from industrialized nations, the unilateral policy commitments of several countries, including Germany, Austria, Denmark, and Australia, to reduce GHGs are significant. For example, the German plan, adopted by the German Federal Cabinet in 1990, calls for a 25% reduction in CO_2 emissions by 2005 compared to 1987 emissions. Meanwhile, the European Community has agreed to stabilize CO_2 emissions at 1990 levels by the year 2000 and is moving to implement measures such as a carbon tax to achieve this goal (Lashof 1992).

The United States

The United States, throughout most of the Reagan and Bush Administrations, was generally opposed to recognizing global warming as a potential problem. Reports from the EPA (1989), OTA (1991), and NAS (1992), however, suggested that the issue be taken seriously, and President Bush's 1991 budget proposed over $1 billion on global climate change research. Nonetheless, action to reduce GHGs was not part of U.S. policy; at the UNCED Conference (Earth Summit) in Rio in 1992, the United States was the major opponent of targets and timetables in the Global Climate Convention Treaty, and successfully prevented their incorporation into the international agreement. The United States then pledged support for the weakened treaty (Stevens 1992). Later in 1992, global climate change surfaced directly and indirectly in Title XVI of the National Energy Act of 1992. This act required the Department of Energy to study the feasibility of stabilizing and reducing GHGs and to assess the NAS (1992) report. It also set up provisions for a Director of Climate Protection, a (voluntary) National Inventory for GHG emissions, and an international global change response fund. Perhaps the most important provision was the emphasis on a least cost energy strategy, which would shift national priorities towards renewable energy sources and efficiency. In 1993, President Clinton announced that the United States would support the global climate convention treaty with strict deadlines and timetables, and that the United States would reduce its CO_2 emissions to 1990 levels by the year 2000 ("Clinton Declares New. . . " 1993). At the present time, it is unclear exactly what additional actions the United States will take domestically to accomplish this objective, although prompt enforcement of the Clean Air Act of

1990 and the National Energy Act of 1992 likely would go a long way towards achieving that goal. The United States, as a member of INC, continues to work on negotiating an international accord on the global warming issue.

The Future: Policy Controversy Today Leads to Consensus Tomorrow???

The nature of the global warming issue, especially uncertainties in the projections for both the magnitude and timing of climate change, and the even greater uncertainties regarding regional climate change and impacts, has provided many policymakers with excuses for inaction. There is also substantial uncertainty on the costs of stabilizing climate change (Nordhaus 1991, 1992; DRI/McGraw Hill 1992; Lashof 1992; NAS 1992; Rubin et al. 1992). This is not only due to uncertainties in climate change and associated impacts, which make costs and benefits difficult to calculate, but also due to inherent difficulties in applying economics to a problem with resources (genetic, natural, cultural) that are difficult to value properly, with widely varying risk perceptions, and with benefits (climate stability for future generations) that accrue to the future (NAS 1992). Unfortunately, the problem of reducing climate uncertainties from GHG emissions may take scientists several decades to solve satisfactorily (Fernau et al. 1993). Until then, policymakers will have to make difficult value judgments based on incomplete information about how much money to spend to mitigate and adapt to climate change. Indeed, there has been substantial debate already regarding appropriate actions to take. It should be noted that reducing GHG emissions to 1990 levels, for example, does not stabilize climate; it would merely let the increase in GHG concentrations to continue (Lashof 1992). It may take an immediate 60% reduction in GHG emissions to achieve climate stabilization (IPCC 1990).

Some U.S. policy studies argue that substantial cuts in GHG emissions, perhaps 10 to 40%, can occur at minimal cost because the primary action for reducing GHGs, improving energy efficiency in many areas of the economy, pays for itself (NAS 1992; Rubin et al. 1992). Others argue that the low costs of some of these actions are illusionary and that resources might better be spent in other areas. The so-called no-regrets strategy argues that many actions to reduce GHGs have significant environmental and economic benefits on their own and ought to be done regardless of their impact on climate change. Others argue that the risks and impacts of climate change are sufficiently large that society may well want to move beyond the no-regrets strategy and expend sufficient money to ensure that large or quick changes in climate do not occur (Schneider 1989; Risbey et al. 1991a,b). Some argue that delay in responding to climate change is acceptable and we should respond as the magnitude of climate change becomes more certain (Schlesinger and Jiang 1991a,b). This approach, however, ignores the fact that most GHGs are associated with other serious environmental issues (see Chapter 12, "World Air Pollution: Sources, Impacts, and Solutions" and Chapter 13, "Stratospheric Ozone Pollution"). There is clearly a great deal of dispute about an appropriate climate change policy at the national level. Yet it is evident that a global approach is needed to effectively deal with climate change. This poses much greater difficulties.

Reaching an effective international accord to limit GHGs and stabilize climate will require accommodation of many conflicting interests. More developed countries, with about 25% of the world's population, contribute the lion's share of GHGs (75%) to the environment at present. Many LDCs, however, such as China and India, are growing rapidly and plan to develop higher standards of living using fossil fuel resources, particularly coal. Other LDCs plan to develop by removing tropical forests for hydropower development and for grazing and agriculture (see Chapter 7, "Hydro-Development"). By 2030, LDCs may account for 40% of GHG emissions (NAS, 1992). There are other countries that may see real or perceived benefits in not responding to climate change. For example, countries with considerable remaining oil resources are opposed to reducing fossil fuel use. Some countries, such as Canada and the former Soviet Union, may benefit from some warming if it enhances grain production in regions that were previously too cold. LDCs appear to be in a double bind: many LDCs do not have the resources to make changes that would reduce GHGs. At the same time, lack of resources makes them less able to adapt to climate change; climate impacts may have the most severe consequences in these areas. Although international agreement to eliminate ozone-depleting CFCs was relatively fast (the ozone hole provided dramatic confirmation of a problem with potentially catastrophic consequences; the costs were much less; relatively few companies produced CFCs), reaching an effective accord on GHGs will certainly tax the abilities of international negotiators.

Conclusions

Climate change poses unprecedented challenges for the human species. Current scientific understanding does not allow accurate predictions of the magnitude, timing, and regional distribution of future climate change, and it may take several decades for scientists to provide much improved accuracy. On the other hand, rapidly increasing GHG emissions are most certainly changing the radiative balance of the Earth, potentially leading to rapid climate change. Substantial change in the Earth's climate would have severe consequences for humans and present ecosystems; moreover, climate "surprises," rapid or unpredicted climate change with potentially disastrous impacts, cannot be ruled out. The present difficulty in attributing the observed temperature record to increasing GHGs, together with the time lag between radiative forcing and climate response, makes it difficult to motivate action; there is no clear and compelling connection between this year's climate and GHG emissions. The long lifetime of most GHGs in the atmosphere, however, means that we may be committed to large climate changes before the change becomes evident. It may take a climate event equivalent to the ozone hole to make people aware of the seriousness of climate change. In principle, the nations of the world are committed to preventing GHG emissions from causing a dangerous change in climate. The difficulty is in translating such a commitment to a workable plan that is accepted and carried out by all nations. Climate change uncertainties, the potentially large consequences of climate change, uncertain costs, and the necessity of a global response all make the climate change problem one of the most difficult that humans have faced.

References

Ardanuy, P. E., H. L. Kyle, and D. Hoyt, "Global Relationships among the Earth's Radiation Budget, Cloudiness, Volcanic Aerosols and Surface Temperature," *Journal of Climate* 5, 1992, pp. 1405–1423.

Austin, J., N. Butchart, and K. P. Shine, "Possibility of an Arctic Ozone Hole in a Doubled-CO_2 Climate," *Nature* 360, 1992, pp. 221–225.

Barrett, P. J. C. J. Adams, W. C. McIntosh, C. C. Swisher III, and G. S. Wilson, "Geochronological Evidence Supporting Antarctic Deglaciation Three Million Years Ago," *Nature* 359, 1992, pp. 816–818.

Brooke, J., "President, in Rio, Defends His Stand on Environment," *New York Times,* I, 4:1, June 13, 1992.

Brooks, H., "Generic Issues Regarding Climate Change," *Joint Climate Project to Address Decision Makers' Uncertainties, Report of Findings*, Washington, DC: Science and Policy Associates, 1992.

Charlson, R. J., S. F. Schwartz, J. M. Hales, R. D. Cess, J. A. Coaxley, Jr., J. E. Hansen, and D. J. Hofmann, "Climate Forcing by Anthropogenic Aerosols," *Science* 255, 1992, pp. 423–430.

Clark, W. C., "Visions of the 21st Century: Conventional Wisdom and Other Surprises in the Global Interactions of Population, Technology, and Environment," K. Newton, T. Schweitzer, and J. P. Voyer (eds.), *In Perspective 2000,* Ottawa: Canadian Government Publishing Center, 1991, pp. 7–32.

Claussen, E., Testimony on the Role of the U.S. Government in the United Nations Negotiations on Global Warming Climate Change, In *Hearing before the Subcommittee on Energy and Power of the Committee on Energy and Commerce, U.S. House of Representatives, 102nd Congress,* March 3, 1992, no. 102–121, Washington DC: U.S. Government Printing Office.

"Clinton Declares New U.S. Policies for Environment," *New York Times,* A, 1:6, April 22, 1993.

DRI/McGraw-Hill, *Economic Effects of Using Carbon Taxes to Reduce Carbon Dioxide Emissions in Major OECD Countries,* Lexington, MA: Standard & Poor's Corporation, January 1992, available from National Technical Information Service (PB92-127562).

EPA (Environmental Protection Agency), *The Potential Effects of Global Climate Change on the United States,* J. B. Smith and D. A. Tirpak (eds.), Report to Congress, Washington, DC: U.S. Environmental Protection Agency, EPA-230-05-89-050, 1989.

Fernau, M. E., W. J. Makofske, and D. W. South, "Review and Impacts of Climate Change Uncertainties," *Futures,* October 1993, pp. 850–863.

IPCC (Intergovernmental Panel on Climate Change), *Climate Change: The IPCC Scientific Assessment,* J. T. Houghton, G. J. Jenkins, and J. J. Ephraums (eds.), Cambridge: Cambridge University Press, 1990.

IPCC (Intergovernmental Panel on Climate Change), *Climate Change 1992. The Supplementary Report to the IPCC Scientific Assessment*, J. T. Houghton, B. A. Callander, and S. K. Varney (eds.), Cambridge: Cambridge University Press, 1992.

Khalil, M. A. K. and R. A. Rasmussen, "The Global Sources of Nitrous Oxide," *Journal of Geophysical Research* 97, 1992, pp. 14651–14660.

Lashof, D. A., Testimony on the Role of the U.S. Government in the United Nations Negotiations on Global Warming Climate Change, In *Hearing before the Subcommittee on Energy and Power of the Committee on Energy and Commerce, U.S. House of Representatives, 102nd Congress*, March 3, 1992, no. 102–121, Washington, DC: U.S. Government Printing Office.

Lehman, S. T. and L. D. Keigwin, "Sudden Changes in North Atlantic Circulation during the Last Deglaciation," *Nature* 356, 1992, pp. 757–762.

Lindzen, R. S., "Some Coolness Concerning Global Warming," *Bulletin of the American Meteorological Society* 71, 1990, pp. 288–299.

MacAyeal, D. R., "Irregular Oscillations of the West Antarctic Ice Sheet," *Nature* 359, 1992, pp. 29–32.

Manabe, S. and R. T. Stouffer, "Century-Scale Effects of Increased Atmospheric CO_2 on the Ocean-Atmosphere System," *Nature* 364, 1993, pp. 215–218.

NAS, National Academy of Sciences, National Academy of Engineering, Institute of Medicine, *Policy Implications of Greenhouse Warming: Mitigation, Adaptation and the Science Base*, Washington, DC: National Academy Press, 1992.

Nordhaus, W. D., "To Slow or Not to Slow: The Economics of the Greenhouse Effect," *The Economic Journal* 101, 1991, pp. 920–937.

Nordhaus, W. D., "An Optimal Transition Path for Controlling Greenhouse Gases," *Science* 258, 1992, pp. 1315–1319.

OTA (Office of Technology Assessment), *Changing by Degrees: Steps to Reduce Greenhouse Gases*, Washington, DC: U.S. Government Printing Office, OTA-O-482, 1991.

Ramaswamy, M. D., M. D. Schwarzkopf, and K. P. Shine, "Radiative Forcing of Climate from Halocarbon-Induced Global Stratospheric Ozone Loss," *Nature* 355, 1992, pp. 810–812.

Raval, A. and V. Ramanathan, "Observational Determination of the Greenhouse Effect," *Nature* 342, 1989, pp. 758–761.

Reinstein, R. A., Testimony on the Role of the U.S. Government in the United Nations Negotiations on Global Warming Climate Change, In *Hearing before the Subcommittee on Energy and Power of the Committee on Energy and Commerce, U.S. House of Representatives, 102nd Congress*, March 3, 1992, no. 102–121, Washington DC: U.S. Government Printing Office.

Rind, D., E.-W. Chiou, W. Chu, J. Larsen, S. Oltmans, J. Lerner, M. P. McCormick, and L. McMaster, "Positive Water Feedback in Climate Models Confirmed by Satellite Data," *Nature* 349, 1991, pp. 500–503.

Risbey, J. S., M. D. Handel, and P. H. Stone, "Should We Delay Responses to the Greenhouse Issue?," *Eos Transactions, America Geophysical Union* 72, 1991a, p. 593.

Risbey, J. S., M. D. Handel, and P. H. Stone, "Do We Know What Difference a Delay Makes?," *Eos Transactions, American Geophysical Union* 72, 1991b, pp. 596–597.

Rubin, E. S., R. N. Cooper, R. A. Frosch, T. H. Lee, G. Marland, A. H. Rosenfeld, and D. D. Stine, "Realistic Mitigation Options for Global Warming," *Science* 257, 1992, pp. 148–266.

Schneider, S. H., "The Greenhouse Effect: Science and Policy," *Science* 243, 1989, pp. 771–781.

Schneider, S. H., *Global Warming: Are We Entering the Greenhouse Century?*, New York: Vintage Press, 1990.

Schlesinger, M. E. and X. Jiang, "A Phased-In Approach to Greenhouse-Gas-Induced Climatic Change," *Eos Transactions, American Geophysical Union* 72, 1991a, pp. 593, 596.

Schlesinger, M. E. and X. Jiang, "Climatic Responses to Increasing Greenhouse Gases," *Eos Transactions, American Geophysical Union* 72, 1991b, p. 597.

Stevens, W. K., "With Climate Treaty Signed, All Say They'll Do Even More," *New York Times,* I, 1:4, June 13, 1992.

Stouffer, R. J., S. Manabe, and K. Ya. Vinnikow, "Model Assessment of the Role of Natural Variability in Recent Global Warming," *Nature* 367, 1994, pp. 634–636.

Wigley, T. M. L. and S. C. B. Raper, "Implications for Climate and Sea Level of Revised IPCC Emissions Scenarios," *Nature* 357, 1992, pp. 293–300.

Winograd, I. J., T. B. Coplen, J. M. Landwahr, A. C. Riggs, K. R. Ludwig, B. J. Szabo, P. T. Kolesar, and K. M. Revesz, "Continuous 500,000-Year Climate Record from Vein Calcite in Devil's Hole, Nevada," *Science* 258, 1992, pp. 255–260.

Chapter 15

Predicting Climate Change: The State of the Art

Kerry H. Cook

Many people today are well aware of the possibility that human activity may be resulting in changes in the climate. This would not have been the case only 10 years ago, but now few days elapse without mention of at least one bizarre-sounding climate problem. We hear about the greenhouse effect, the ozone hole, acid rain, African drought, or El Niño regularly on television, radio, and in the newspapers. This paper examines how scientists study climate and climate change to understand these kinds of problems, and how they try to predict their occurrence and their future development. The main goals of this article are to enhance understanding of how these issues are approached, how advanced our science is in dealing with them, and in what directions the science of climatology should develop to meet the needs of a society faced with the very real possibility of climate change.

Definition of Climate and Climate Change

To start thinking about how the climate changes, we should first define climate. When we talk about the climate of a particular area, we are referring to what the weather is like on average over time scales of many years. For example, we might say that the climate of the southwestern United States is "hot and dry." So we are using our perception of the average atmospheric temperature near the Earth's surface and precipitation rates to define *climate*. In general, when we worry about the climate changing we are concerned with the temperature of the atmosphere near the ground getting colder or hotter, or maybe an increase or decrease in the amount of rainfall.

When we study the climate, however, we have to be concerned with more than the thin layer of the atmosphere near the surface. The climate is a complex system that can be represented as five subsystems that are constantly interacting with each other—the atmosphere, the oceans, the cryosphere, land surfaces, and the biosphere. The cryosphere includes sea ice as well as the snow and ice on land surfaces. The biosphere subsystem consists of the terrestrial and marine biota, and we can also include the increasingly important human influences on climate. Figure 15.1 shows some of the connections between the atmosphere and the other subsystems of the climate system; each subsystem also interacts with the others.

No subsystem of the climate can change significantly without causing changes in the other subsystems. A large change in one subsystem will change how that subsystem interacts with the others, and these changes can, in turn, affect the first subsystem, feeding a response back and modifying the original change. This is a useful way to think about how climate changes. We say that the climate system "responds" to a given change with some degree of "sensitivity." The climate system's sensitivity to a given change depends on how each subsystem is affected.

To illustrate how climate change occurs, imagine what would happen if the Sun were to become dimmer. The direct effect might be that the land and ocean surfaces would become colder by an amount that is directly related to the magnitude of the decrease in sunlight. But this cooling would have consequences in the other climate subsystems. In the cryosphere, for example, we might expect an increase in the extent of ice and snow. This would make the surface brighter than it was with the sun full strength. When the surface is brighter, more sunlight is reflected back to space than before, and this warmth never enters the climate system. The surface may then become even colder, and this could lead to more ice and snow, an even brighter surface, and still colder temperatures. So, in the end, the cooling that occurs due to the dimming of the Sun may be larger than originally expected because the total climate system was sensitive to the change and reacted in a way that made the cooling more severe. This kind of reaction is called a *positive feedback* because the original tendency—cooling due to a decrease in the amount of sunlight—is amplified by how the climate reacts.

Consider another, perhaps simultaneous, reaction that the climate may have to a dimming of the Sun. What if the climate somehow responded to the change in the Sun by a decrease in cloudiness. In this case, more sunlight would get through to the surface, and the cooling caused directly by the decreased sunlight would be opposed. This type of reaction is called a *negative feedback,* because the original tendency is de-amplified, or suppressed, by how the climate reacts.

In this example, the decrease in the incoming sunlight is called the external *forcing* for the climate change. It is a factor that changes outside the climate system and, in this case, it is unchanged by the climate's response, like a catalyst in a chemical reaction. Climatic change forcing functions may also be internal to the climate system but, whatever the forcing, the sensitivity of the climate to a particular forcing is essentially the sum total, or net result, of all the positive and negative feedbacks.

"Natural Climate Change"

Although we suspect that human activity is changing the climate, we know that the Earth's climate can change without human influence. There have been cold periods, the Ice Ages, when extensive land surfaces in the Northern Hemisphere were covered by massive sheets of ice. The most recent ice age was at its peak around

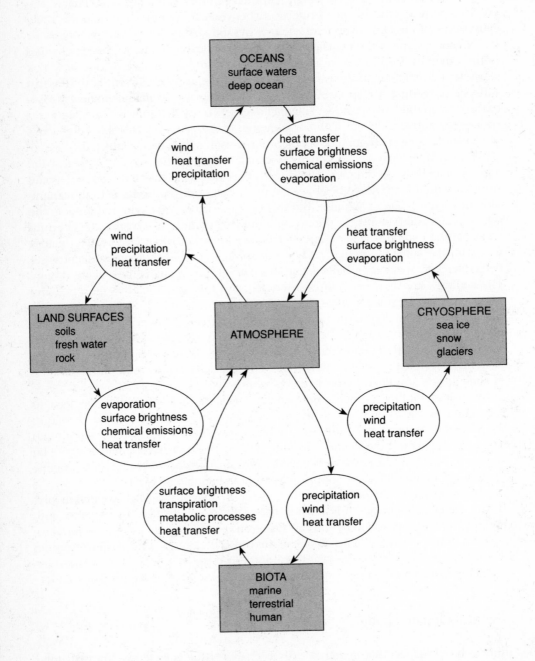

Figure 15.1. Schematic illustration of the components of the climate system, with some of the important ways in which they interact with the atmospheric component indicated.

18,000 years ago, but there have been many glacial periods in the Earth's history. The last 5 million years of the climate's history is characterized by alternating ice ages and warmer periods, or interglacials, such as the one we are in now. There have also been periods in the climate's history that were warmer than the present, for example, 65 to 200 million years ago during the Mesozoic Era when dinosaurs were alive.

Studying past climates can teach us a lot about climate change, both in terms of magnitudes and rates of climate change. There are still many unknowns concerning the climate's oscillation between the ice ages and interglacials, but there is some new information that is helping to solidify our ideas. But before discussing some of this new information, consider what might be the cause of the "natural" climate change.

The most widely theorized cause of the ice ages concerns changes in the amount of sunlight coming into the climate system. This has nothing to do with the output of energy from the Sun, which varies on much longer time scales, but with the relative positions of the Sun and the Earth. The Earth's orbit around the Sun is not fixed—it changes in a known and regular way over time periods of tens of thousands of years. There are three changes thought to be important for climate change. First, the eccentricity of the orbit—the elongation of the orbital oval— changes with a period of about 100,000 years. Second, the time of year when the Earth is closest to the Sun changes. Right now, the closest approach (perihelion) is in December, when it is winter in the Northern Hemisphere. You might imagine that it would make a difference if we were closest to the Sun in the summer. In the Northern Hemisphere, summers might be warmer and winters colder. Perihelion varies with a period of about 20,000 years. The third variation in the Earth's orbit is the tilt of the axis of rotation. Today, the Earth spins on an axis that is tilted 23.5° from the perpendicular to its plane of revolution about the Sun. This tilt changes with a period of 41,000 years.

It has long been supposed—and there is evidence to support this idea—that these changes in the Earth's orbit and the accompanying changes in sunlight (solar forcing) cause the ice ages. The idea is called the Milankovitch theory of climate change after an early proponent of the concept (Berger et al. 1984). There is also evidence that does not fit exactly with the predictions of the Milankovitch theory, and some climatologists assign more control of climate change to the climate system, even on these very long time scales.

A lot of our knowledge about other stages in the climate's history comes from cores. Cores are tubes of rock or ice that have been drilled out of bedrock or glaciers. The deeper the core, the farther back in time that is represented. There is one recent ice core that is particularly interesting because of its length. It is called the Vostok core because it was drilled near Vostok in Eastern Antarctica by the Soviet Antarctic Expedition (Barnola et al. 1987). This core is very long—over 2 km. The ice at the bottom of the Vostok core formed 160,000 years ago, and so we now have a continuous record from Antarctica that spans two ice ages and two interglacial periods. The top of the core represents the present interglacial period. Farther down, the core samples ice throughout the last ice age, which ended around 15,000 years ago and began some 110,000 years ago. The core goes back further still to the next interglacial at around 120,000 years, and well into the previous ice age that ended around 140,000 years ago.

The Vostok core has been analyzed to give time series of certain climate variables back to 160,000 years. One of these is the Antarctic surface temperature, which can be calculated with some accuracy from the details of the chemical composition of the ice. Another time series that has been constructed is the carbon dioxide (CO_2) concentration of the atmosphere. Even though regular monitoring of the atmospheric CO_2 concentration began in 1958, nature has been saving air samples for us through the ages. When snow packs down to become ice, air is trapped in the structure. In a permanently frozen place, like Vostok, the ice keeps piling up over thousands and thousands of years. The Vostok core has brought 160,000-year-old ice with samples of the atmosphere to the surface for analysis. The Vostok core shows that glacial periods are 6–10°C colder than the interglacials in Antarctica, and that the CO_2 concentration during glacial periods is significantly lower than its concentration during interglacial periods. In other words, high CO_2 concentrations seem to be naturally associated with warm climate states. This connection is intriguing, especially in the light of our current concerns about increasing atmospheric CO_2 levels as discussed below.

If the Milankovitch theory of the ice ages is correct, we expect to see periodicities in the geological records from the Vostok core that are the same as the variations in the orbital parameters, namely, 20,000, 41,000, and 100,000 years. The Vostok surface temperature record does indeed show each of these periods, although the 20,000 year period is weak. High resolution analysis indicates that the temperature first changes before the CO_2 concentrations, not as a result of the CO_2 changes. Thus, it is not likely that the orbital parameters induce a change in the climate through control of the CO_2 concentration; but perhaps CO_2 variations amplify the climate's response to changes in the orbital parameters.

Climate Models

One way to study how climate changes is to compare today's climate with past climates, as with the Vostok core discussed above. Such comparisons assure us that the climate's history is characterized by large changes on time scales of thousands of years, and that we are not dealing with a static system when we study climate. It is very difficult, however, to piece together a complete picture of climates that occurred so long ago. Cause and effect can be blurred in the records because they usually do not offer the kind of resolution in time that is needed to determine which change came first. It is difficult to determine the actual physical processes that might have occurred to bring about the observed climate change.

Another way to study climate is with computers. We know the mathematical equations that describe how fluids move and how they react to external forces. These equations are called the *fluid dynamics* equations, and they can be used to study air as well as water. They serve as the basis for a mathematical description of the atmosphere and the oceans. To see how this works, imagine a volume of air or water, which we sometimes call a *parcel*. The state of a parcel is defined by its temperature, three-dimensional velocity, density, and moisture content. The fluid dynamics equations evaluate the forces that are acting on the parcel conspiring to make it move in a certain direction. If we also consider an equation to account for all of the heat sources acting on the parcel (a thermodynamic equation), we can also calculate if the parcel is going to get colder or warmer.

In this type of description of the atmosphere or ocean, there are a lot of forces and heat sources to be taken into account. They are interdependent, and they depend on the condition of the parcel; but if these equations are translated into a computer language, a computer program can keep track of every force and heat source at every time and place. That is a climate model. It is a computer program that sums all the forces and heating acting on each volume of air or water, and calculates how the parcel is obliged to move and if it gets hotter or colder. If the parcel is air, as in an atmospheric model, we also have to keep track of how much water vapor is contained in the parcel to predict rain and snow. In an ocean model, the salinity of the water is important. In the most advanced climate models today, complex ocean and atmospheric models are linked together and allowed to interact just as the real atmosphere and ocean climate subsystems interact (Washington and Parkinson 1986).

Climate models serve as laboratories for experimenting with the climate, since we cannot tamper with the real climate—at least not on purpose. One kind of experiment involves generating two solutions of the equations. The first experiment is called the *control* experiment. The climate model is set up to represent today's climate. For example, the concentration of carbon dioxide and ozone in the atmosphere, the positions of the continents, and the amount of energy reaching the Earth from the Sun each season are all set at values that describe these "boundary conditions" for the present-day climate. Then, usually starting with motionless, isothermal conditions, the fluid dynamics equations are solved repeatedly, advancing (or integrating) the model climate from the initial conditions to a new stable climate state. This climate state is compared with the observed climate to see if the model climate is reasonable.

The results of the control experiment, and not the observed (real) climate, are the basis of comparison for the second solution of the equations. For the second experiment, one condition in the climate model may be changed. To study the example above, concerning a dimming of the Sun, the amount of energy coming into the climate system from the Sun would be decreased. Another experiment that is often performed with climate models is to increase the atmospheric CO_2 concentration. Whatever the nature of the experiment, the computer integrates the fluid dynamics equations again after the modification and, if the climate model is sensitive to the Sun getting dimmer, the temperature, wind, and precipitation fields will be different—a "new" climate state. The model climate from the second solution is then compared with the control climate to see where significant changes occurred; the difference is called the *equilibrium response* to increased CO_2. These results are useful for exploring theories about how and why climate changes occur, or for testing mechanisms hinted at in observations. Subsequent experiments can help to clarify these ideas.

The most complete climate models today include complex representations of the atmosphere and ocean, and often biospheric components as well. The development of coupled atmospheric-ocean models has made a second kind of climate change experiment possible. Since the mass of the oceans is much greater than the mass of the atmosphere, the oceans respond to forcing much more slowly. Without the realism of the ocean's slow response times in a climate model, one cannot get a realistic estimate of the time dependence of climate change from a model. With the oceans included, experiments with models can be done with more realistic forcings. For example, coupled atmosphere-ocean models are currently being forced by gradual, more realistic, atmospheric CO_2 increases to provide information about rates of climate change, not just magnitudes of change as from the equilibrium experiments.

The perspective afforded by realistic time dependence in climate models can make a significant difference in how we perceive the threat of climate change, since we really must respond to rates of climate change. Figure 15.2 (top) shows the equilibrium surface air temperature change (annual mean) that occurs when the atmospheric CO_2 concentration is doubled in a climate model. Note that tropical temperatures increase by about 3°C, and that temperature changes at higher latitudes in the Northern Hemisphere are larger, with over 5°C warming over portions of North American and Eurasia. At high southern latitudes, the warming is also enhanced.

Also shown in Figure 15.2 (bottom) is the time-dependent response to doubling CO_2. Rather than being given enough time to completely adjust to increased CO_2 levels, this figure represents a snapshot of the temperature field at the time of CO_2 doubling in a model with gradual CO_2 increases. Significant differences occur. Warming in the tropics has reached about 80% of the equilibrium response, but over the North Atlantic Ocean and in the Antarctic, temperature changes are much less than the equilibrium changes. The differences between the equilibrium and time-dependent simulations are due in large part to the fact that some regions take longer than others to respond to the CO_2 forcing.

What are the limitations of climate models? We cannot predict the climate, or even tomorrow's weather for that matter, with complete accuracy, so there must be some problems. The first is our lack of knowledge about the climate system. To improve the atmospheric part of a climate model, for example, we need more basic research on cloud formation and precipitation processes. Also, we have concentrated on the atmosphere and ocean climate subsystems in our models, and the three other components of climate—land surfaces, biosphere, and the cryosphere—are treated much more schematically. It is possible that the interactions of these subsystems need to be treated with more detail at this stage, or the next stage, in the development of climate models.

The second source of inaccuracy in climate models is due to the limitations of the present generation of computers. In writing the fluid dynamics equations in a form that the computer can deal with, the world is represented by a grid. Imagine taking a huge screen and wrapping it smoothly around the globe near the surface. Then take a second screen and position it above the first, suspended at a specified height in the atmosphere. Place a third screen about the second, and so on through the depth of the atmosphere and down through the ocean. This is analogous to the mathematical construction of a computer model. At every intersection—where wires cross on the screen—the computer solves the fluid dynamics equations for a

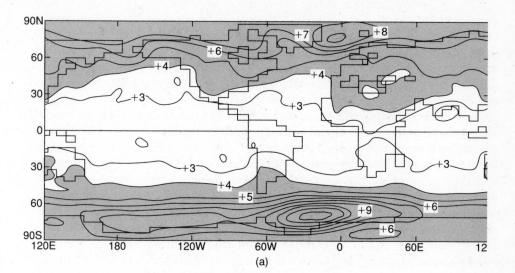

(a)

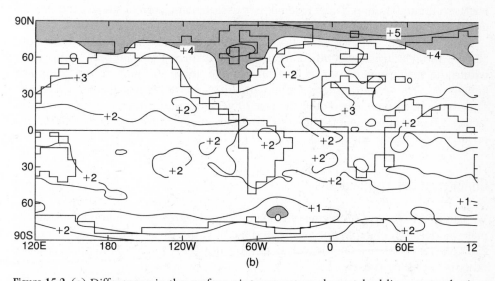

(b)

Figure 15.2. (a) Differences in the surface air temperature due to doubling atmospheric CO_2 in a equilibrium climate model experiment. (b) Differences in the surface air temperature at the time of CO_2 doubling in a climate model with gradual CO_2 increases. Contour intervals are 1°C, and warming over 4°C is denoted by shading. (From Manabe, S. et al., "Transient Response of a Coupled Ocean-Atmosphere Model to Gradual Changes of Atmospheric CO_2," *Journal of Climate* 4, 1991, pp. 785–818.)

parcel at that location. We assume that the parcels in between the grid intersections behave the same as the parcels at the intersections. The distance between the grid intersections is called the *horizontal resolution* of the model. The distance between the vertically stacked grids is the vertical resolution. Resolution varies from model to model and increases as more powerful computers are made available. A typical horizontal resolution is a few hundred miles, and about 10 vertical levels are used through the lower atmosphere (troposphere). In a model with coarse resolution, the assumption that the parcels in between the grid intersections act like the parcels at the intersections can be a very bad assumption—like assuming the climate in Florida is the same as the climate in Georgia. As the grid size is decreased, that is, the model resolution is increased, this assumption of homogeneity between grid points is potentially less dangerous. The model's resolution is determined by considering computer limitations, given the complexity of the equations to be solved. In other words, since these models are run on the most advanced computers, resolution can be increased only by giving up some physical complexity in the equations.

To impart an understanding of our current capabilities, Figure 15.3 shows how well the outlines of the continents and topography are commonly resolved in climate models. The top figure is from a lower resolution model, and represents models used to study climate and climate change through most of the 1970s and 1980s. While scientists still find that this coarser resolution is useful, many current studies and predictions use higher resolution models, as shown in the lower part of the figure.

It is easy to discuss the limitations of these models, but it is important to note climate models are designed to treat climate change as accurately as possible within the limitations of our lack of knowledge about the system and computer limitations. These models are among the most complex models in existence. They have been developed over the last 20 years by choosing the most important factors involved in a given climate change problem and representing them appropriately in the model.

Climate Model Applications

We can address many problems of climate change with some confidence using climate models. Other problems are more difficult at the present time. Below are two examples of problems. One is well suited to studying with climate models, and the other is not at the present time.

African Drought

Recent years have brought devastating drought conditions to large parts of Africa. Are these droughts a natural feature of climate in this region or an indication of climate change? If the climate in that region is changing, why? Is it natural or due to human activity? (see Chapter 10, "Desertification," for an in-depth discussion on this topic).

Rainfall records in northern Africa show that persistent droughts may be part of the climate there. There are periods of drought indicated in these

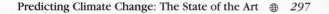

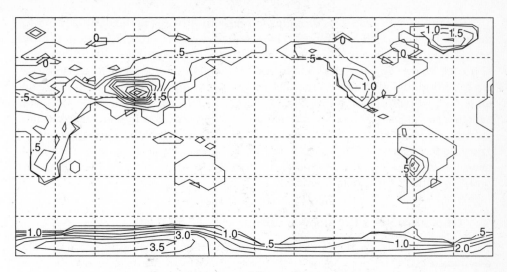

(a)

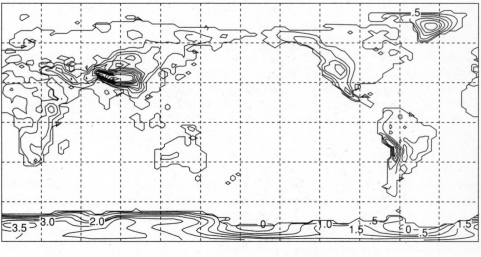

(b)

Figure 15.3. Continental outlines and topography as resolved by two versions of a climate model. Contours are drawn for every 0.5 km of surface elevation. See text for explanation.

records, but none is as long lasting as in the 1970s and 1980s. The region most affected is the Sahel, a marginal climate zone between the Sahara desert to the north and the Sudan to the south. By simply looking at the climate record for this region it is not possible to conclude whether a human-induced drying trend exists (Nicholson 1985).

The problem of recurring droughts in northern Africa is particularly difficult for the present state of climate models. Foremost is the ever-present resolution problem. The Sahara Desert is in the far north of the continent with less than 100 mm of rain each year. The Sahel and Sudan regions, which are the areas at greatest risk, are transition regions between the desert and wetter regions to the south. They are long, narrow bands, not compatible with the model's rectangular grid. Also, precipitation is one of the most highly parameterized processes in general circulation models (GCMs); that is, it is not treated explicitly, but with substitute equations on grids much larger than the space scales of the actual physical processes. Approximate treatment of surface hydrological processes also play a role in making this problem difficult for climate models.

"Greenhouse" Climate Warming

Climate models are better suited for handling more global-scale problems. The *greenhouse effect* is a popular name that refers to a global warming due to the presence of certain trace gases in the atmosphere. An entire issue of *Scientific American* is devoted to the many aspects of global warming (Clark 1989; Schneider 1989), and a scientific assessment of climate change has been published by the Intergovernmental Panel on Climate Change (IPCC 1990, 1992). The greenhouse warming occurs only in the lower part of the atmosphere where our weather occurs (the troposphere). The layer above the troposphere—the stratosphere—cools when greenhouse gases increase. Warming is instigated by the so-called greenhouse gases but, as discussed below, the human releases of these gases are not directly responsible for all the warming. The response of the climate system amplifies the effect (see Chapter 14, "The Challenge of Global Climate Change," for further discussion of this topic).

To begin exploring this issue, first consider what determines the temperature of the atmosphere near the surface of the Earth. The atmosphere is largely heated from heat released by the surface, and not by directly absorbing sunlight. If the Earth were a black[1] planet without greenhouse gases in its atmosphere, it would absorb an amount of sunlight determined only by its size, its distance from the Sun, and the brightness of the Sun. In that case, the temperature near the surface would be about 60°F (33°C) colder than the observed value. Our greenhouse-active atmosphere causes the surface air temperature of the Earth to be warmer than this "black body temperature," making life (as we know it) possible. The warming is accomplished by the greenhouse effect, which causes heat in the atmosphere to be redistributed so that more is held near the surface.

Greenhouse gases are largely invisible to the shortwave energy from the Sun, which passes through a layer of atmosphere containing greenhouse gases just as sunlight passes through the glass ceiling of a greenhouse. When this energy strikes the surface of the Earth, a fraction of it is absorbed and the surface warms. Warm surfaces give off long-wave energy, and it is this form of energy that greenhouse gases absorb. As the long-wave energy travels upward into the atmosphere, a por-

---

[1]An ideal "black body" absorbs all the radiation incident on it, regardless of wavelength.

tion of it is absorbed by the greenhouse gases, and then re-emitted in all directions. Some of this energy is "turned around" and the net effect is a trapping of heat near the Earth's surface. (This part of the process is unlike a greenhouse, which traps heat by keeping the air warmed at the surface from rising). When the concentrations of greenhouse gases increase, the trapping increases and less energy escapes to the upper atmosphere and space. But the direct trapping of long-wave energy is only part of the story—the response of the climate must be taken into account.

CO_2, methane, and other "greenhouse gases" are minor components of the atmosphere and they are not present in sufficient quantities to determine the atmosphere's thermal structure by themselves. When the concentrations of these gases increase, however, a positive feedback occurs that amplifies the direct warming. As the temperature of air increases, it is able to hold more water in the vapor phase. So, a small direct warming due to an increase in a trace gas concentration warms the atmosphere near the surface. More water evaporates from the warmer surface, and some of this extra water can be held in the warmer atmosphere without condensing out. This increase in the primary "natural" greenhouse gas—water vapor—causes additional warming and thus amplifies the human-induced warming. This is a positive feedback, as defined above.

The water vapor response is a strong positive feedback, but it is not the only response from the climate system. The ice albedo/temperature feedback process discussed above also occurs in response to the warming, so that we expect the high latitude warming to be larger when CO_2 increases. A number of other feedbacks are also possibly important, namely, the response of clouds and the biosphere. The ocean's response is crucial because it mostly determines the reaction times of the climate system and, to a large extent, the amount of CO_2 in the atmosphere.

Testing Climate Models

Since even the most advanced climate models are incomplete, we need to explore the question of how well climate models can predict future climates. With a host of processes missing, or treated very simply, and coarse resolution in both time and space, how can we believe the predictions from climate models? There are a number of ways that we can test these models, and the results are encouraging.

One way to build confidence in climate models is to apply them to the climates of other planets and see how well they can replicate the observed structures in these planets' atmospheres. The same set of equations that we use to describe the atmosphere and oceans of the Earth (the fluid dynamics equations) can be modified to describe the atmospheres of the planets. This has been done with some success for Jupiter and Venus, for example, and adds to our confidence in global climate models. But we do not know many of the details about the atmospheric circulations of the other planets and they test only for atmospheres, since Earth has the only oceans known in the solar system.

Another way to test climate models is to see how well they can replicate the observed features of the atmosphere and the oceans. The models are not made to treat a particular day, like weather models, but a model's average climate should be similar to the observations on the average. We are developing long series of observations to make a good average for comparing with climate models, along with other uses. But some quantities are still poorly observed, especially over the oceans.

A good replication of the present-day average climate by a climate model is a necessary condition for a trustworthy climate model, but it is not sufficient for climate change applications. The physical processes that are important in maintaining the climate may be different from the processes that are important in changing the climate.

Every year the climate goes through large changes. We can test the models' ability to treat climate change by seeing how well they can replicate the changes that we observed in the climate from one season to the next. The seasonal variations in climate are larger than any changes that we expect to occur due to human activity, at least in the near future. Climate models are competent in mimicking the changing seasons. When the amount of sunlight specified in the climate model is changed to represent seasonal solar forcing, the temperature and precipitation rates in the model change in a realistic way. There are some minor— and not so minor—differences, but overall the agreement is good (Washington and Parkinson 1986).

There are some limitations of this test of the models' ability to handle climate change. First of all, the response to only one kind of forcing is tested, namely, solar forcing. Also, while seasonal variations may be as large, or even larger than, the amplitudes associated with changes in the climate, the time scales involved are very different. The component subsystems of climate each have characteristic time scales, which means they take different amounts of time to respond to a forcing. The atmosphere can respond very quickly, but the deep oceans take many centuries to fully react. The upper surfaces of the ocean or shallow areas near land, on the other hand, have reaction times on the order of months because there is less water mass involved. Seasonal climate change is too fast to fully involve the slower reacting components of the climate system. The deep oceans do not "see" the changing seasons.

A third way to validate climate models takes very long time scales into account. Model climates can be compared with other climates of Earth's history (Barron 1984; Kutzbach and Guetter 1986; Broccoli and Manabe 1987). One remarkable data set describes the climate of the last ice age, which was at its height 18,000 years ago. We seem to learn more about the "Last Glacial Maximum" every year. An effort to put together a coherent picture of this climate state was begun in the early 1970s when a group of scientists from a variety of disciplines set out to "reconstruct" the Ice Age Climate (CLIMAP Project Members 1976). From this effort, we have a good idea about the temperature of the sea surface, heights and locations of the glaciers, and what plants lived on the continents. More recent studies have added to and changed the original picture of the ice age climate, including the fact that the CO_2 concentration was two-thirds of its present-day value. Information about this time is also being extended into the period of the deglaciation, when the climate made the transition from the last ice age to the present interglacial.

Detecting Climate Change

The ultimate test of climate models is the observation of a climate change as predicted by a model. Have we detected climate change due to human activities? An examination of the average global temperature of the atmosphere near the ground over land surfaces for the last 100 years shows a warming trend over that period amounting to about 1°F. But the warming trend is not smooth; temperatures do not get predictably higher every year (Figure 15.4). From 1880 until 1925, the average temperature increased about three-quarters of a degree Fahrenheit. But between 1925 through the mid 1970s, the increase in temperature leveled off, and maybe even some cooling occurred on the global average. But a definite warming has resumed as of the 1980s.

Simply observing the warming trend does not prove that it is due to human activity. There are a number of considerations in making this connection. One problem is that the global temperature, as well as any other measure of climate, has its own natural variability. In other words, the global mean temperature in one year can easily be 1° warmer or cooler than the previous year with no help from human activity. It is a problem of signal-to-noise ratio. What the recent data indicate is that a signal—a long-term warming trend—is emerging from the noise. But does it necessarily hold that this signal is due to human activity? The suspected culprits are the "greenhouse gases." These compounds have various human sources. In addition to the natural sources, CO_2 is released into the climate system by the burning of coal and other fossil fuels. Other greenhouse gases have their origins in refrigerants and cleaning fluids, for example (see Chapter 14, "The Challenge of Global Climate Change" for further discussion of this topic).

An increase in the CO_2 concentration of the atmosphere has been observed and amounts to about 25% since late in the last century. The concentrations of other greenhouse gases are also known to be increasing (Graedel and Crutzen 1989). But can we say with confidence that their increase is causing the increased temperature? That is difficult at this time. The only thing we can do is to look carefully at the temperature observations and compare them with the climate model predictions. If the magnitude of the observed temperature change, as well as the geographical distribution and the timing of the change, agree with the predictions from experiments with climate models, then we might be able to establish a causal relationship between the greenhouse gas increase and the temperature increase.

A number of climate models have been applied to the problem of the climatic effects of increasing CO_2, and they are probably the most popular application of these models. The model predictions do not agree with each other in many of the details of what a warming forced by CO_2 increases would look like, but they do indicate that the increase in the global mean atmospheric temperature should be more than the currently observed increase given a 25% increase in CO_2. Does this mean that the models and observations are in disagreement? Not necessarily, because these model predictions do not take into account that the climate's response to a CO_2 increase would probably lag behind the atmospheric CO_2 increase itself. In other words, the climate cannot change as fast as the CO_2 concentration is changing.

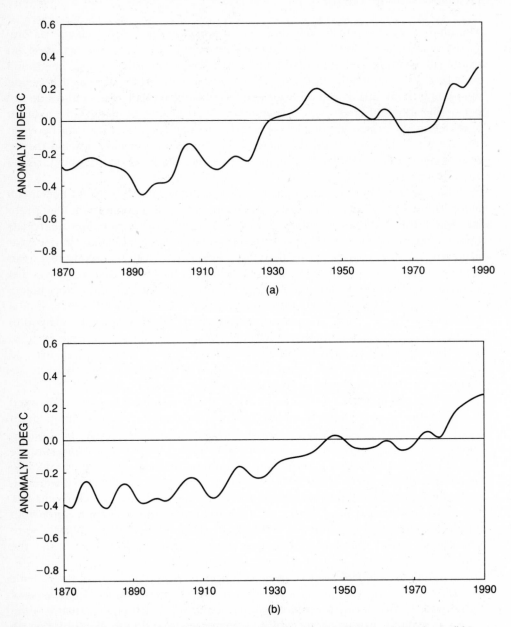

Figure 15.4. Air temperatures over land surfaces for the Northern (a) and Southern (b) hemispheres from one study (Jones, 1988). Values are expressed as differences from the average temperature in the 1951–1980 period. (Taken from IPCC [Intergovernmental Panel on Climate Change]), *Climate Change: the IPCC Scientific Assessment,* J. T. Houghton, G. J. Jenkins, and J. J. Ephraums (eds.), Cambridge: Cambridge University Press, 1990. Original data from: Jones, P. D., "Hemispheric Surface Air Temperature Variations: Recent Trends and an Update to 1987," *Journal of Climate* 1, 1988, pp. 654–660.)

The reason for a lag between the forcing (the CO_2 increase) and the response is due to the ocean subsystem of the climate. You may have noticed that the ocean is colder than the air in the late spring, but warmer than the air in early fall. The ocean has a longer "memory" than the atmosphere, and in the spring the ocean is still "remembering" winter temperatures. The huge mass of the world oceans will slow down the climate system's response to any forcing, CO_2 included. Since the model that made the predictions did not currently take this lagging effect into account, the observed CO_2 and temperature increases are not inconsistent, which means that the connection is neither proved nor disproved (see Chapter 14, "The Challenge of Global Climate Change," Chapter 13, "Stratospheric Ozone Depletion," and Chapter 12, "World Air Pollution: Sources, Impacts, and Solutions," for discussions of other aspects of this topic).

As far as the magnitude and timing of the warming is concerned, the jury is still out on whether the warming is due to human activity. What about the geographical distribution of the change? Does the observed change look like the predicted change? For example, models predict that the CO_2-induced warming will be higher at higher latitudes. One of the reasons for this is the ice albedo–temperature feedback loop. This process does not occur in the tropics and subtropics, of course, because there is no snow to begin with. But at higher latitudes, you might expect an amplification of the warming. So far, this has not been shown to be the case for the observed warming. However, a recent study has shown that differences in cloud optical thickness along the tropic to polar gradient may be causing a more uniform pattern of global warming (Tselioudis et al. 1993).

It is important to note that even if the warming trend continues or accelerates into the next century, we will not be able to say for certain that the warming is being caused by CO_2 increases, although our confidence in statements of this kind could increase as the signal becomes clearer and the models improve. It will be up to the public and our policymakers to be informed on these issues and decide, with input from the scientific community, when it is probable that the CO_2 increase is causing the temperature increase. Beyond that, whether we actually do anything about the trend is a whole other issue. That will depend on whether the warming has more negative impacts on the world's people than reducing the burning of fossil fuels.

Conclusions

As we have seen, all climate problems are not created equal. We can study and explain some kinds of problems with more confidence than others. Some climate problems can be studied, at least to a first approximation, by considering only one or two subsystems of the complete climate system. Climate change has historically been the realm of meteorologists and, more recently, oceanographers. But, as the science of climatology advances, scientists from a broader range of scientific disciplines, including biology, geology, and hydrology, are involved. A global perspective with an interdisciplinary approach is developing rapidly, and we have great hopes of refining our predictions of future climate.

References

Barnola, J. M., D. Raymond, Y. S. Korotkevich, and C. Lorius, "Vostok Ice Core Provides 160,000-Year Record of Atmospheric CO_2," *Nature* 329, 1987, pp. 408–413.

Barron, E. J., "Ancient Climates: Investigation with Climate Models," *Reports on Progress in Physics,* 47, 1984, pp. 1563–1599.

Bell, M. and M. J. Walker, *Late Quaternary Environmental Change,* New York: John Wiley & Sons, 1993.

Berger, A., J. Imbrie, J. Hays, G. Kukla, and B. Saltzman (eds.), *Milankovitch and Climate: Understanding the Response to Climate,* Boston: D. Reidel, 1984.

Broccoli, A. J. and S. Manabe, "The Influence of Continental Ice, Atmospheric CO_2, and Land Albedo on the Climate of the Last Glacial Maximum," *Climate Dynamics* 1, 1987, pp. 87–99.

Clark, W. C., "Managing Planet Earth," *Scientific American* 261 (3), 1989, pp. 46–54.

CLIMAP Project Members, "The Surface of Ice-Age Earth," *Science* 191, 1976, pp. 1131–1137.

Graedel, T. E. and P. J. Crutzen, "The Changing Atmosphere," *Scientific American* 261(3), 1989, pp. 58–68.

Hansen, J. and S. Lebedeff, "Global Surface Temperatures: Update through 1987," *Geophysical Research Letters* 15, 1988, pp. 323–326.

IPCC (Intergovernmental Panel on Climate Change), *Climate Change: the IPCC Scientific Assessment,* J. T. Houghton, G. J. Jenkins, and J. J. Ephraums (eds.), Cambridge: Cambridge University Press, 1990.

IPCC (Intergovernmental Panel on Climate Change), *Climate Change 1992: The Supplementray Report to the IPCC Scientific Assessment,* J. T. Houghton, B. A. Callander, and S. K. Varney (eds.), Cambridge: Cambridge University Press, 1992.

Jones, P. D., "Hemispheric Surface Air Temperature Variations: Recent Trends and an Update to 1987," *Journal of Climate* 1, 1988, pp. 654–660.

Kutzbach, J. E. and P. J. Guetter, "The Influence of Changing Orbital Parameters and Surface Boundary Conditions on Climate Simulations for the Past 18,000 Years," *Journal of Atmospheric Sciences* 43, 1986, pp. 1726–1759.

Lorius, C., J. Jouzel, D. Raynaud, J. Hanien, and H. Letreut, "The Ice-Core Record: Climate Sensitivity and Future Greenhouse Warming," *Nature* 347, 1990, pp. 139–145.

Manabe, S., R. J. Stouffer, M. J. Spelman, and K. Bryan, "Tranient Response of a Coupled Ocean-Atmosphere Model to Gradual Changes of Atmospheric CO_2," *Journal of Climate* 4, 1991, pp. 785–818.

Nicholson, S. E., "Sub-Saharan Rainfall 1981–1984," *Journal of Climate and Applied Meteorology* 24, 1985, pp. 1388–1391.

Schneider, S. H., "The Changing Climate," *Scientific American* 261(3), 1989, pp. 70–79.

Tselioudis, G., A. A. Lacis, D. Rind, and W. B. Rossow, "Potential Effects of Cloud Optical Thickness on Climate Warming," *Nature* 366, 1993, pp. 670–672.

Vinnikov, K. Ya., P. Ya. Groisman, K. M. Lugina, and A. A. Golubev, "Variations in Northern Hemisphere Mean Surface Air Temperature Over 1881–1985," *Meteorology and Hydrology* 1, 1987, pp. 45–53 (in Russian).

Washington, W. M. and C. L. Parkinson, *An Introduction to Three-Dimensional Climate Modeling,* Oxford: Oxford University Press, 1986.

Chapter 16

Disaster Revisited: Bhopal and Chernobyl— What Are the Lessons?

Michael R. Edelstein

T he catastrophic mid-1980s environmental disasters at Bhopal and Chernobyl represent worst-case outcomes of our two most hazardous areas of recent technological enterprise—the chemical and nuclear industries. In neither case was the release of toxic agents deliberate, yet both of these "industrial accidents" resulted from the deliberate application of complex technology. The employment of these technologies entails known but abstract risks that prior to the concrete events at Bhopal and Chernobyl were generally viewed as an acceptable part of modern life.[1]

After describing the Bhopal and Chernobyl disasters in some detail, this article explores some key differences and similarities between these events before posing two questions: "What have we learned from chemical and nuclear disaster?" and "What have we failed to learn from these incidents?"

Overview of the Two Tragedies

The 1984 Bhopal Disaster

The yellow-white cloud of mixed chemicals that blanketed residential areas of Bhopal, India (Figure 16.1) on December 3, 1984 killed thousands and injured hundreds of thousands of panicked victims. It also dramatically underscored the hazards associated with the rapid diffusion of chemical technology across the globe as had no previous incident.

Background to the Bhopal Disaster

In the 1960s, during the Green Revolution, the use of pesticides in India increased dramatically. In 1969, Union Carbide built a plant to blend pesticides in the

[1]For an in-depth analysis of the Bhopal and Chernobyl disasters see Edelstein (1991a,b).

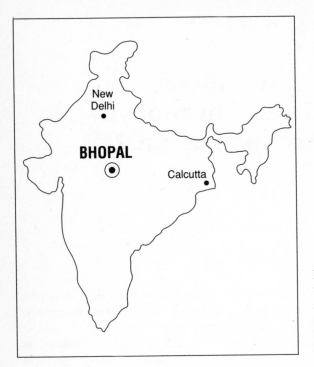

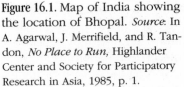

Figure 16.1. Map of India showing the location of Bhopal. *Source*: In A. Agarwal, J. Merrifield, and R. Tandon, *No Place to Run,* Highlander Center and Society for Participatory Research in Asia, 1985, p. 1.

transportation hub of Bhopal, India, the 1,000 year old capital of Madhya Pradesh (Weir 1986). The Bhopal plant was principally owned by UCIL, an Indian Union Carbide subsidiary primarily engaged in battery manufacture (Weir 1986). While local officials opposed this location of the UCIL plant, UCIL had powerful friends in the state of Madhya Pradesh and in India; they employed former government officials at high pay and maintained close relationships to current government officials (Shrivastava 1987). Despite the lack of industrial infrastructure in Bhopal, the plant's location there reflected the general pressure by the Indian government for industrialization (Shrivastava 1987). In fact, the original plant was constructed on five acres of government land rented for under $40 (U.S.) an acre with taxes included (Weir 1986).

In 1974, UCIL was licensed to manufacture carbaryl-based pesticides; production commenced in 1977. With increased competition and a decline in use of pesticides in the 1970s, industries such as UCIL sought to save money by manufacturing rather than importing their component compounds. As a result, UCIL began to make five pesticide components, including MIC (methyl isocyanate), used in the synthesis of Sevin and other carbamates. Because a financial crisis suffered by Indian farmers in the 1980s further reduced demand for Sevin, the Bhopal plant never produced more than half its annual capacity of 5,000 tons of Sevin. By 1984, it was producing less than 1,000 tons (Weir 1986). The loss of profitability made the plant less important to Union Carbide. There was less oversight by the parent corporation, diminished career prospects for employees of UCIL's agricultural products division, lowered employee morale, and the resulting high turnover led to the departure of 80% of the employees trained in MIC manufacture in the United States (Shrivastava 1987).

There was a corresponding decline in plant safety. Instrumentation was poor and easily led to misunderstandings. Employees checked off inspection forms that they could not understand. Operators had inadequate safety training, while managers and plant workers had little information regarding potential hazards. Furthermore, the number of operators manning the MIC unit was cut in half between 1980 and 1984, from 12 to 6 and, significantly, the maintenance supervisor was cut from two shifts. As a result, no one was responsible during those shifts for ensuring that a device was installed so that flushing water did not enter the MIC tank (Shrivastava 1987). Because there was no reliable way to detect MIC, workers were told that eye irritation was the first sign of MIC exposure. This information engendered a belief that MIC was only an eye irritant and not lethal (Weir 1986; Shrivastava 1987; Bogard 1989). Workers learned to detect MIC leaks by the same method used the night of the disaster—namely when their eyes burned and watered. Contingency plans for a major accident were never developed (Ramaseshan 1985). In addition, there were few federal employees dealing with environment—most of whom were concerned with deforestation. Due to cost, there was little pollution monitoring or control. Additionally, there was an absence of community watchdogs (Ramaseshan 1985).

Two-thirds of Bhopal's surging population were unemployed. Besides the neighborhoods that had predated the UCIL factory, squatter settlements were allowed to develop in the plant's shadows. Unsafe housing conditions in the teeming Moslim neighborhoods were tolerated by the Hindu government. Water was polluted by sewage and there was much illness. Modern conveniences, such as telephones, were absent. Ironically, its neighbors saw UCIL as a harmless manufacturer of "medicine" for plants (Alvares 1986; Morehouse and Subramaniam, 1986; Shrivastava 1987).

The Accident

It was shortly after midnight on December 3, 1984 when a control room operator at the Union Carbide pesticide plant noticed that a tank storing MIC was overheating and that the pressure in the tank was rising. He and his co-workers failed to stop the reaction after trying every strategy that they had been trained to consider. In panic, they fled the plant (Shrivastava 1987).

Immediate Causes of the Accident

Although Union Carbide later attempted to blame the accident on sabotage, it is clear that the Bhopal accident resulted from simultaneous failure in design, technical subsystems, safety development, managerial decisions and operating procedures that could not possibly be the result of sabotage. In an immediate sense, the Bhopal accident involved a complex chain of unanticipated interactions among multiple technological failures. At the time of the accident, MIC at UCIL in Bhopal had been stored for more than two months in the only MIC holding tank that was in service at the plant. The tank was in use despite the fact that it lacked positive nitrogen pressure, a condition that allowed "trimmer" (a plastic formed from the chemical reaction of MIC and water) to build up. When late on December 2, 1984, an attempt was made to use water to flush this trimmer out of pipes, the water was inadvertently mixed directly with 20 tons of MIC in the tank. Metallic impurities introduced by operation at inadequate pressure facilitated the reaction of the MIC with the water. Further adding to the reactivity of the MIC was the fact that MIC is unstable when not stored at low temperature; yet the refrigeration unit for the tank had been shut off by UCIL to save energy costs and the alarm thresholds had been reset to tolerate warmer temperatures, making them less sensitive. When workers noticed that water was not escaping from a drain during the flushing, their manager, who had just been transferred to the plant and had no experience with MIC, ordered the flushing to continue, at least until after an upcoming scheduled tea break (Ayers and Rohatgi 1987; Shrivastava 1987; Bogard 1989).

By the time that the control room discovered the problem, there was little that could be done to prevent a release. All major safety systems were inoperative, including the scrubbers and the vent flare apparatus needed to detoxify gas emissions. The 100 foot high vent pipes on the tower above the MIC tank were beyond the reach of emergency sprinklers that were futilely deployed. The spare overflow tank car was full of MIC, not empty. Unmitigated, MIC gas, foam and liquid at 400°F under 180 psi escaped through an open safety valve during the next two hours (Morehouse and Subramaniam 1986; Ayers and Rohatgi 1987; Shrivastava 1987; Bogard 1989). Along with some 54,000 pounds of unreacted MIC, some 26,000 pounds of various reaction products were released, many of them highly toxic. Among these were lighter gases, such as carbon monoxide, that were trapped beneath heavier gases and, thus, unable to disperse. These gases joined MIC as a cause of death and injury (Morehouse and Subramaniam 1986).

Consequences of the Accident

Warning alarms designed to alert nearby residents of problems at the UCIL plant had been shut down at the time of the disaster (Shrivastava 1987). Thus, awareness of the accident that occurred early on December 3, 1984 came directly from the yellow-white cloud of gas that escaped from the UCIL plant and spread out over the sleeping city (Weir 1986). People living nearby were awakened from their sleep "coughing violently, with eyes burning as if chilies had been thrown into them" (Agarwal et al. 1985). Unlike more well-to-do residents warned by telephone to escape and likely to have available transportation, the poor were forced to flee from the toxic cloud on foot (Bhagat 1985). Some 200,000 residents from the area surrounding the plant fled. While some 3,000 people died, overall, as many as 300,000 people were physically injured by the chemical release. Reports suggested that as many as 85% of the nearby residents suffered moderate to severe illness. Nearly 95% suffered respiratory symptoms, 90% eye problems, half had gastrointestinal symptoms, and just under half had neuromuscular symptoms. Some 60,000 were severely disabled and another 40,000 mildly disabled (Morehouse and Subramaniam 1986; Shrivastava 1987). Gynecological illnesses involving shortened menstrual cycles and loss of lactation were widely evidenced. Significant pregnancy problems were encountered—of 2,700 recorded pregnancies, there were 400 reported abortions, 52 stillbirths, 132 deaths shortly after birth, and

30 malformed babies. These health problems had an important significance in a society where the social role of women was so directly tied to their reproductive capacity, threatening to stigmatize female Bhopal victims within the larger community (Shrivastava 1987). Ninety percent of the victims came from the adjacent slum and working class neighborhoods of Chola, Jaiprakash Nagar and Tila Jamalpura (Bhagat 1985).

Beyond persistent physical consequences of the accident, victims of the release suffered a variety of psychosocial effects. Not only were victims put through an intensely stressful event, but many had lost friends and relatives, were displaced from their homes and lost their livelihood, and many thereafter were reduced to beggary. The neighborhoods surrounding the UCIL plant were gripped by continuing panic events, brought on by the fear that another toxic release was occurring. Between December 16 and 19, when the remaining MIC was neutralized, 200,000 residents fled in mass panic. Results of this fear were evident in survivors' emotional states as well.

The accident additionally destabilized the economic well-being of already impoverished survivors. Thousands of primary and secondary jobs were lost with the closing of the UCIL plant. Government food relief sabotaged local grain markets (Shrivastava 1987). Neighborhoods around the Bhopal plant were devastated by the loss of residents who had participated actively in local commercial and social life. In the Jayaprakash Nagar neighborhood alone, more than a quarter of the 7,000 inhabitants died (Morehouse and Subramaniam 1986). Many injured neighborhood men, previously dependent upon physically demanding manual labor, could no longer work. Women faced strenuous household chores such as gathering wood and cooking over smoky stoves despite injuries that affected respiration and their eyes. In many cases, households were deprived of wage earners, requiring the formation of new extended family groupings. The net effect of these changes was a dramatic fall into debt on the part of survivors (Raghunandan 1986; Shrivastava 1987).

These effects were not offset by minimal awards through government assistance, from a legal settlement with Union Carbide or from the courts (Associated Press 1989a, 1990a,c; Morehouse 1994). The Indian government undertook 90% of the relief, rescue and rehabilitation efforts in the wake of the accident, yet 85% of the affected population never received any assistance. Not surprisingly, the more powerful residents received more relief more quickly. For the poor, the very form of relief created a dilemma. Lacking bank accounts in which to deposit government relief checks, the victims were often exploited by local money handlers (Shrivastava 1987).

Accordingly, it is not surprising that the Bhopal disaster resulted in a dramatic loss of public trust in government. Not only had government helped to create the conditions that led to the disaster, but their remedies in the aftermath were ineffectual. Equally disturbing, government failed to adequately provide accurate information in the wake of the accident. For example, residents were incorrectly assured that local water and vegetables were safe to consume. The government minimized the health effects of the accident and denied medicine that may have alleviated some long term consequences (Morehouse and Subramaniam 1986; Shrivastava 1987). Through the Bhopal Gas Disaster Bill, the government became the sole legal representative of victims (Shrivastava 1987). That role led eventually to the $470 million dollar settlement with Union Carbide branded as a sellout by many opponents (Associated Press 1989a; Morehouse 1994).

The 1986 Chernobyl Disaster

On Saturday April 26, 1986, the Chernobyl 4 reactor, located 80 miles north of Kiev, exploded. For 4 days, a variety of fissionable materials were released to the atmosphere. These included the short-lived but dangerous iodine-131, with a half-life of 8 days, and the longer lived cesium-137, with a 30 year half-life (Broadbent 1986; Flavin 1987). The exiled Russian biochemist Medvedev (1990) notes that large amounts of radioactive strontium-90 were also dispersed. Radioactive particles were sent high into the sky by the temperature of the Chernobyl blaze, just missing Kiev and its 2.4 million people, but blanketing large portions of three Soviet Republics (Figure 16.2). This plume of radioactivity was then blown by the wind into Scandinavia, where it was capped by a temperature inversion. The air plume was later carried to the south over Central and Eastern Europe. In all, 20 countries, some over more than 2,000 kilometers away received radiation from the accident (Figure 16.3), spreading west to the British Isles, north to the Arctic, and south to Greece (Broadbent 1986; Flavin 1987).

Compared with the loss of 15 curies of radiation at the 1979 American Three Mile Island accident, recent estimates for Chernobyl suggest that about 185 million curies were released (Travis 1994). The Chernobyl accident has caused more than 30 immediate deaths, left more than 200 acute radiation victims, contaminated

Figure 16.2. Contaminated areas in the vicinity of Chernobyl. The shaded regions have contamination levels of 1–5 curies of radiation per km². *Source*: Eijgenraam, F., "Chernobyl's Cloud: A Lighter Shade of Gray," *Science* 252, 1991.

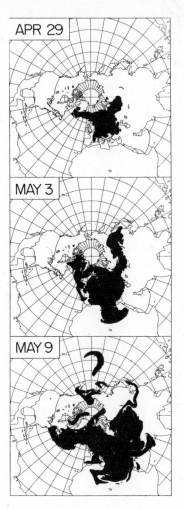

Figure 16.3. Distribution of cesium-137 released from Chernobyl as of the indicated date in 1986. The shaded areas indicate calculated airborne concentrations greater than 3 fCi/m³, based on estimates of the source strength. *Source:* Davidson, C. I. et al., "Radioactive Cesium From the Chernobyl Accident in the Greenland Ice Sheet," *Science* 237, 1987.

thousands of square kilometers and resulted in the temporary burial of some 500,000 m³ of radioactive wastes (Schroeder 1990). Long-term cancer risk throughout north and west Europe has been enhanced (Reason 1987). Disruption across a wide swath of the globe also includes economic, social, and political impacts.

The Causes of the Accident

The Chernobyl 4 reactor opened in 1984, as did the other three 1,000 megawatt reactors at the nuclear complex. Chernobyl 4 was developed from a 1954 design that lacked complex automatic emergency devices. Operation procedures for the reactor recognized that there was a major design flaw—allowing a runaway nuclear reaction at low power settings—so that operation below 20% capacity was forbidden. To improve the safety of plant operation, experiments were planned to test a device to keep the emergency core cooling system of the reactor operating in case of an off-site power failure. These experiments required that the reactor operate at reduced power for repeated trials (Reason 1987). At midnight on Friday, 25 April 1986, the plant was on its way downward toward a 25% capacity target. When the operator undershot the setting, however, power dropped to 1% of capacity, eventually being stabilized at only 7%. Thus, while the tests were supposed to occur above the reactor's danger zone, in fact, they proceeded well within it (Reason 1987).

This was just the first of the many steps leading to the accident, however. To simulate the emergency core cooling system, which was shut off, and to cool the reactor during the experiment, eight pumps were started, even though protocols called for no more than six to be used at one time. The presence of additional water meant that too many neutrons were absorbed to sustain power; to offset the problem, additional control rods were withdrawn. The plant supervisor decided to continue the experiment with only six control rods inserted, half the number called for by safety requirements. This problem was exacerbated when the operators further increased water flow, and then had to override automatic shutdown devices to keep the plant operating. When it was finally realized that the plant was overheating, an attempt was made to "scram" the reactor, to drop all the control rods in order to stop the reaction. By then the excessive heat had so warped the rods that they jammed. Chernobyl's fate was sealed (Reason 1987).

At 24 minutes after 1 a.m. on April 26, 1986, May Day was prematurely greeted with a spectacular fireworks display as two explosions occurred back-to-back at Chernobyl 4, ripping off the 1000-ton roof and starting 30 fires in the area (Figure 16.4). While exterior fires were put out during the first day, the graphite core continued to burn for several days (Reason 1987). Hundreds of fire fighters and nuclear workers fought the five-story-high blaze that threatened to engulf a neighboring nuclear reactor, as well (Associated Press 1987b). After an initial explosion of radioactive material from the plant, emissions dropped dramatically, but then built back up as residual fission products decayed. By the ninth day of the accident, the daily release of radioactive material nearly equaled the initial release (Anspaugh et al. 1988). Essentially, what happened was that the initial effort to smother the fire under 5,000 tons of sand, clay, lead and dolomite contributed to a new build up of heat that caused a second meltdown. In the end, 1,000 kg of plutonium, as well as other contaminated materials, were buried in a 32-foot deep trench dug around the plant and encased in waterproof concrete (Figure 16.4). This containment, however, failed to meet international regulations requiring safe disposal for 10,000 years, in actuality lasting only a few years (Medvedev 1990).

Impacts Within the Former Soviet Union
The worst fallout from the accident fell on 10,000 km² of the former Soviet Union: 70% in Byelorussia, now Belaras, (with nearly 2½ million people affected) and the remainder divided between the Russian Republic (with 700,000 people affected) and in the Ukraine (with more than 4 million people affected) (Schroeder 1990). It is estimated that at least 200,000 Ukrainians and Byelorussians and thousands of Russian cleanup workers were exposed to significant amounts of radiation, along with the cost of 1,000 immediate injuries, 31 immediate deaths and some 250 long term deaths as of Spring, 1990 (Barringer, 1986;

(a)

(b)

Figure 16.4. Unit 4 of the Chernobyl nuclear power plant: (a) after two explosions and a fire; (b) encased in reinforced concrete in May 1989. *Source:* Medvedev, Z. A., *The Legacy of Chernobyl,* New York: W.W. Norton, 1990. (From ITAR-TASS/SOVFOTO.)

Marshall, 1986; Associated Press 1990a).[2] Childhood leukemia was reported in Pripyat, Chernobyl, and Kiev (in Russia and the Ukraine) and lip, mouth, and other cancers in Byelorussia. Malformed cattle were observed in areas affected by contamination. A Russian woman may have spoken the collective mind when she asked, "My daughter recently got married. What kind of grandson will I have?" (Associated Press 1989c). By 1992, the area north of the reactor yielded a rate of thyroid cancer in children under age 15 that was 80 times the normal rate (Associated Press 1992b). Ukrainian doctors have identified an immune deficiency disorder that they term *Chernobyl AIDS* (Carothers 1991).

[2]Note that one well placed observer, Vladimir Chernousenko, director of the 18-mile exclusion zone surrounding the reactor complex, has claimed that there were 10,000 additional short term deaths due to Chernobyl (Associated Press 1991). Another source claims that 3,166 residents of the 18-mile zone surrounding the Chernobyl complex died in 1990 alone (Scripps Howard 1992a).

A Church World Service team visited extensively with medical personnel in Byelorussia and Moscow in July 1990. They reported that in the short run, the principal radiation-related disease evident was thyroid pathology due to iodine–131 fallout. This symptom of radioactive exposure could have been avoided had the government distributed iodine immediately instead of initially covering up the accident. They further reported discoveries of cardiovascular disease and weakened immune systems. Spontaneous abortions are said to have doubled. Voluntary abortions have risen due to malformed fetuses. Of particular concern are genetic effects. Children born in 1987–88 have been found to have more problems than those born in 1986. The birth rate has dropped because many young couples have left the area and those remaining are afraid to bear children. A particularly vulnerable group is the 600,000 liquidation workers, many of them young army conscripts ordered to work at Chernobyl and now dispersed to their homes (Schroeder 1990).

While it is not known how much radiation exposure accumulated, the impacts of the accident are clearly not reflected in the modest early death toll. Unlike the much studied external radiation exposures of Japanese in Nagasaki and Hiroshima, exposed Russians have received continuing internal exposures from eating cesium-contaminated food, from drinking cesium-contaminated milk and water, and from living in contaminated buildings and on contaminated land. As a result, the real health effects will only be known in the long run, reflecting not just long-latency disease, but also the fact that the exposures themselves are continuing. Physicians are monitoring 160,000 children exposed to more than 30 rem of radiation. Overall, as much as 70% of the work of medical teams in the impacted area is given to the effects of Chernobyl, with only the remaining 30% directed to normal medical care (Schroeder 1990). Projections by American scientists range from an estimate that between 1,000 and 7,500 new cancers will develop over the next 50 years and that a total number of between 50,000 and 250,000 Chernobel-related deaths will occur within the former Soviet Union (Barringer 1991).[3]

The Chernobyl accident caused extensive contamination to the land and water, with serious resulting dilemmas. The 10-km radius around Chernobyl has been made into an ecological research reserve, with the surrounding 30 km, known as "the zone," abandoned. Meanwhile, a concrete-lined trench some 32 meters deep was dug around the Chernobyl plant to hold contaminated organic matter, such as topsoil, leaves, and manure, as well as water (Schroeder 1990). An additional 800 shallow burial pits around the reactor also hold waste materials. These sites allow for radioactive dust to escape and are expected to leach contamination to the groundwater (The Economist 1991; New York Times 1991b). Perhaps the most serious concern, however, is that the 300,000-ton sarcophagus hastily built to contain the remains of the reactor is no longer airtight, is leaking and is structurally unsafe and requires replacement (The Economist 1991; Stone 1992).

[3]One dissenting set of findings came from a 1990 review by international scientists. They found considerably lower exposure to radiation that had been calculated by Soviet scientists. Furthermore, the actual levels, they concluded, often did not justify evacuation of residents because exposures were high enough that remaining in the contaminated area was not expected to account for any appreciable additional risk (Eijgenraam 1991; International Advisory Committee 1991).

Over a wide swath of the three republics, farmland cannot be used for a minimum of 30 years, and in some areas it may be contaminated for millennia with strontium-90 and plutonium. Fearing the spread of radioactivity through erosion, proposals have been made to plant trees or grass on the fields. However, such steps have been blocked by a concern that fires might spread contamination bioaccumulated in the plants. A massive portion of the marshland in Pripyat, known as the "Polesye" swamp, is contaminated. Representing the largest swamp in Europe, the Polesye is particularly likely to retain radioactivity over time. Radioactivity washing into the bottom sediments of area waterways threatens important fisheries. In nearly 800 inhabited areas, wells had to be sealed, roofs replaced, buildings washed, soil removed, dust suppressed and contaminated areas covered by asphalt, wood chips, or new earth. The city of Kiev required a new water supply (Schroeder 1990; The Economist 1991; Scripps Howard 1992b). Five years after the accident, Chernobyl officials were continuing to refrigerate 30 railroad cars full of contaminated beef for which no disposal site had been identified (New York Times 1991b). The load of beef had been rejected with much rancor in Georgia and had become somewhat the equivalent of the Long Island garbage barge that could find no welcome.

Many residents have continued to live in areas that they no longer trust to be safe. A visitor from Kiev told the author in early 1993 that local fish could not be consumed, water quality was uncertain, whole areas of the city were "hot," and that leaking radioactivity from the poorly contained Chernobyl 4 plant still wafted unpredictably over Kiev. Indeed, many residents have been left to live in areas that are contaminated, either because testing for radioactivity has been so slow and inadequate or because no relocation sites were available. At Gomel, for example, an inhabited area was discovered to be contaminated 3 years after the accident. And in Byelorussia, some 175,000 people, a third of them children, live in virtual quarantine on the 7,000 km² most heavily contaminated by the accident (Schroeder 1990).

Such continuing exposures from the environment magnify the effects of the original exposures at Chernobyl, as the Director of the Institute of Nuclear Medicine near Minsk explained:

> The people don't have a normal diet because they can't eat the food they produce. The majority of the people in this region used to be dependent on forest products such as mushrooms and berries which they would gather in the forest. No longer can they either eat or sell their products. How can a farmer understand the danger of milking his cow? Contamination goes from the soil to the grass, to animals, and to humans. It creates the paradox that internal dosages are often higher than the level of radiation in areas of rigid control. Only now, the authorities are trying to deliver "clean" foodstuffs to the people.

The official described children whose contamination was entirely caused by their swimming in contaminated water and playing in fields where there were ever increasing levels of radiation (Schroeder 1990).

Relocation of victims has proven to be a major social consequence of the Chernobyl accident. Beginning 2 days after the accident, some 135,000 people were evacuated in 1,100 buses from within an 11-mile-radius zone around the Chernobyl plant. The accident forced the building of new villages to house the dislocated, who may never be able to return home. The Soviet town of Pripyat and surrounding farmland may have to be abandoned for decades. Nearby forests must be razed and topsoil removed and buried. The area of contamination has increased in the 4 years after the accident, as radiation was carried by ground and surface water, and as winds spread radioactive dust and smoke from forest fires. With 135,000 people already relocated from their homes by 1990, Soviet officials expected that another 200,000 would require relocation, some living as far as 200 miles from Chernobyl. Some officials and scientists predicted that 4 million people would need to be re-settled in all (Associated Press 1988; Scripps Howard News Service 1990).

While billions were spent to construct new housing for the relocated, some residents were returning to their homes and farms by avoiding checkpoints (Associated Press 1988; Scripps Howard News Service 1990). A visiting Church World Service team in 1990 witnessed Byelorussian women tending gardens in otherwise abandoned villages. Officials explained that some farmers chose to remain on their land even if it meant dying from the radiation. The team was informed that 1,500 people refused to leave the 30 mile zone around Chernobyl itself. This attachment to place was augmented by the difficulties encountered in newly created settlements. The disruption of family, job, and place involved in resettlement has rarely been remedied through relocation (Schroeder 1990).

The Church World Service team heard reports of increased alcoholism and psychological disturbance among Chernobyl victims. They identified increasing stress-related illness. Suicide has been such a common response that the Orthodox church has been forced to assist the victims' families by giving dispensation for burial in church graveyards. Particularly vulnerable have been the 600,000 soldiers and workers who have participated in the liquidation efforts. Also affected psychologically are children (Schroeder 1990). Byelorussian researchers have confirmed the increase of stress-related disorders; 38% of the subjects in one study were affected by depression, illusions, or the constant intrusion of radiation into daily conversations (Rich 1991).

An independent review by non-Soviet scientists participating in the International Chernobyl Project concluded that these high levels of psychological stress exceeded the actual risk from the accident. Such stress was not viewed as irrational "radiophobia," as many Soviet observers had dismissively claimed, but was instead considered to be a legitimate secondary impact of the accident blamed on confusion due to government handling of information, on disruptive relocation efforts, on politicians who exaggerated the issue, and the fact that so many people were exposed to unknown amounts of radiation (International Advisory Committee, 1990; Eijgenraam 1991; Ginzburg and Reis 1991).

A final psychosocial impact of the Chernobyl accident was a fundamental loss of trust in others, particularly in government. Chernobyl victims have been stigmatized and shunned by others fearing the transmission of contamination. But, more seriously, they have confronted the realization that their government denied them timely and truthful information and effective help. People were allowed to continue eating contaminated meats and produce, some of which was even sold in state stores (Medvedev 1990). People were allowed to live in areas exceeding international radiation guidelines (Schroeder 1990). The Chernobyl complex was declared safe at the end of emergency action and other reactors at the complex were allowed to continue operation (Associated Press 1987b). The total result of government intervention was what one observer termed "the fatal triad—incompetence, secrecy and bureaucratic squabbling" (The Economist 1991, p. 20). Beyond government, distrust extended to physicians who, lacking specialized expertise, adequate equipment, and medicine, are unable to provide certainty in the midst of ambiguity. Impressively, 399 of 400 patients at the Center for Radiation Studies in Kiev reported not believing what their doctors told them (The Economist 1991).

The legacy of distrust and inaction underscored the Chernobyl accident as a failure of the Soviet system. Chernobyl was the central catalyst for political and economic reform within the Soviet Union, beginning with the move toward openness or *glasnost* begun with Gorbachev's pronouncements of the people's right to know on the 18th day after the accident (Medvedev 1990; The Economist 1991). The Soviet government even went so far as to admit gross failures in the management of the disaster and to state an intent to pursue criminal charges against former officials (New York Times 1991b). The inadequate response of the central government forced the affected republics into taking a more critical and independent course. Rukh, the Ukrainian independence movement, was formed in the wake of the accident (The Economist 1991). The Ukrainian government vowed to close the entire Chernobyl complex by 1993 (New York Times 1991a). Periods of democratic experimentation and also ethnic violence followed. The republics were moved to succession from the Soviet Union, in no small part by the aftermath of Chernobyl (Barringer 1991).

Indeed, the past and continuing costs of the Chernobyl accident have been staggering. The accident entailed financial impacts on an unprecedented scale. Direct financial losses from the accident were estimated at between $3 and $5 billion. By 1990, the cleanup of Chernobyl had already cost $13 billion and was expected to cost another $26 billion (Flavin 1987; World Watch 1990). Medvedev (1990) estimates that the costs from this one accident have exceeded the Soviet Union's entire investment in nuclear power. Carothers (1991) cites the projection of the former head of the Supreme Soviet's Environmental Committee that the accident will eventually cost $400 billion. The reality of the breakup of the Soviet Union, however, has meant that the new independent Republics have little energy and few resources to deal with the impacts of Chernobyl. Ironically, even less aid is available now to ensure relocation, safe environments and needed medical care than was available in the short run aftermath of the disaster (Barringer 1991; Schroeder, personal communication, 1993).

Impacts Outside the Former Soviet Union

Radioactive fallout from Chernobyl blanketed Europe and the entire Northern Hemisphere, as well as the former Soviet Union, with a mixture of some fifty radioactive isotopes. Some 1,500 deaths are expected over the next 30 years due to exposure to short-lived iodine-131 resulting principally from milk consumption. The real long term health threat, however, is due to the more persistent cesium-137

(Marshall 1986; Flavin 1987; Goldman 1987). A comprehensive study done by the U.S. Department of Energy estimated the health consequences from cesium-137 by modeling total global fallout from Chernobyl. As a result, it was estimated that while 1 million curies of cesium-137 fell in the USSR, another million blanketed the rest of Europe, and still another million fell throughout the Northern Hemisphere. The total fallout was seen to be only slightly less than that from all previous atmospheric weapons tests combined. Using an estimate of 2.3 fatal cancers per 10,000 person-rem of radiation exposure, the study estimated that there will be 39,000 extra cancer deaths due to the accident, most outside the former Soviet Union (Marshall 1986, 1987; Goldman 1987).[4] Not surprisingly, there has been considerable international concern over the health effects from the incident. It is estimated that some 100 million people may have changed their diets in the wake of the accident in order to avoid contaminated foods (Flavin 1987). Another indicator of concern involved the abortion of fetuses conceived at the time of or after the accident. For example, the Associated Press (1987a) reported estimates by Greek doctors that 2,500 additional abortions were performed in 1986 as the result of the Chernobyl accident. This amounts to almost one quarter of Greek pregnancies during that time. In the same vein, the International Atomic Energy Agency estimated that between 100,000 and 200,000 abortions occurred in all of Western Europe in the wake of the accident.

Several countries tallied their costs from the Chernobyl accident for submission to the Soviet Union for reimbursement. Particularly hard hit were Eastern European countries that lost hard currency from the loss of crop exports (Flavin 1987). Czechoslovakia placed severe restrictions on consumption of contaminated milk; Poland—the most severely contaminated country outside the Soviet Union—and Hungary restricted both cattle grazing and milk consumption. Poland also distributed iodine tablets to millions of children. Hungary was the only socialist country to reimburse farmers for losses, paying out some $10 million. Meanwhile, in the West the Austrian government destroyed large amounts of vegetables from the Salzburg and Innsbruck areas, compensating farmers some $80 million. West Germany destroyed vegetables, grass, and hay in the southern part of the country, paying farmers about $100 million in compensation. Large amounts of vegetables were also destroyed in France, and milk and vegetable sales were restricted in northern Italy. The United Kingdom banned sales of 3 million lambs and sheep, paying some 5 million pounds as compensation to herders. Massive amounts of hay, milk and reindeer meat were discarded in Sweden (Schroeder 1990).

Additional costs of the Chernobyl accident have resulted from the stigma associated with products from areas where fallout was extensive. Grain products and other items from Russia, Scandinavia and Western Europe came to be identified as dangerous. Exemplifying such stigma was a controversy within Mexico over the purchase by the government of thousands of tons of contaminated milk imported from Ireland. The government claimed that the milk had been de-

---

[4]Barringer (1991) cites much higher estimates: between 50,000 and 250,000 deaths within the former Soviet Union and a similar loss in other countries.

stroyed, whereas the environmental group responding to the issue charged that the milk was distributed under a subsidy program to the poor (Cox News Service 1988). In a similar vein, a major controversy occurred in Africa over the belief that contaminated foods were being imported from Europe. In another instance, the West German government stopped shipments of highly contaminated powdered milk from leaving port for Egypt and Angola. The Angolan shipment was supposedly being sent in response to requests for hunger assistance. The African press depicted such alleged importation of foodstuffs contaminated by Chernobyl as another "killer waste" sent from the industrialized nations. A Nigerian paper charged that (Brooke 1988) "The Chernobyl nuclear facility disaster happened thousands of miles away from Africa. But Europe is making sure that Africans will make up the bulk of victims."

Perhaps the region outside of the Soviet Union that was most heavily impacted by the Chernobyl accident was the tundra area of northern and western Europe, particularly the Swedish provinces of Vasterbotten and northern Jamtland. The contaminated areas there encompassed some 55,400 km^2 and a human population of 250,000. The local economy was based upon reindeer herding, as well as fishing, hunting and berry picking (Broadbent 1986). This entire food chain was disrupted due to radioisotopes released from the Chernobyl accident. Central Norway and Sweden were particularly affected by cesium-137, with its 30-year half-life. The cesium-137 was taken up by the reindeer's primary food, lichen, which acts as a sponge for radiation because it gets nutrients from air rather than soil. As a result, while there are disputes over how long it will take before the lichen can safely be consumed by reindeer and the level of contamination at which reindeer is safe for human consumption, there is agreement that much of the indigenous food in this part of the tundra area will not be edible for at least an entire generation (Stephens 1987).

The Lapps (or Sami), an indigenous people, have herded reindeer in the tundra since the last ice age (Broadbent 1986). A typical family would consume between 10 and 20 reindeer per year. Throughout the northern area where the Sami live, the traditional separation of the herds has been a key social and economic event for centuries. The Sami culture was so dependent upon reindeer that personal identity often correlated with the herd—a diminished herd meant a loss of one's own substance (Stephens 1987).

In the wake of the Chernobyl accident, a gradual increase of one or two hundred cancer deaths is predicted for the Sami over 40 years. Indirect health effects due to contamination of food have attracted even more attention. Much of the impact stemmed from the effect on reindeer due to consumption of contaminated lichen, but there were other impacts on people of the north. As a result of the accident, herders were forced to keep their milk calves indoors for a month and feed them uncontaminated fodder, a difficult commodity to acquire. Then, within two months of the accident, the government warned against the consumption of reindeer meat, fearing the bioaccumulation of radioactivity in the meat. Indirectly, this warning also affected the sale of meat from animals slaughtered prior to the accident because it caused the collapse of markets for reindeer products. Then, a month later, mass slaughter of the reindeer was ordered; carcasses were buried in pits (Broadbent 1986). Fish were also quickly contaminated, partially because the food chains in northern lakes are simple, with rapid trophic turnover. Within a period of weeks, warnings were escalated from limiting consumption to once or twice per week to twice a year. People whose diets heavily depended on fish were forced to buy food for the first time in their lives. Similar limitations occurred with berry picking and elk hunting. Thousands of tourists cancelled their visits to the area, in part spurred by erroneous press reports of thousands of deaths in Sweden (Broadbent 1986).

In these ways, Sami families became dependent upon purchased food. Although governments paid for the lost meat, the cultural significance of losing the herd could not be mitigated. The Sami identity and sense of well-being was damaged. Parents worried that their children would lose the will to become reindeer herders. This concern was expressed at a meeting of Sami leaders (Broadbent 1986, p. 237):

> Reindeer and the Sami form a culture. The reindeer herding family is the bearer of this culture. The reindeer can be replaced but it is very difficult to replace a family which has left reindeer herding. Can one replace a lost symbol?

Similarly, a Sami woman explained (Stephens 1987, p. 38):

> This is not just a matter of economics, but of who we are, how we live, how we are connected to our reindeer and each other. Now we must buy everything. Thread, material, food, shoes are now all different things, when they used to be part of one thing. . . . Even if there comes a time when we can eat the reindeer again, it may be too late to pass the knowledge of how to take and use niestti [meat from one's herd] on to our children.

Human stress resulting from the herd impacts included anguish over the devaluing of the reindeer, the intangible nature of the accident and its effects, and a resulting sense of helplessness. For many, their trust in experts was shaken by conflicting warnings about the safety of consuming reindeer meat. Some Sami ignored the restrictions and continued to consume reindeer (Broadbent 1986). Others fearing contamination reported worry about health effects. Sami spoke of their concern for their children and unborn generations. Some women had precautionary abortions. Their safe environment had changed fundamentally, as one Sami reported (Stephens 1987, p. 38):

> It seems sometimes that things have become strange and make-believe. You see with your eyes the same mountains and lakes, the same herds, but you know that there is something dangerous, something invisible, that can harm your children, that you can't see or touch or smell.

Current estimates of the duration of the problem suggest that in 25 years there will still be significant contamination present in Lapland. The question is how to keep Sami children living in the tundra until reindeer herding becomes once again possible? The problem is not limited to a small area either. Perhaps 80% of Swedish reindeer were affected. Furthermore, in the wake of the accident, the larger socio-political environment promises to exert influences on the Sami that may be indirectly as disruptive as the radiation from Chernobyl was directly. Sweden, currently 50% dependent upon nuclear power, has decided to phase out its 12 reactors by 2010, a decision hastened by the Chernobyl disaster. A strong emphasis upon alternative energy had already made hydropower central to the nation. Three quarters of the remaining hydro capacity is in the north. Plans for an additional 100 hydro-electric plants are in the works. The result is likely to be a further disruption of native economy (Broadbent 1986). Additionally, in the wake of Chernobyl, there is increasing pressure by development interests to move into the north, already impacted by large NATO bases (Stephens 1987).

It will remain to be seen whether Sami culture can persist under these circumstances. Beyond the direct impacts, secondary impacts on the economy and on the development of hydropower promise to further test the region.

Conclusions

The Chernobyl accident illustrated the potential for nuclear accidents to have major cross-boundary impacts. Since the inception of the nuclear enterprise, the "fallout" from this technology has been evident. But it took the Chernobyl accident to demonstrate with certitude that the "peaceful" atom could still extract a frightful price.

Some Key Differences and Similarities Between Bhopal and Chernobyl

The Scope of Impact

The physical impacts of Bhopal and Chernobyl were different in scope. While its secondary impacts rippled across the globe, Bhopal was a proximate disaster. Its direct physical effects were on the adjacent populations and environment. Bhopal can be considered a global disaster only in the sense of the role of a multinational corporation, and of the potential for parallel localized disaster because of the broad diffusion of chemical technology that alarmed people worldwide. In contrast, Chernobyl had important distant physical effects. While those immediately around the plant were acutely impacted, radioactive poisons were released to regional, super-regional, and cross-global atmospheric paths of circulation. As a result, its direct impact was global in scope. Although comparatively restricted and controlled, nuclear threat has also been widely dispersed through civilian and military uses. In short, beyond the local level, chemical risk has been spread by diffusion of chemically based technologies, whereas radioactive risk is spread by transport, winds and waters, as well as by technological diffusion.

Onset of Impact and its Invisibility

Environmental hazards, such as those unleashed during the acute toxic disasters at Bhopal and Chernobyl, often work in invisible ways that makes definitive linkage of cause and effect impossible (Vyner 1987). The immediate health impacts of Bhopal upon its victims left little doubt about their cause. In contrast, with the exception of the 31 near term deaths, much of the health effect from Chernobyl is delayed and, therefore, of more easily obscured causality. Charges of "radiophobia" against Chernobyl victims reflect the ease with which doubt can be raised over latent health effects.

Duration of Threat

While the principal external threat in Bhopal existed only so long as the cloud of gas swept across the community, the duration of threat from Chernobyl persists due to environmental rebound—once altered, the environment became an agent of continuing contamination of considerable but unknown duration—as well as from the continued chronic leakage from the Chernobyl site itself. It took days for the acute event at Chernobyl to end, but the duration of the toxic release will persist for generations.

The threat of continued internal threat, however, exists for victims of both disasters. Victims face the prospect that their bodies were changed in permanent and evolving ways. Beyond impacts that were immediately evident, as with the widespread incidence of blindness among survivors of the Bhopal incident, long latency diseases, such as cancers, were unleashed, as were consequences that will emerge only in the next generation: teratogenic effects, impacts on in-utero children, and mutagenic impacts, chromosomal changes.

Responsibility for the Accident

Who was at fault for the Bhopal and Chernobyl accidents? The issue of accountability, responsibility and blame is complex in the wake of major environmental disasters. But the options are somewhat clarified when we make several distinctions. First, are we blaming people or technology? Second, are we allocating responsibility for causing the technological failure that precipitated the disaster or for the failure to remedy the consequences of the disaster once it has been unleashed?

Blaming People

After Bhopal, Union Carbide sought to assume responsibility while shifting blame elsewhere. Even as the corporation's president Warren Anderson flew personally to India, the corporation pointed the finger of blame at UCIL, its Indian subsidiary. Anderson's action underscored the diffusion of responsibility inherent in multinational corporate arrangements. Meanwhile, UCIL's attempt to blame worker sabotage is in itself instructive, both underscoring the potential security problems found with such facilities and the ease of using sabotage as an excuse. The Bhopal accident involved a certain degree of government responsibility as well, given questions about how UCIL came to be sited and about its allegedly poor regulatory monitoring. The Indian government took a strong posture of accepting responsibility while it sought to shift blame to Union Carbide, therefore employing the same tactic adopted by Union Carbide itself. Given their comparable positions in this matrix of blame, it may not be surprising that India and Union Carbide were able to settle the matter in a manner that accomplished maximum damage control at the eventual expense of the actual victims.

Meanwhile, the Bhopal accident contributed to yet another level of blame. One excuse widely offered for the Bhopal accident was the supposed inability of Third World workers to safely operate complex modern technological facilities. In this view, which combines blaming of the victim with western chauvinism, the Bhopal accident was less a failure of technology or of multinational corporate structures than of the presumably technically illiterate workers employed at the plant. The "Third World blame" tactic serves a further aim of suggesting that such accidents are unlikely to occur in First World settings, helping to diffuse the claims of those blaming industry and taking Bhopal as a portent of future disaster.

Yet, the Chernobyl accident involved workers from the modern world who similarly were unable to control their industry. Much as with Bhopal, the Chernobyl accident also points to the social component of technological systems as the root of major catastrophe. There are major questions about human decisions leading up to the accident. The tests involved a collaboration between the plant operators, who had little experience with non-normal operation of the reactor, and outside electrical engineers who knew nothing about reactors. Not only did operators violate rules and miscalculate the reactor's response at every step of the experiment, but formal central approval for the tests was never received, leaving plant management responsible (Reason 1987).

Even the chemical industry in the developed world is hardly free of serious accidents, even if none were so catastrophic as Bhopal. Witness, for example, the Seveso, Italy dioxin contamination that resulted from a 1976 accident at a Hoffmann—La Roche facility (Edelstein 1991a). If accidents such as this can happen in industrial nations, one must look to the technology and to the manner of employment for an explanation to the accident, not to some simplistic difference between highly developed and less developed nations.

Blaming Technology

In the wake of the respective accidents, much attention was directed to technology as the cause of the accident. It was suggested that the UCIL plant was not built to the specifications of an American chemical factory. Particular attention was paid to the ways that the Chernobyl plant differed from the American "light water" nuclear reactors. Design features were blamed, notably the reactor's graphite core and lack of full containment. But the blame of technology for these disasters runs even deeper than the claim that the technology was inadequate. What if such disasters are inherent risks of modern technology?

Bhopal and Chernobyl (and for that matter, Seveso and Three Mile Island) involved the simultaneous failure of multiple facets of complex technological systems. While there are significant differences between chemical and nuclear industries, the similarities between these technological systems may be even more salient. Both were tightly coupled systems, in that an error in one component offered the possibility of causing complex and hard-to-anticipate effects on other subsystems. In other words, there was relatively little margin of error. For example, the Chernobyl reactor operators had only a short time, minutes at most, to correct the system once the accident began. In contrast, the operators of the American Three Mile Island (TMI) #2 reactor had hours to stop an accident during the near meltdown of the plant in 1979 (Ahearn 1987). Challenging this minimal tolerance for error were what Perrow (1984) calls normal accidents, the small errors and failed or unavailable mitigations that occur routinely and are bound to happen, but that can combine and interact to produce something decidedly abnormal. Operator error was grafted onto technological and organizational errors to produce a disastrous synergy. Despite elaborate control systems, both accidents evolved from physical dynamics that were only controllable under normal conditions. Under hybrid and uncertain circumstances, they were beyond control. Yet such abnormal circumstances are actually probable occurrences given the combination of tight coupling with normal accidents.

Complacency and the Blind Belief in Technical Infallibility

Neither of these accidents was supposed to happen. They were unintended. Quite to the contrary, the respective technologies were surrounded by a belief in their safety, a confidence that nothing too dire could occur. Assurances of safety were presented in the form of mitigations and safeguards intended to offset any hazard. But, as Bogard (1989) argues, mitigations are merely a relabeling of hazards, a linguistic packaging that makes risk acceptable. In this view, a safety system is merely an acknowledgment that a hazard exists combined with a denial that it will occur. One must believe that such mitigations will work in order to accept the inherent risk of the technology. Such beliefs become in themselves a hazard, as they contribute to a confidence in the plant's infallibility that grows over time. As a result, workers become slack in upkeeping the mitigations, unleashing the hazards inherent in the perceived infallible facility. At Bhopal, numerous mitigative technologies existed as ideas, but few as reality.

At Chernobyl, the mitigative rules were perhaps well known, but deliberately violated. Reason (1987) similarly notes that Chernobyl and the American Three Mile Island nuclear accidents were similar in that operators thoroughly believed that an accident of such magnitude could not possibly happen to them. Their blind confidence invited complacency. In both cases, the operators debilitated safety systems deliberately and evidenced insufficient understanding of how their plants operated. In both cases, as at Bhopal, the accident that could not happen happened!

The Import of Psychosocial Impact and Environmental Stigma

Together Bhopal and Chernobyl changed the lives of millions of people for the worse, affecting individuals, family, neighborhood, community, government, and the entire society (Edelstein 1988). The Indian disaster claimed by far the larger number of immediate lives and injuries, with Chernobyl promising to eventually more than catch up. But these disasters also establish the importance of psychosocial impact that stems both from the physical consequences of toxic disaster, as well as from the social effort to remedy the disaster. In both disasters, physical injury has taken people formerly in good health and caused them to see themselves as unhealthy victims. While surviving Bhopalese were mostly "free" to return to their homes in the wake of the accident (they had little choice, having no place else to go), Chernobyl has forced the possibly permanent relocation of hundreds of thousands of people. In Bhopal, there are reports of people afraid to return to their homes; at Chernobyl, some fled while others have refused to leave their homes and communities. In both cases, the bonds of home, community, and place that may have existed prior to the accident have been fundamentally altered if not destroyed.

A related impact at Bhopal and Chernobyl was the occurrence of "environmental stigma," or the devaluing of people, places, and products associated with the hazard. Chernobyl tainted the physical environment in the persistent manner frequently found as well with toxic chemical releases. At Bhopal, even if no residues of MIC remained in the environment, it was still indelibly tainted by the accident in the minds of residents. After Chernobyl, an invisible maze of radioactivity persisted, giving a concrete reason to fear the environment, as well. When the environment thus comes to be viewed as harboring invisible dangers, it ceases to be a safe place for people to be. Nature is rendered un-natural. The home is discounted as a sanctuary. In the slums of Bhopal, fear of recurrence added to the complex adjustments needed to regain some approximation of normalcy after the accident. And, particularly for those whose lives are closely tied to the cycle of nature, as with the Swedish Lapps, whose reindeer herding society was destroyed by the Chernobyl accident, the Russian peasants living on contaminated farms, or for that matter, the native and western residents of Alaska impacted by the subsequent Exxon Valdez accident, the results of contamination were disastrous. Such changes to the environment further affect the view people hold of themselves. They are now "victims," harmed by the failure of others; they are likely to be angry and even vengeful against those who caused the accident or allowed it to occur. The inability to trust government, technology, and environment can be disconcerting, particularly because they all point to the victims' increased vulnerability. In such instances, blaming of victims is often noted (Edelstein 1988).

What We Have Learned

Both the Bhopal and Chernobyl accidents were unthinkable. That they occurred involves an important demonstration of the moral, legal, organizational, and physical limits of technology. What are some of the lessons of these disasters?

The Vulnerability of Private Corporations

Would (and should) major technological failure place a limit on the spread and conduct of multinational corporate control of the planet? The Bhopal disaster, much as the subsequent Exxon Valdez accident, raised fascinating implications for the liability of multinational corporations for environmental disaster. In the wake of the disaster, Union Carbide deflected actual blame to UCIL, its Indian subsidiary, while seeking to publicly assume moral responsibility for the disaster. An effort to humbly accept responsibility fell on its face when Union Carbide chairman Warren Anderson was arrested after he flew to India immediately after the accident (Shrivastava 1987). Meanwhile, despite the effort to appear like it was acting responsibly, Union Carbide fought aggressively in court to minimize its financial liability. At stake was the potential for setting a precedent affecting multinational responsibility worldwide, as Morehouse and Subramaniam noted (1986, p.1):

> The disaster at Bhopal confronts us with a critical choice. If on one hand, justice is done to the poor and hapless victims . . . [it] will require a substantial portion, if not all, of the company's assets. An unambiguous message will thus be delivered to hazardous industries the world over that they can no longer give the quest for profit priority over human life. If on the other hand, Union Carbide is allowed to settle the case for an amount that does not significantly diminish its assets, the opposite message will be conveyed: that these industries may continue business as usual.

In fact, the settlement by the Indian government was widely contested and criticized, suggesting collusion between two agents both holding responsibility for the disaster. And, even after the settlement, future Indian government action continued against Union Carbide. For example, in the spring of 1992, an Indian court sought to extradite Warren Anderson, Union Carbide's former CEO, to face homicide charges for an alleged 4,037 fatalities due to the Bhopal accident (Associated Press 1992a).

The threat of both civil and criminal liability has challenged the historic independence of multinational corporations. Furthermore, in the wake of the accident, Union Carbide and others came under greater scrutiny and have faced more stringent regulatory requirements and citizen action. Yet, one would conclude that the consequences of disaster have not diminished the global control of multinational corporations. If Bhopal proved a test of this control, then it can only be concluded that the corporations won. At the same time, the equation will never be quite the same.

The Vulnerability of Government

The Chernobyl incident shook the status quo in the Soviet Union in a way that perhaps no other single event could. Government was caught committing unforgivable lies. The massive social implications of the disaster could not be managed. Chernobyl provided a forum for open doubt and criticism of government on new ecological grounds. It provided an issue in which three of the most powerful republics—Russia, Ukraine, and Byelorussia—were forced to take independent action to protect citizens in the face of central government procrastination and deceit. That the central government could control hazardous facilities within a republic's boundaries was a major issue for dispute. Overall, Chernobyl was the watershed event leading to the breakup of the Soviet Union (Medvedev 1990).

Secrecy in the Soviet Union was perhaps to be expected, but the similarity between information management during the Chernobyl incident and western nuclear disasters is telling. Full case studies of the American Three Mile Island and the British Windscale disasters would reveal general parallels to the Soviet management of Chernobyl (Flavin 1987; Dickson 1988). Likewise, in the wake of Chernobyl, a new era of openness regarding America's military nuclear legacy has unveiled a complex history of deception and control. Chernobyl, therefore, not only called attention to the closed nature of government control over technology behind the "iron curtain" but to the similar way that such technology has been handled in so-called democracies as well.

In fact, both Bhopal and Chernobyl pose fundamental questions about democracy. Neither sector of industry has been subject to the democratic principles by which nations such as the United States operate. Protected as "individuals," industrial corporations have enjoyed an amazing "freedom" to operate. Proprietary information is protected. Citizens have little access and control over the chemical or nuclear enterprises, despite a growing right–to–know because it is "not their business." Key decisions affecting nations and the globe are made in corporate board rooms according to the principles of profit maximization. On its part, the nuclear industry has been able to operate outside the normal checks and balances of open government because of the intimacies of the military–industrial complex and because civilian nuclear power is shielded by authorities that provide private entities with the rights of government. In fact, authorities often enjoy more independence than do government agencies.

Ironically, in light of the shakeup that Chernobyl produced in the Soviet Union, there has been no comparable challenge to governance as usual in the United States. Perhaps it is because no single large catastrophe has yet occurred on the scale of Chernobyl, we are somewhat free to continue our complacency and denial until confronted by an anomaly so great that we are forced to face reality. While thousands of revelations of contamination across the United States have shaken the trust of corporations and of government, and a large toxics rights social movement is evident, environmental disaster is yet to have a similar effect on the operative democracy compared to its revolutionary effect in the Soviet Union. For us, the environment is not quite yet a political issue. We do not see its implications in the black and white way that it appeared in a totalitarian state. Perhaps we all still have too much investment in the polluting system to rebel. In any case, we have not yet been forced to tackle the basic characteristics of the United States that make it a polluting state. In a final irony, the rush to capitalism as the answer to the problems of the new Soviet Republics may deliver to the multinational corporate world the very control that has been wrested from the totalitarian state. The right mix of market, government and citizens as watchdogs is yet to be concocted.

The Vulnerability of National Boundaries to Contamination

Along with oil dependence (Pirages 1978), cross-boundary pollution has become an important issue affecting the relationship of nations. That a nuclear or chemical plant can use the air or waters to undermine the quality of life of neighboring nations has been firmly established, spurring the development of models of how these issues are to be settled. In the wake of Bhopal and the subsequent transboundary Sandoz incident in Europe, legislation was tightened in some nations to ensure corporate responsibility (Mahon 1987). Likewise, well-publicized incidents of the most developed countries dumping their wastes in the least developed countries has contributed to a climate of increasing scrutiny around the world.

Chernobyl illustrated how a nuclear accident in one country can affect others. This realization had special significance for Europeans, where 110 reactors lie within 62 miles of national borders. The resulting concern has led to friction between Denmark and Sweden, Austria and West Germany, France and its neighbors, Ireland and the United Kingdom, and in Asia, between China and Hong Kong. Chernobyl has led to agreements regarding cross-boundary contamination covering obligations to warn other countries and provision of assistance to the stricken country (Flavin 1988).

A further frontier between nations requires attention to the cross-boundary impacts of chronic air pollutants, such as acid deposition precursors, as well as cumulative contributions to global environmental degradation (see Chapter 12, "World Air Pollution: Sources, Impacts, and Solutions"; Chapter 13, "Stratospheric Ozone Depletion"; and Chapter 14, "The Challenge of Global Climate Change"). Serious work on these issues is now underway, although national self-interest remains a serious roadblock, as illustrated at the Rio de Janeiro conference in 1992.

The Bhopal and Chernobyl accidents reflect the crossing of a different kind of boundary as well. Both disasters involved the machinations of the developed First and Second Worlds, but impacted peoples of the Third and Fourth Worlds. Although generally termed *developing* in the language of modernity, relatively few Third World peoples share in the wealth or other benefits created by development. Rather, Third World peoples live at the edges and off the scraps of modern society. In contrast, indigenous "Fourth World" peoples are directly dependent upon nature for their entire way of life and do not participate in modern life. Global technological disaster is not only visited upon those invested directly in the modern world. Rather, none are spared. And, to the extent that Bhopal signifies the shift of risky enterprise away from the developed world, that disaster signifies as well the inequity and bias of exported risk.

The Vulnerability of Technology in an Era of Risk Consciousness

In the United States, Bhopal was a contributing influence to the addition of "right-to-know" provisions in the new Superfund amendments. Activists, particularly in Asia, have organized in memory of Bhopal's thousands of victims. The United Nations has emerged as a key player in negotiations over global environmental protection. The growing climate of scrutiny and regulation worldwide, however, has led multinational corporations to work through the developed nations to establish protective free trade agreements that may limit the restrictions a given government might bring to bear. The era of the hegemony of the multinational corporation has not lapsed. Thus, while Bhopal has led to reforms in chemical production and use, it has not led to a deep questioning of the Green Revolution and the overall costs of chemical dependency. Nations did not shed their chemical dependencies in the wake of Bhopal.

In contrast, the Chernobyl accident brought the far-flung implications of nuclear disaster to the world's attention, posing a systemic critique of this technology in combination with the unsolvable problem of nuclear waste disposal. Much as the Three Mile Island accident helped to sensitize even pronuclear communities to the dangers of nuclear power (Shelton 1982), the Chernobyl accident has made it even harder to believe that nuclear power is safe. This scrutiny could not help but leave a mark on the nuclear industry, most evidenced by the negative turn of Europeans toward nuclear power (Associated Press 1988). In the aftermath of the Chernobyl disaster, Sweden was joined by Italy, Austria, Greece, China, Brazil, and the Philippines, along with other nations, in abandoning nuclear power (Flavin 1987, 1988). The concerns of these nations were summarized by Austria's foreign minister when he addressed the 1986 meeting of the normally pronuclear International Atomic Energy Agency (IAEA) (Flavin 1987, p. 76).

> For us the lessons from Chernobyl are clear. The Faustian bargain of nuclear energy has been lost. It is high time to leave the path pursued in the use of nuclear energy in the past, to develop new alternative and clean sources of energy supply and, during the transition period, devote all efforts to ensure maximum safety. This is the price to pay to enable life to continue on this planet.

The accident was a key nail in the coffin of an already dying industry. While a few countries remain highly dependent upon nuclear power, few new reactors are in the offing worldwide and many operating facilities are closing (Flavin 1987, 1988). Despite these signs of a nuclear demise, it may be premature to write off nuclear energy. There continues to be an establishment dedicated to the nuclear enterprise. It remains to be seen whether the nuclear industry can offset Chernobyl's stranglehold by promoting "safer" reactors that will be "greenhouse" friendly.

What We Have Failed to Learn

It may take additional tragic disasters to get action on nuclear and chemical production safeguards. The question remains, does the tragedy have to be repeated in every locale, or can people learn from the experience of others to be vigilant against such hazards?

In the wake of the respective accidents, much attention in the United States was drawn to the question, "Could such a thing happen here?" Union Carbide and the nuclear industry made similar arguments aimed at public reassurance. Just as chemical corporations explained that a Bhopal accident was not possible at American MIC plants, such as the Union Carbide complex in Institute, West Virginia, the Department of Energy reassured Americans in the wake of Chernobyl that U.S. reactors barely resemble the uncontained graphite-core Soviet plants. Much as pronuclear supporters interpreted the Three Mile Island accident as demonstrating how safe nuclear power is, both Bhopal and Chernobyl were configured to boost, or at least not damage, the industries involved.

Such mind games illustrate the depth of our psychological denial of the fragility of the technologies underpinning modern consumer-oriented society. Ayers and Rohatgi (1987) suggest that the Bhopal accident demands the testing of underlying technological assumptions, including the assumption that complex technologies can be sufficiently understood to be safely designed, that such systems can be operated by human workers safely, that government can regulate such systems effectively, and that legal "fault" can be determined after a system failure. Similarly, Reason (1987) notes that the "it couldn't happen here" response that followed Chernobyl is in itself an example of the same kind of rationalization generally evident when major technological accidents are not anticipated. His analysis of records of U.S. and British reactors found little basis for such reassurance:

> . . . there was nothing unique about the Chernobyl accident. The principal ingredients are common to nearly all major system disasters.

Not that such lessons are entirely missed by the general public. Polling suggests that many people no longer seek to bear the risk of technology; instead they have come to doubt it (Milbrath 1984; Dunlap et al. 1993; Olsen et al. 1993). Public opposition to potentially hazardous facilities suggests that such doubts are often reflected in action. But skepticism about technology, and of the corporations and government agencies responsible for safe operation, has not as yet translated into an approach for assuring safety. After Bhopal and Chernobyl, some compromises and sacrifices had to be made by the respective industries as more information came to light about both MIC production and nuclear power in the United States. Union Carbide was forced to strengthen safety features on its American MIC plants after a string of embarrassing minor accidents occurred during a period of heightened public scrutiny. And the DOE closed the Hanford N Reactor in Washington State, America's only graphite core reactor. These steps were, in themselves, mitigations for the chemical and nuclear industries, sacrifices necessary in order to limit a more widespread and critical evaluation. In fact, while much technology has come to be distrusted, there is yet absent a fundamental shift in beliefs and values necessary to address our use of risky technologies, even if Bhopal and Chernobyl have helped to move us toward this shift. We are caught in the contradiction that, while we do not want the risk, we still believe that risky technology is a requisite for the good life.

Technology Transfer Implications

Working during the period that bridged the Bhopal and Chernobyl disasters, the World Commission on Environment and Development prepared its influential report published under the name *Our Common Future*. Perhaps the most important concept placed into common usage by this report has been the notion of *sustainable development,* defined by the commission as

> . . . development that meets the needs of the present without compromising the ability of future generations to meet their own needs. (World Commission on Environment and Development 1987)

The Bhopal and Chernobyl disasters bear on the concept of sustainability in several important ways. First, Bhopal and particularly Chernobyl have caused harm to the next generations, directly by affecting living children and indirectly by altering the teratogenic and mutagenic environments. Second, Chernobyl has altered the habitability of the environment over a multigenerational time span. Third, both Bhopal and Chernobyl illustrate the inevitability of catastrophic accidents in complex technological systems. As such systems age, are dispersed to widespread locations without adequate supervision, or come to be ignored in complacency, the prospect of future disaster is assured. Finally, the Bhopal disaster forces a general questioning of the practice of modernization in the developing world, forcing one to question whether technology transfer is merely a polite way for describing the shift of hazards from developed to developing nations.

Bhopal's disaster directly questions the assumptions behind modernizing the Third World along the model of the West. It raises the moral issues in the actions of multinational corporations to engage in the neocolonial exploitation of others for their personal gain. It suggests that technological initiatives, such as the Green Revolution, have actually had a great secondary cost not foreseen in the optimism of promoters. It underscores the lose/lose situation that developing countries have encountered due to their need for hard currency. Finally, it suggests the need for a sustainable approach to development that is not so closely tied to industrial chemicals and nuclear energy.

In the guise of progress, we have too often allowed for technology transfer to be merely the acquisition of risks by those least able to handle them. Hazardous facilities and products are exported in the absence of a corresponding infratechnology and without the expertise to evaluate the consequences of using the technology (Weiss and Clarkson 1986). True inequities emerge in such considerations. Bogard (1989) points to the basic dilemmas of choice that underlie accidents such as Bhopal and that select the poor to be the victims. In his view, once a technology is defined as needed—in this case the need was for pesticides to support agriculture to feed growing populations—people proceed to become dependent upon the technology. This dependence influences them to tolerate catastrophic potential and to redefine hazards as mitigations. Thus, in India the risk of pesticides was discounted while their use was seen as a cure for hunger and a solution for unemployment, overpopulation, and economic dependency. As India became increasingly dependent upon the use of pesticides to feed masses of people, it was left without other options, as this technology became continually more ineffective due to increasing insect tolerance to the treatments. They were stuck.

Conclusions

In many respects, the disasters at Bhopal and Chernobyl point to the collision between the post-World War II miracle of technological and economic growth and the physical parameters and realities of our Earth. Perhaps more than any other single incidents during the 1980s, Bhopal and Chernobyl were forces for the reevaluation of the acceptable risk of technology. Along with Love Canal, they called attention to the plight of contamination victims. Along with the Exxon Valdez oil spill in Alaska, these disasters called attention to the human impact on our environmental surroundings. Along with the Challenger space shuttle accident, they haunt our underlying belief that we can solve any problems with advanced technology.

The disasters at Bhopal and Chernobyl united the technological props of modern society with the lives of citizens who previously were only remotely connected to such technology in the course of their daily lives. Contamination, once released, transformed reality for these victims from relative security to disaster, signaling the very failure of their social systems. Individuals and families, neighborhoods, and communities were now united by one common experience, the shared exposure to the toxins. After exposure, the once tangential technology became central to their lives. Life itself was now defined by the exposure and its resulting physical, economic, social, and psychological effects.

Bogard (1989) notes that the language we use to refer to such events is often imprecise. He prefers the Greek sense of tragedy in describing the suffering caused by unintended outcomes, recognizing that like Chernobyl,

> Bhopal was more than a mere accident, disaster or catastrophe. Each of these descriptions minimizes the problem of human agency and intention and thus refuses to address directly the issue of responsibility. . . . a tragedy, in contrast, emerges out of a complex of confused and misguided intentions, many of which may be honorable in themselves but when forged to the actual chain of events produce the worst possible outcome . . . no one desired such an outcome.

With the tragedy of our times symbolized by Bhopal and Chernobyl—the specter of failed technology and suffering victims—we are left to question whether such fate awaits us all. Or, do we, in fact, have choices?

References

Agarwal, A., J. Merrifield, and R. Tandon, *No Place to Run: Local Realities and Global Issues of the Bhopal Disaster,* Knoxville, TN: Highlander Research and Education Center, 1985.

Ahearn, J., "Nuclear Power After Chernobyl," *Science* 236, 1987, pp. 673–679.

Alvares, C., "A Walk through Bhopal." *In* D. Weir, *The Bhopal Syndrome: Pesticide Manfacturing in The Third World*. Penang, Malaysia: International Organization of Consumers Unions, 1986.

Anspaugh, L., R. Catlin, and M. Goldman, "The Global Impact of the Chernobyl Reactor Accident," *Science* 242, 1988, pp. 1513–1515.

Associated Press, "Greek Abortions up Due to Fear of Chernobyl's Effects," *The (Middletown, N.Y.) Times Herald Record,* November 2, 1987a, p. 24.

Associated Press, "Soviet People Still Bear Scars of Chernobyl," *The Times Herald Record,* April 19, 1987b, p. 64.

Associated Press, "Chernobyl Evacuees Sneak Home," *The (Middletown, N.Y.) Times Herald Record,* April 17, 1988.

Associated Press, "$50 a Month Going to Bhopal Victims Heirs," *The (Middletown, N.Y.) Times Herald Record,* April 22, 1989a, p. 30.

Associated Press, "Mass Exodus of Chernobyl Area Advised," *The (Middletown, N.Y.) Times Herald Record,* July 30, 1989b.

Associated Press, "Cancer, Horror Tales Fallout for Chernobyl," *The (Middletown, N.Y.) Times Herald Record,* February 16, 1989c.

Associated Press, "Chernobyl Victims Get New Benefits," *The (Middletown, N.Y.) Times Herald Record,* April 8, 1990a, p. 40.

Associated Press, "Strikes, Telethon Mark Chernobyl Anniversary," *The (Middletown, N.Y.) Times Herald Record,* April 27, 1990b, p. 4.

Associated Press, "2000 Bhopal Gas Victims Rally for Justice," *The (Middletown, N.Y.) Times Herald Record,* June 30, 1990c, p. 33.

Associated Press, "Senior Scientist Says Chernobyl Killed Up To 10,000 People." *The (Middletown, N.Y.) Times Herald Record,* April 15, 1991, p. 32.

Associated Press, "Bhopal Court Seeks Extradition of ex-CEO," *The Times Herald Record,* March 28, 1992a, p. 50.

Associated Press, "Chernobyl Cancer Rate Mushrooms," *The (Middletown, NY) Times Herald Record,* September 3, 1992b, p. 18.

Ayers, R. and P. Rohatgi, "Bhopal: Lessons for Technological Decision-Makers," *Technology in Society,* 9, 1987, pp. 19–45.

Barringer, F., "Chernobyl Disaster Leaving a Long Trail of Lists," *The New York Times,* September 6, 1986, p. 2.

Barringer, F., "Five Years Later: Chernobyl: The Danger Persists," *The New York Times Magazine,* April 14, 1991, pp. 28–39, 74.

Bhagat, D., "A Night in Hell," *In* L. Surendra (ed.), *Bhopal: Industrial Genocide?,* Hong Kong: Arena Press, 1985.

Bogard, W., *The Bhopal Tragedy: Language, Logic, and Politics in the Production of a Hazard,* Boulder, CO: Westview Press, 1989.

Broadbent, N.D., "Chernobyl Radionuclide Contamination and Reindeer Herding in Sweden," *Collegíum Antropologicum* (Zafgreb, Croatía), 10, 1986, pp. 231–242.

Brooke, J., "After Chernobyl, Africans Ask If Food Is Hot," *The New York Times,* January 10, 1988, p. 15.

Carothers, A., "Children of Chernobyl." *Greenpeace Magazine,* January/February 1991, pp. 8–12, 27.

Cox News Service, "Mexico Accused of Distributing Tainted Milk," *The (Middleton, N.Y.) Times Herald Record,* January 31, 1988, p. 85.

Dickson, D., "Doctored Report Revives Debate on 1957 Mishap," *Science* 239, 1988, pp. 556–557.

Dunlap, R., G. Gallup, Jr., and A. Gallup, "International Public Opinion Toward the Environment," *Impact Assessment* 11, Spring 1993, pp. 3–26.

Dunlap, R., G. Gallup, Jr., and A. Gallup, "International Public Opinion Toward the Environment," *Impact Assessment* 11, Summer 1993, pp. 113–144.

Economist, "What Chernobyl Did: Not Just a Nuclear Explosion," April 27, 1991, pp. 19–21.

Edelstein, M. R., *Contaminated Communities: The Social and Psychological Impacts of Residential Toxic Exposure,* Boulder, CO: Westview Press, 1988.

Edelstein, M. R., "Bhopal and the Spectre of Third World Chemical Disaster," *In* W. J. Makofske, H. Horowitz, E. F. Karlin, and P. McConnell (eds.), *Technology, Development and the Global Environment,* Mahwah, NJ: Institute for Environmental Studies, Ramapo College of NJ, 1991a, pp. 116–144.

Edelstein, M. R., "Chernobyl and the Spectre of Nuclear Power Accidents," *In* W. J. Makofske, H. Horowitz, E. F. Karlin, and P. McConnell (eds.), *Technology, Development and the Global Environment*, Mahwah, NJ: Institute for Environmental Studies, Ramapo College of NJ, 1991b, pp. 145–167.

Eijgenraam, F., "Chernobyl's Cloud: A Lighter Shade of Gray," *Science* 252, May 31, 1991, pp. 1245–1246.

Flavin, C., "Reassessing Nuclear Power," In *State of the World 1987: A Worldwatch Institute Report on Progress Toward a Sustainable Society*, New York: W.W. Norton, 1987, pp. 57–80.

Flavin, C., "How Many Chernobyls?," *World Watch* 1, 1988, pp. 14–18.

Ginzburg, H. M., and E. Reis, "Consequences of the Nuclear Power Plant Accident at Chernobyl," *Public Health Reports* 106, 1991, p. 38.

Goldman, M., "Chernobyl: A Radiobiological Perspective," *Science* 238, 1987, pp. 622–623.

International Advisory Committee, International Atomic Energy Committee, "The Radiological Consequences on the USSR from the Chernobyl Accident: Assessment of Health and Environmental Effects and Evaluation of Protective Measures," International Atomic Energy Agency, April 23–27, 1990.

Mahon, J. F., "Managing Toxic Wastes—After Bhopal and Sandoz," *Long Range Planning* 20, 1987, pp. 50–59.

Marshall, D., "The Aftermath of Chernobyl," *Science* 233, 1986, pp. 1141–1143.

Marshall, D., "Calculating the Cost of Chernobyl," *Science* 236, 1987, p. 658.

Medvedev, Z., *The Legacy of Chernobyl*, New York: Norton, 1990.

Milbrath, L., *Environmentalists: Vanguard for a New Society*, Albany, NY: SUNY Press, 1984.

Morehouse, W. "Unfinished Business: Bhopal Ten Years After," *The Ecologist* 24(5), 1994, pp. 164–168.

Morehouse, W. and M. A. Subramaniam, *The Bhopal Tragedy: What Really Happened and What it Means for American Workers and Communities at Risk*, Preliminary Report for the Citizens Commission on Bhopal. New York: Council on International and Public Affairs, 1986.

New York Times, "Chernobyl Still Raining Sickness," *The (Middletown, N.Y.) Times Herald Record*, November 3, 1991a, p. 53.

New York Times, "Soviets Admit Failed Cleanup of Chernobyl," *The (Middletown, N.Y.) Times Herald Record*, February 8, 1991b, p. 12.

Olsen, M., D. Lodwick, and M. Olsen, *Viewing the World Ecologically*, Boulder, CO: Westview Press, 1993.

Perrow, C., *Normal Accidents*, New York: Basic Books, 1984.

Pirages, D., *The New Context for International Relations: Global Ecopolitics*, North Scituate, MA: Duxbury Press, 1978.

Raghunandan, D., "Ill-Effects Persist: Report of a Survey," *Economic and Political Weekly*, February 22, 1986, pp. 332–334.

Ramaseshan, R., "Government Responsibility for Gas Tragedy," Reprinted from *Economic and Political Weekly*, December 15, 1984, in Asian Regional Exchange for New Alternatives, *Bhopal: Industrial Genocide?*, Hong Kong: Arena Press, 1985.

Reason, J., "The Chernobyl Errors," *Bulletin of the British Psychological Society* 40, 1987, pp. 201–206.

Rich, V., "USSR: Chernobyl's Psychological Legacy," *The Lancet* 337, May 4, 1991, p. 1086.

Schroeder, K. (ed.), "Chernobyl: The Intangible Catastrophe Continues," Report of the World Council of Churches/ Commission on Inter-Church Aid, Refugee and World Service Team Visit to Moscow and Byelorussia from June 23 to July 4 1990, New York. World Council of Churches, August 17, 1990.

Scripps-Howard News Service, "Contamination from Chernobyl Spreading Through Soviet Union," *The (Middletown, N.Y.) Times Herald Record,* April 23, 1990.

Scripps-Howard News Service, "'Only God Knows' What Chernobyl Wrought," *The (Middletown, N.Y.) Times Herald Record,* April 26, 1992a, p. 49.

Scripps-Howard News Service, "Chernobyl's Sick, Dying Find Solace Coming Home," *The (Middletown, N.Y.) Times Herald Record,* April 26, 1992b, p. 49.

Shelton, D., "Emergence of Community Doubts at Plymouth, Massachusetts, *In* D. Sills, C. P. Wolf, and V. Shelanski (eds.), *Accident at Three Mile Island: The Human Dimensions,* Boulder, CO: Westview Press, 1982, pp. 83–84.

Shrivastava, P., *Bhopal: Anatomy of a Crisis,* Cambridge, MA: Ballinger Publishing, 1987.

Stephens, S., "Lapp Life after Chernobyl," *Natural History,* December 1987, pp. 33–40.

Stone, R., "West May Help Re-Entomb Chernobyl," *Science* 258, November 20, 1992, p. 1295.

Travis, J., "Inside Look Confirms More Radiation," *Science* 263, 1994, p. 750.

Vyner, H., *Invisible Trauma: The Psychosocial Effects of the Invisible Environmental Contaminants,* Boston: Lexington Books, 1987.

World Commission on Environment and Development, *Our Common Future,* New York: Oxford University Press, 1987.

World Watch, "Cleanup of Chernobyl," 3/4, July–August 1990, p. 6.

Weir, D., *The Bhopal Syndrome: Pesticide Manufacturing in the Third World,* Penang, Malaysia: International Organization of Consumers Unions, 1986.

Weiss, B. and T. Clarkson, "Toxic Chemical Disasters and the Implications of Bhopal for Technology Transfer," *The Milbank Quarterly* 64, 1986.

Chapter 17

Gaia, a New Look at the Earth's Surface

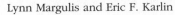

Lynn Margulis and Eric F. Karlin

As one studies the Earth and the continuum of ecosystems that cover its surface, many anomalies become apparent. It is the only planet known to harbor life. In a world of finite resources, life has flourished for at least 3.5 thousand million years. In a solar system where the Sun has become significantly hotter over time, the Earth's surface temperature has remained relatively constant. Oxygen gas is found in only trace amounts in the atmospheres of other planets, yet it is the second most abundant gas in the Earth's atmosphere. Clearly something quite different is happening on Earth, a process that has allowed the sustenance of life for over 3.5 billion years. Has life passively "adapted" to a changing environment for over 3.5 billion years? Although most scientists assert that the Earth's surface is controlled primarily by physical and chemical processes, it seems more likely to us that life has had a major role in maintaining and regulating the surface of the planet.

The Gaia hypothesis, developed about 20 years ago by James E. Lovelock (1979, 1988), states that life, as the sum of all interacting organisms, shapes and maintains certain aspects of the Earth's surface environment: reactive atmospheric gases, temperature, alkalinity, etc. The name *Gaia*, which refers to the ancient Greek goddess of Mother Earth, is the hidden word in the *Geos* satellite, the supercontinent Pan*gea*, and the sciences of *ge*ology and *ge*ography. It was suggested by the late English novelist William Golding (author of *Lord of the Flies*) in response to Lovelock's request for a "good four-letter" word to focus the attention of his scientific colleagues on the whole Earth.

The basic tenet of the Gaia hypothesis is that life plays a significant role in regulating the global environment. It is a physiological view, rather than a strict physico-chemical view, of the global environment (Margulis and West 1993). What we usually consider to be the physical environment, that is, the atmosphere, the oceans, and the land, is really part of a planetary ecosystem connected through space (by atmospheric and hydrospheric interaction) and through time (by evolution). The Gaia hypothesis sees temperature and aspects of atmospheric composition being actively maintained by a vast, complex network of interacting ecosystems. Gaia concepts reject traditional views of the inert physical environment. The environment is not apart from and independent of living systems. From a Gaian perspective, the rocks, the air, the biota, and other components of the Earth's surface are parts of a continually changing global ecosystem. If we regard the Earth's surface, life enmeshed in its surroundings, as a holocoenotic system, our attitude toward the planet dramatically changes. Especially in light of the extensive modifications to the global environment by human activities, the implications of a gaian view differ significantly from those of the typical Judeo-Christian view that suffuses our culture (Sagan and Margulis 1993).

The Gaia hypothesis is less of a hypothesis than it is a radical change of scientific orientation (Joseph 1990). The term *Gaia* generates visceral emotional response in certain scientists. Insight into this malaise via the philosophical discussion of "Gaia" theory at the American Geophysical Union, the first debate on Gaia in "polite scientific society," is provided by Sagan (1988).

Evidence from Comparison of Planets

The Gaia hypothesis was born of the international space program in the last three decades; it comes from the comparison of Earth with Mars, which is farther from the Sun and Venus, which is closer. Both Russians and Americans have landed probes on these planets. We can state definitively that Mars is cold: so cold that carbon dioxide ice (dry ice) forms polar ice caps. Unlike on Earth, ultraviolet light passes largely unfiltered through the thin Martian atmosphere and penetrates to the planet's surface. Venus, on the other hand, is extremely hot. The atmospheric temperature far exceeds the boiling point of water. Venus is dry, acidic, and has a dense atmosphere. At present, both planets are entirely uninhabitable.

Whereas the atmospheres of both Mars and Venus are carbon dioxide rich and very dry, the Earth's is carbon dioxide poor, moist, enjoys a moderate temperature, and contains huge quantities of that very reactive gas: oxygen (Figure 17.1). All three planets are believed to have had very similar atmospheres at one point in time (Hunten 1993). We can only understand what happened to remove the carbon dioxide from the atmosphere of the Earth and why so much free atmospheric oxygen is present if we recognize the relevance of one of the most important evolutionary achievements in the history of the planet: the origin in cyanobacteria of oxygen-releasing photosynthesis. These bacteria and their descendants, algae and plants, remove carbon dioxide while they produce oxygen. The Earth's atmosphere is a dynamic system, and for the past 3.5 billion years it has been extensively altered by the activities of life, most importantly, oxygenic bacterial photosynthesis (Schopf 1978).

Manifestations of Gaia

Global Temperature

The baseline data indicate that the Earth's temperature has not changed greatly throughout the existence of life. If oceans had entirely frozen or, on the other hand, had boiled, at any time in the history of the Earth, it would be detectable in the geologic record. At no time was there either extensive ocean freezing or ocean boiling,

Figure 17.1. The planet Earth.
Source: NASA, 25 October 1975.

which attests to the relatively narrow range of temperatures enjoyed by the planet throughout the history of life. During the most severe glaciations, the global mean surface temperature was only about 8°C cooler than at present. The evidence of this drop in temperature abounds: continents were scoured as glacial tillite formed, glacial erratics litter the landscape, continental landmasses isostatically rebound, and extensive ecosystem disruption occurred in formerly glaciated areas. Similar hints tell if the temperature has risen: hot water precipitated minerals, higher sea levels, and fossils of coastal redwood (*Sequoia sempervirens*) in Alaska and of Gondwana flora in Antarctica. We conclude from reading the geological record that the mean temperatures remained within 10° or 20°C of what they are now.

Although the global temperature has been relatively stable, astronomers posit that stars similar to our Sun increase in luminosity as they age. Because they simply track the increase in solar luminosity, the temperatures on Mars and Venus rose over time. The Earth's temperature should have also progressively become warmer, but apparently it has not. A planet that maintains a relatively stable temperature in the face of the tendency of the whole solar system to increase in temperature is very strange, this resistance is inexplicable by chemistry and physics alone. To understand global temperature stasis, the regulatory abilities and properties of life must be considered.

Critics of the Gaia hypothesis say that purely physical-chemical systems alone maintain the atmospheric CO_2 balance and regulate planetary temperature. They miss the crucial role of life, however, in the greenhouse effect. Incoming solar radiation heats the Earth's upper atmosphere and the Earth's surface. The energy absorbed at the Earth's surface is reradiated back to space as infrared radiation. The absorption of the outgoing infrared radiation by carbon dioxide, water vapor, and other so-called greenhouse gases heats the lower atmosphere in the process called the *greenhouse effect*. Changes in concentration of any greenhouse gas may lead to changes in the temperature of the lower atmosphere. Increases in the concentration of atmospheric carbon dioxide would lead to global warming; by contrast, one way for the Earth to maintain a relatively constant atmospheric temperature as the sun increases in luminosity is by reduction of carbon dioxide concentration. A primary way by which atmospheric carbon dioxide decreases is by the formation of carbonate sediments. Much of the carbon dioxide dissolved in ocean water is chemically transformed into carbonates (CO_3^-) and bicarbonates (HCO_3^{2-}), and these, in turn, can be precipitated as calcium carbonate ($CaCO_3$). The absorption of CO_2 by the oceans is a physical and chemical process, enhanced in rate and extent by photosynthesis and other biotic processes that uptake carbon dioxide. The precipitation of calcium carbonate is also a biological activity. The chalky remains of protoctists (especially coccolithophorids and foraminiferans) and animals (clams, snails, and corals) are obvious examples. Also significant, but less obvious, is calcium carbonate precipitation by bacteria. An oceanic sink for carbon thus exists, as carbon dioxide is removed from the atmosphere and deposited as carbonate sediments on the ocean floor.

Critics claim that the autoregulatory ability of the environment, a fundamentally physical/chemical system, would exist if life were removed. The Gaia hypothesis states that autoregulation, associated with living systems, would disappear in the absence of life. Gaia protagonists posit that without life, the Earth's environment would revert to simple interpolation between that of Mars and Venus.

Nitrogen and Gaia

Much attention is given to the biologically significant gases, oxygen, and carbon dioxide, but the major gas in our atmosphere is nitrogen. The relative concentration of atmospheric nitrogen to total CO_2 on the Earth is similar to that on Venus and Mars. Most of the Earth's CO_2, however, has been removed from the atmosphere and is located, instead, in carbonate rocks, dispersed reduced carbon in shale, and to a lesser extent fossil fuels. If we were able to return all of the carbon from the rocks, sediments, and soil to the atmosphere, the relative nitrogen content of the atmosphere would decline from 78% to only 2% or 3%. This indicates that the present atmosphere has a high nitrogen concentration because nitrogen remains whereas CO_2 is effectively removed. Is the nitrogen content of the atmosphere regulated at all by biotic activities?

The atmospheric nitrogen reservoir is drained by nitrogen fixation, a process in which gaseous nitrogen reacts with oxygen to form nitrate or with hydrogen to form amines and other organic compounds. Nitrogen fixation results in the transformation of nitrogen gas into solid form, and occurs via both abiotic and biotic processes. Biotic pathways provide roughly 90% of the nitrogen fixed on a global basis (Delwiche 1970). Nitrate is a stable compound and very little is converted by abiotic processes back into nitrogen gas. But nitrate, a nutrient for bacteria, algae, and plants, is used to produce amino acids. Some bacteria, in addition, under anoxic conditions transform nitrate into gaseous nitrogen (a process called *denitrification*) and thus return nitrogen to the atmosphere. Without this life, nitrate would accumulate in great quantities on the land surface and in the oceans. Nitrate is thought to be abundant on Mars, for example. On a lifeless Earth, nitrate would be expected to accumulate and atmospheric oxygen would be dramatically reduced to trace quantities. No one doubts that the atmospheric content of even such a chemically intractable gas as nitrogen is regulated by living systems (Figures 17.2 and 17.3).

Understanding Gaia Better

The Gaia hypothesis suggests that ecosystem interactions on a global scale have maintained an environment suitable for life over long periods of time. Obviously, however, the environment has not remained constant—entirely fixed—over the past 3.5 billion years. When life first started there was no, or very little, atmospheric oxygen. Oxygen gas is highly reactive, and when it first appeared in the atmosphere most microbes could not tolerate exposure to it. Yet today oxygen is a requirement for nearly all large life on Earth and a large fraction of the microorganisms. Some claim a contradiction: Gaia establishes environmental guidelines permitting the existence of life, but these environmental guidelines change. If Gaia is a self-regulating system, what is the optimum end point toward which it is working? Botanists would say the best environment for plants might be an atmosphere with less oxygen and more carbon dioxide than at present. Zoologists, on the other hand, claim that the current levels of oxygen and carbon dioxide are suitable for marine and land animal populations.

But there is no contradiction: the observation that the details of regulation have changed through time does not mean that regulation does not occur. The problem of teleology (purposeful activity) disappeared when proponents of Gaia stopped talking about what is "optimal" as if Gaia, like a committee, could plan in advance. Rather Gaia, like life itself, is a natural phenomenon—the result of interacting species. The nub of it is that there are between 5 and 50 million species alive today, each with unique environmental optima. That each has different needs and produces different exudates drives essential ecosystem processes. All organisms require water, but some thrive in dry habitats, while others must be submerged in an aquatic environment. All exchange gas, but some are destroyed by exposure to oxygen, whereas others absolutely need it in high concentrations. The solar radiation, temperature, pH, gas concentration, and other environmental optima of different species vary; many are even mutually exclusive. If they did not vary, the refuse of one organism would not serve as the food of another. If all organisms produced identical wastes, whether solid (feces, wood, feathers), liquid (urine, slime, pine sap), or gaseous (carbon dioxide, methane, oxygen), ecosystems would grind to a halt. A diversity of environmental requirements is essential for the gaian phenomenon of environmental modulation by nutrient cycling. The wastes of one species are utilized as resources by a series of others.

Environmental regulation is a result of the complex interactions of a vast interacting network of organisms. A brief and simplified review of how Gaia might respond to a gain in solar luminosity illustrates this concept. As the increase in solar luminosity proceeded, the Earth's surface temperature would be expected to warm incrementally over time. As all organisms have the capacity to tolerate a

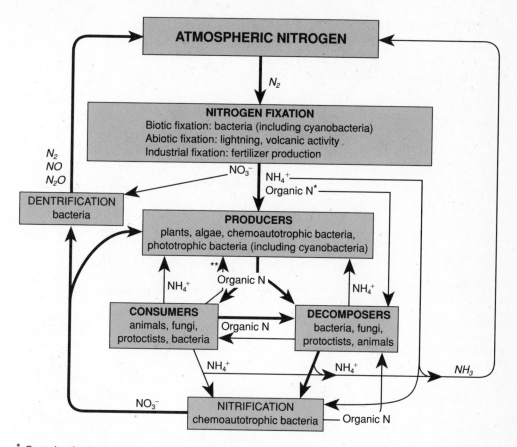

Figure 17.2. The nitrogen cycle. The thick arrows indicate the primary route of nitrogen. Gaseous compounds are italicized.

* Organic nitrogen: organic compounds that contain nitrogen (i.e. amino acids, nucleic acids, ATP, NADH)

** Carnivorous plants such as sundews (*Drosera*), bladderworts (*Utricularia*), and pitcher plants (*Sarracenia*) obtain a significant portion of their nitrogen via the capture and digestion of small animals and protoctists.

range of environmental conditions, a slight increase in temperature could initially be handled by behavioral changes and acclimation. Consider the response of humans as the air temperature increases. If one's temperature goes up the first response is behavioral. Excess clothing is shed, or a person moves into the shade. But as one reacts behaviorally, the body also responds physiologically. Sweating begins, blood is shunted toward the surface vessels of the skin, the heart beat rises, and so on. If the change in temperature persists, becoming seasonal, the response to the elevated temperatures may be periodic and hormonal. The shedding of hair by dogs in the spring and the ability of many plants to withstand

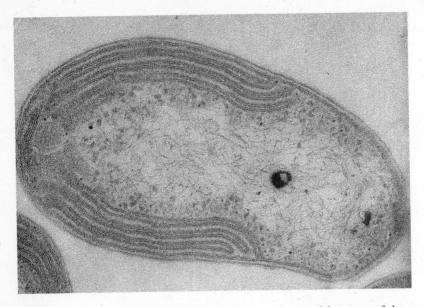

Figure 17.3. Electron micrograph of *Nitrobacter winogradskyi,* one of the many bacteria associated with the nitrogen cycle. This autotrophic species obtains energy to produce organic compounds by oxidizing nitrite (NO_2^-) to nitrate (NO_2^-). (From Watson, S. W., *International Journal of Systematic Bacteriology* 21, 1971, pp. 254–270.)

freezing temperatures in winter but not in summer exemplify physiological expressions of genetic potential (acclimation) to temperature change. But as solar luminosity increases and local environments continue to heat up, behavior and physiological change will not suffice. Survival would depend upon either migration to cooler habitats or innovative genetic responses to the increasing temperatures. The rate of climate change would be slow enough to provide time for evolution to occur. In those cases where neither option is possible, the death of individuals would ultimately lead to population and species extinctions. Organisms that survive would be likely to produce offspring that possess the capability to tolerate the elevated temperature levels.

If global temperature were to rise significantly, increased temperature would eventually become a major variable in the process of natural selection in all ecosystems. Life on Earth would be faced with strong selection pressure to change with the warming climate; eventually, each species would become extinct or evolve the toleration of the new environmental conditions. A multitude of evolutionary pathways would be developed by organisms in response to increased temperature: changes in color, efficient heat exchange systems (larger ears, loss of blubber, highly compound plant leaves), and more heat-tolerant physiology (e.g., C_4 and CAM photosynthetic pathways, more heat-stable proteins) provide a small number of examples. Biological processes are also generally affected by a change in temperature (i.e., rates of photosynthesis and respiration, rates of growth and development). As organisms undergo change in response to warmer temperatures, their impacts on the environment also change.

The sum of the environmental modifications brought about by all life on Earth as it evolves, adjusting to warmer climates, would also change the Earth's environment. If these changes involved significant reductions in greenhouse gases (brought about by increased rates of calcium carbonate precipitation in the oceans, photosynthetic rates in excess of respiration, altered rates of methane emission), then the Earth would warm minimally. Those portions of the planet that are warmed by solar radiation could also be modified. Prior to the presence of atmospheric oxygen, most of the solar ultraviolet radiation intercepted by the Earth passed through its atmosphere and reached the planet's surface. Not only did this severely limit where life could exist on Earth, it also added to the heat budget of the planet's surface. As oxygen and ozone absorb portions of the ultraviolet spectrum (UV-B and UV-C), the appearance of these two gases in the atmosphere allowed for atmospheric filtering of ultraviolet radiation. As levels of oxygen and ozone increased, the absorption of UV-B and UV-C would occur at higher and higher elevations. At present, most of the UV-B and all of the UV-C are absorbed not at the Earth's surface, but in the stratosphere (see Chapter 13, "Stratospheric Ozone Depletion"). Thus the inhabited surface of the Earth has lost a heat source and the stratosphere has gained one. This oversimplified example shows how temperature could be modulated by life. Such responses developed by billions of individual members of millions of different species are the collective gaian mechanism. Gaia is a very complex assemblage.

The Earth's environment continually changes. No perfect environmental optimum exists—any given organism at any time alters its surroundings. In the winter, the temperature inside beehives in Minnesota may be 70°C warmer than the ambient air. While the humidity of the ambient air is only 5–10%, inside the hive it can be 95%. We do not call the air alive, but it is obvious that the air inside is modified dramatically by the activity of the bees. The bees are energetically pumping their wings and patching up holes. They expend energy from last summer's glucose to maintain the hive's environmental conditions. What is happening in the hive is a microcosmic example of what occurs, involving all organisms, at the planetary level.

Although we see a single bee, one flying hymenopteran buzzing about from flower to flower, we know that a bee is composed of several billion organisms living together in intimate association. Like all animals, bees are composites, mobile microbial communities. The study of any animal reveals the presence of other associated species. Indeed, even the cleanest people harbor mites in their eyebrows and intestinal bacteria. We each maintain a field of underarm bacteria that would amaze us if seen magnified on the screen of a scanning electron microscope. We speak of ourselves, and other animals, as single organisms, and yet in reality we are all more or less integrated interacting communities of many types of organisms. What appears to be the purposeful behavior of a single individual is really the self-maintaining behavior of symbiotic complexes, in the sense that if certain interactive metabolism and behavior ceases, the complex fails to persist and disintegrates. We call this literal disintegration *death*. The same sort of self-maintaining behavior we observe in the individual animal can be loosely extrapolated to a larger scale. On a vast level it leads us to look at the complexity and interaction of all life as resulting in a dynamically stable global surface environment.

Gaia as a Superorganism?

If we accept that life regulates the global environment, we must ask how? The super strong Gaia hypothesis, mistakenly believed by many to be the Gaia hypothesis, is that all life on Earth is part of one *global superorganism*. This terminology has been proffered, by both proponents and critics, and debated since the inception of the Gaia hypothesis. Claiming Gaia as a single "Earthorganism" is counterproductive; the global superorganism model is not ecologically valid. No single organism utilizes exclusively its own wastes as resources. We deny that Gaia is a superorganism, and reject the super strong hypothesis.

This formulation, that all life on Earth forms one global superorganism, as the basic tenet of the Gaia hypothesis has been a major stumbling block to the seriousness of attention paid to gaian ideas by the scientific community. The idea that plant communities might be superorganisms, championed by Clements (1936), was hotly debated by ecologists. But Gleason (1926), Whittaker (1951, 1953), and McIntosh (1967) clearly established that the superorganism concept was invalid. Thus the resurrection of the superorganism by the super strong Gaia hypothesis struck a sensitive nerve in many ecologists. This was unfortunate, because ecologists, with their focus on interdisciplinary research and knowledge of ecosystems, should be among the first of scientists to understand the full ramifications of the Gaia hypothesis.

The most important aspect of the Gaia hypothesis is that life plays a major role in the regulation of the global environment. As it is not essential that a superorganism exist in order for life to have a profound impact on the planet, the superorganism concept is a secondary issue. Rejection of the superorganism concept does not impact the central ecological concepts presented, or implied, by the Gaia hypothesis:

1. All organisms depend upon members of different species for survival (food source, gas delivery, waste removal, etc.).
2. The abiotic environment is an essential component of ecosystems.
3. Every ecosystem depends upon other ecosystems for long-term sustainability, and none is truly isolated from material and energy exchange; the global environment is composed of interacting ecosystems.
4. Ecosystems regulate their environment to varying extents and also change over time, in response to changes in their environment (May 1978).
5. The complex continuum of ecosystems that cover the Earth's surface play a significant role in the development and maintenance of the planet's surface environment, with microbes having the most extensive role in chemical cycling and energy conversion. The Earth's surface environment results from the interaction of the planetary regolith and atmospheric gases with life.

The first four ideas are central concepts in modern ecological theory and are not unique to Gaia. They are used to develop the fifth concept, which forms the foundation of the Gaia hypothesis. Although accepted by many ecologists in principle, much work remains before it is fully established in ecological theory.

The inherent simplicity of the superorganism analogy is one reason that this terminology is employed in descriptions of the Gaia hypothesis. "Superorganisms" are readily grasped by the public, by politicians and by scientists unfamiliar with ecology. A second reason to include it is that it provides a context for social change—people generally treat living beings differently than they treat nonliving objects. Unfortunately, it is all too easy for the superorganism analogy to be seen as the central point of the Gaia hypothesis and we thus discourage its use. Although eradicating the superorganism terminology will make the Gaia hypothesis more palatable to many scientists, it also results in the loss of an effective mechanism to transmit basic gaian concepts to the public. As ecosystems form a continuum across the surface of the Earth (McIntosh 1967), we propose that the term *Gaia* could be used to refer to the largest ecosystem of all, the grand sum of all ecosystems, namely, the biota in the biosphere. The biota, atmosphere, rocks, and water would then be perceived as being a vast global ecosystem comprised of interacting smaller ecosystems, all with a role in Earth's environmental regulation. More precisely defining Gaia in terms familiar to ecologists would accomplish much of what the superorganism analogy does without the exacerbating terminological quagmire.

Testing the Gaia Hypothesis

Scientists and others question whether it is possible to test the Gaia hypothesis. Unfeasible and insidious, a definitive proof would follow the destruction of all life on Earth and subsequent documentation of the consequences to the Earth's surface. The Gaia hypothesis predicts that if life were eradicated, the Earth would revert to conditions comparable to those of Mars and Venus. Carbon dioxide would accumulate in the atmosphere, oxygen would be depleted, hydrogen would escape, water would be lost, and surface temperature, in following the solar curve, would begin to increase.

We can, however, test these predictions in a less devastating manner. We can make sealed glass enclosures in which environments are created with and without life. Professor Clair Folsome and Carmen Aguilar-Diaz at the University of Hawaii enclosed small ecosystems in glass flasks and measured oxygen levels in them. When life was absent in a flask and no oxygen introduced at the beginning, no gaseous oxygen developed. But when another glass flask, containing "Hawaiian mixture," that is, unidentified living algae, cyanobacteria, and many other microbes, was placed in light, the oxygen content of the flask air increased. Not only that, it fluctuated for a while before leveling off. The concentration of oxygen gas was maintained between limits for the 3 months it was monitored. These closed systems, fluctuating but regulated, provide suggestions for the Earth system as a whole. Unfortunately, Professor Folsom died before this work, described in Sagan (1990), was completed.

The largest experiment of this kind (named Biosphere II, in analogy to Biosphere I in which we live) is a project taking place at Sunspace Ranch in Oracle, Arizona. South of Tucson, it involves a 3.15 acre greenhouse, the largest enclosed ecosystem in the world. The structure is an expandable bubble that is supposedly closed off from any material exchange with the outside world; that is, the air, soil, and water are recycled. A number of ecosystems were introduced into Biosphere II, including a rain forest, a marsh, and an agricultural system (Sagan 1987; Kelly 1990). Eight people, "biospherians," were sealed into the Biosphere II system for a 2 year stay (September 1991 to September 1993). External researchers monitored and analyzed the changes that occurred inside. Significant changes included an increase in carbon dioxide, a surprisingly large decrease in oxygen, and the loss of a number of species, including two-thirds of the 300 insect species that had been introduced to Biosphere II. To survive, the biospherians used about 20% of reserve food supplies. Such clues indicate that the human-designed complex of ecosystems found in Biosphere II was not sustainable. Unfortunately, a number of problems associated with the management of the facility and a lack of scientific integrity meant that only minimal scientific results were obtained (Watson 1993; Appenzeller 1994).

Other approaches to the study of Gaia are currently taking place. One is the formation of oceanic and lacustrine manganese nodules. Experimenters place geochemically and geologically insignificant quantities of specific microbial poisons, such as antibiotics, into the nodules, or their milieu, and then observe if the rate of precipitation of manganese is the same with or without microbial life. Such experiments allow us to distinguish processes developed and controlled by life. Another approach is to identify the elements required by life (i.e., carbon, nitrogen, sulfur), as well as those that are avoided because they are poisonous, and ask questions about their distribution on the Earth. The distribution of gold by physical and chemical processes is very different from its biotically influenced distribution (Mossman and Dyer 1985). A large amount of work needs to be done to test Gaia, yet the outline of life's dramatic effect on global element distribution is clear (Lapo 1987).

Sulfur, for example, is a "gaian element." Required by all cells as a component of the sulfur-containing amino acids (cysteine and methionine), sulfur is abundant in the ocean but very limited on land. Since transfer through the rock cycle is not fast enough for the needs of life, the Gaia hypothesis requires that massive ways of moving sulfur from the ocean to the land exist. Therefore, based on gaian thinking, one expects to find a significant sulfur flux via the fluid phase (either through water or the atmosphere). As the net flow of liquid water is from land to the oceans, the movement of sulfur to land is far more likely to be through the atmosphere. This gaian reasoning led Lovelock to seek sulfur-containing gases to fulfill these needs. He discovered the major sulfur transporter was not hydrogen sulfide, as claimed, but dimethyl sulfide and dimethyl sulfoxide, gases produced by marine phytoplankton (Charlson et al. 1992). This serves as a specific example of science generated by the Gaia hypothesis.

Gaia: A Special Relationship with Humans?

As a relatively young species, humans have not been present long enough to play an important role in gaian development. Have humans recently developed a special place in Gaia, its continued existence or its destruction? An all-out nuclear war would destroy industrial centers, major cities, and greatly alter ecosystems. Global warfare would most likely eliminate Beethoven, analytical geometry, and the construction of particle accelerators. Our actions even in peace time, significantly impact the planet. Pollution, habitat degradation, ecosystem destruction, and resource depletion threaten the existence of a significant percentage of the species on Earth, including ourselves. We are not, however, in a position to destroy Gaia (Sagan and Margulis 1993).

During the Proterozoic Eon, beginning 25 hundred million years ago, gaseous oxygen first appeared in significant quantities (Schopf 1978; Cloud 1988). Exposure to the increasing levels of atmospheric oxygen caused naturally occurring deposits of uranium oxides to undergo spontaneous nuclear fission, resulting in enormous nuclear explosions in what is now Gabon in central Africa. Although similar to what we can do with our weapons technology, life survived these prehuman disasters. The same life that survived the Cretaceous-ending meteorite can certainly endure other environmental ravages.

We are presently worried about atmospheric carbon dioxide because it has increased from about 280 ppm (0.028%) in 1860 to about 350 ppm (0.0350%) at present, an increase of 25%. This rapid change in carbon dioxide has the potential to significantly change the global environment and thus severely disrupt present human societies and many ecosystems (see Chapter 8, "The Loss of Biodiversity," and Chapter 14, "The Challenge of Global Climate Change"). But it will not extinguish life on Earth. Atmospheric oxygen went from 0.0000001% about 3 billion years ago to the present 21%—eight orders of magnitude (Schopf 1978). Life and ecosystems have dramatically changed, for example, the extinction of many anaerobic species preceding the proliferation of aerobic species. Life, however, itself has never been extinguished: Gaia persists (Barlow and Volk 1989).

Because we best relate to objects within our size range, adult humans actually comprehend ecosystems much like kindergarten children do. We are more readily aware of the machinations of elephant-sized beings than of bacteria or protoctists. We find it difficult to comprehend urban blight the size of a continent. Consequently we tend to believe that the more noticeable organisms are the most significant. This then naturally leads to a belief that the destruction of these organisms and their associated ecosystems signals destruction of all life on Earth. The major components in gaian atmospheric and sediment maintenance, however, are microbes, not animals and plants (Figure 17.4). What we tend to perceive as scum, crust, turbid water, floc, and beach foam, in fact, are thriving microbial ecosystems. The crust in Figure 17.4a yields a sample like that shown in Figure 17.4b. Electron microscopy reveals tightly packed phototrophic bacteria like those shown in Figure 17.5.

Animals and plants exchange CO_2 and O_2 with the atmosphere and hydrosphere. But O_2 and CO_2 gas exchange are activities completely represented in the microbiota. Although the processes of photosynthesis and respiration are found in both prokaryotes (bacteria) and eukaryotes (protoctists, fungi, animals, and plants), it was in the prokaryotes that these biological processes evolved. Nitrogen from the atmosphere is fixed by metabolic virtuosities only found in prokaryotes. Nitrification, denitrification, reduction of sulfate to sulfide, and the oxidation of organic sulfur are also unique to prokaryotes. So whereas the prokaryotes perform all the basic biochemical transformations, animals and plants accomplish only a very small subset. If all animals and plants were driven to extinction, the speed of ecosystem processes would no doubt slow down and their quality would change, but the basic planetary surface processes would remain intact. If microorganisms were to be lost, then all life would be lost; the articulations of the biosphere would be unhinged. It is the microbes, not the animals or plants, that would survive even the worst of the environmental disasters that humans impose on the planet.

We can dramatically alter present ecosystems and even destroy ourselves. The question facing humans is not if we will eliminate life from the planet Earth. Rather, it is what kind of planet do we want to inhabit and leave for our descendants? Do we want a world with a diversity of life—forests, grasslands, and coral reefs—a world where all humans have access to basic necessities and a healthy environment? Or do we want a world where plants and animals have been largely eliminated and in which sufficient resources to meet the needs of humanity are absent? Our problem is that even if we concur on an optimal world, given our growth rate and expansionist tendencies, it will be very difficult to implement such an agreement. Whether or not our decisions are conscious, our actions impact the future. In the ferocity of our quest to develop and prosper, we tend to exterminate the ecosystems needed for our own survival. Gaia existed for over 3.5 billion years without humans, and it can certainly exist in our absence. The acknowledgment that life plays a significant role in regulating the global surface environment and the incorporation of that knowledge into our everyday lives represents an essential step towards ensuring the persistence of our species.

Interdisciplinary Discourse

The Gaia hypothesis has been refined as communication between its proponents and some geophysicists has developed. Geophysicists and climatologists recognize that many of the phenomena they study are at least in part biological; even worse, they are microbiological and protoctistological. Geologists generally are learning that they cannot fully understand their subject until they study microbes, including protoctists. Biologists experienced a similar revolution when they were told by physical scientists that they had to study chemistry in order to understand biology. Although there was initial resistance, eventually biologists embraced chemistry. Two generations ago, biologists did not study chemistry because there was no chemistry in biology. Today, most geologists do not study biology because they are unaware of its relevance. The Gaia hypothesis indicates that biology and geology intertwine in great mutual relation and that courses in microbiology and atmospheric chemistry are essential to Earth system science. It also instructs biologists that evolutionary theory developed in the absence of knowledge of meteorological, geological, and geochemical sciences is inadequate. The gaian viewpoint is not popular because research scientists and teachers want to continue business as usual. There is resistance now and there will be much more; it may be a generation or so before gaian analysis is fully integrated into "ordinary science."

(a)   (b)

Figure 17.4. A *Microcoleus* microbial mat ecosystem in the Sippewissett, salt marsh (Massachusetts). *Microcoleus* mats develop in intertidal zones separated from the open ocean by barriers such as volcanic rocks, mangrove swamps, or sand dunes as at Sippewissett. (a) The *Microcoleus* mat covers the ground on the right side of the picture. (b) The distinct layering of the microbial communities is evident in this hand-held sample of the mat. The darker layer just below the surface is dominated by *Microcoleus cthonoplastes,* a cosmopolitan sheathed filamentous cyanobacteria. The light colored layer just below it is dominated by purple phototrophic bacteria (e.g., *Chromatium* and *Thiocapsa*) like those in Figure 17.5.

But progress is being made. Even without wide acceptance, the Gaia hypothesis has encouraged thought about ecosystems at a global level and facilitated increased awareness that life plays a significant role in maintaining the planet's environment. Scientists now realize that many problems have to be studied with an interdisciplinary approach at a global level. They are far more open to the possibility that life has a powerful influence on the Earth's environment, and they mount experiments with this in mind. For example, complex computer programs called general circulation models (GCMs) are used to predict climate change, but most treat the Earth as if it were a dead planet; the influences of life are often ignored. Recently, however, two ecologists (Colin Prentice and Ian Woodward) have crafted GCMs that include the interactions of ecosystems with the global climate system, and testing of these new GCMs is currently taking place (Baskin 1993). Bacteria have been found to have a role in the chemical weathering of subsurface rock, and they may also influence the flow of groundwater (Appenzeller 1992; Hiebert and Bennett 1992). The oxidation of ferrous iron by anoxygenic photosynthetic bacteria has been proposed, in addition to the oxidation brought about by oxygenic photosynthetic bacteria, as a major process in the generation of the extensive iron ore deposits (Widdel et al. 1993). And the large drop in atmospheric levels of CO_2 that occurred in the Paleozoic Era has been most strongly linked to an increase in solar luminosity and the rise of vascular plants, not to tectonic processes (Berner 1993).

Conclusions

The Gaia hypothesis represents a paradigm shift. The resistance to a changing paradigm is different from simple hypothesis testing. Many assumptions are simultaneously affected. Lovelock is fond of saying that the 1974 Gaia hypothesis (Margulis and Lovelock 1974) has not been controversial; it has just been ignored. Although Gaia has generated science literature, both technical and popular, for 25 years (Margulis and West 1993), it has only recently gained some notice and reaction. When Darwin argued that humans had evolved from anthropoid ape ancestors, that was controversial. People understood it, they hated it, and they fought it. But the Gaia hypothesis is not yet even controversial in this sense because people do not yet know and understand it. In contrast with evolutionary theory, which has primarily influenced only the scientific discipline of biology, acceptance of the Gaia hypothesis will dramatically affect many fields of science. The acceptance of a gaian viewpoint will alter our behavior toward each other, the other animals, the plants, the microbiota, and the abiotic environment. As with Darwinian evolution, the comprehension of Gaia theory will be accompanied by cultural and even religious responses. We concur with Thompson's (1981) assertion that gaian concepts are entering our awareness first as global music and the graphic arts. Hopefully, we

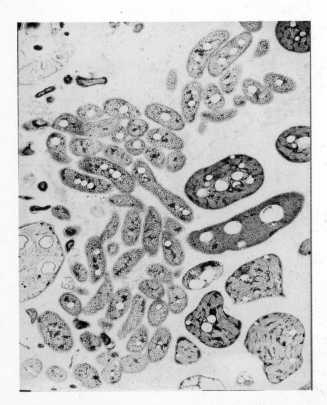

Figure 17.5. Electron micrograph of purple phototrophic bacteria on whose vesicles photosynthesis—transformations of atmospheric carbon dioxide to organic compounds—occurs. Taken directly from a natural sample, unidentified. These are most probably members of the Chromatiaceae. (Courtesy of Lynn Margulis. Picture taken by David G. Chase.)

will stop viewing the natural world merely as an exploitable resource and come to understand that human survival and the quality of our lives are intimately intertwined with Gaia. Ultimately, gaian theory will allow us to comprehend that we are but one thread, among many billions, of the ancient, complex fabric of life.

References

Appenzeller, T., "Deep Living Microbes Mount a Relentless Attack on Rock," *Science* 258, 1992, p. 222.

Appenzeller, T., "Biosphere 2 Makes a New Bid for Scientific Credibility," *Science* 263, 1994, pp. 1368–1369.

Barlow, C. and T. Volk, "Open Systems Living in a Closed Biosphere: Implications for the Gaia Debate," *Biosystems* 23, 1989, pp. 371–384.

Baskin, Y., "Ecologists Put Some Life Into Models of a Changing World," *Science* 259, 1993, pp. 1694–1696.

Berner, R. A., "Paleozoic Atmospheric CO_2: Importance of Solar Radiation and Plant Evolution," *Science* 261, 1993, pp. 68–70.

Charlson, R. J., S. E. Schwartz, J. M. Hales, R. D. Cess, J. A. Coakley, J. E. Hansen, and D. J. Hofmann, "Climate Forcing by Anthropogenic Aerosols," *Science* 255, 1992, pp. 423–430.

Clements, F. E., "The Nature and Structure of the Climax," *Journal of Ecology* 24, 1936, pp. 252–284.

Cloud, P. E., *Oasis in Space,* New York: W.W. Norton, 1988.

Delwiche, C. C., "The Nitrogen Cycle," *Scientific American* 223(3), 1970, pp. 136–146.

Gleason, H. A., "The Individualistic Concept of the Plant Association," *Bulletin of the Torrey Botanical Club* 53, 1926, pp. 7–26.

Hiebert, F. K. and P. C. Bennett, "Microbial Control of Silicate Weathering in Organic Rich Ground Water," *Science* 258, 1992, pp. 278–281.

Hunten, D. M., "Atmospheric Evolution of the Terrestrial Planets," *Science* 259, 1993, pp. 915–920.

Joseph, L., *Gaia: Growth of an Idea,* New York: St. Martins Press, 1990.

Kelly, K., "Biosphere II," *Whole Earth* 67, 1990, pp. 2–13.

Lapo, A. V., *Traces of Bygone Biospheres,* MIR Publishers: Moscow and Synergestic Press: Oracle, AZ, 1987.

Lovelock, J. E., *Gaia: A New Look at Life on Earth,* New York: Oxford University Press, 1979.

Lovelock, J. E., *The Ages of Gaia: A Biography of Our Living Earth,* New York: W.W. Norton, 1988.

Margulis, L. and J. E. Lovelock, "Biological Modulation of the Earth's Atmosphere," *Icarus* 21, 1974, pp. 471–489.

Margulis, L. and O. West, "Gaia and the Colonization of Mars," *GSA Today* 3, 1993, pp. 277–280, 291.

May, R. M., "The Evolution of Ecological Systems," *Scientific American* 239(3), 1978, pp. 160–175.

McIntosh, R. P., "The Continuum Concept of Vegetation," *Botanical Review* 33, 1967, pp. 130–187.

Mossman, D. J. and Dyer, B. D., "The Geochemistry of Witwaterrsand-type Gold Deposits and the Possible Influence of Ancient Prokaryotic Communities on Gold Disolution and Precipitation," *Precambrian Research* 30, 1985, pp. 303–319.

Sagan, D., "Biosphere II: Meeting Ground for Ecology and Technology," *The Environmentalist* 7, 1987, pp. 271–281.

Sagan, D., "What Narcissus Saw: The Oceanic 'I'/Eye," *In* J. Brockman (ed.), *The Reality Club 1*, New York: Prentice Hall, 1988, pp. 247–266.

Sagan, D., *Biospheres: Metamorphosis of Planet Earth*, New York: McGraw-Hill, 1990.

Sagan, D. and L. Margulis, "God, Gaia and Biophilia," *In* E. O. Wilson and S. Kellert (eds.), *Biophilia*, Washington, DC: Island Press, 1993.

Schopf, J. W., "The Evolution of the Earliest Cells," *Scientific American* 239(3), 1978, pp. 110–138.

Thompson, W. I., *The Time Falling Bodies Take to Light*, New York: St. Martin's Press, 1981.

Watson, T., "Can Basic Research Ever Find a Good Home in Biosphere 2?," *Science* 259, 1993, pp. 1688–1689.

Whittaker, R. H., "A Criticism of the Plant Association and Climatic Climax Concepts," *Northwest Science* 25, 1951, pp. 17–31.

Whittaker, R. H., "A Consideration of Climax Theory: the Climax as Population and Pattern," *Ecological Monographs* 23, 1953, pp. 41–78.

Widdel, F., S. Schnell, S. Heising, A. Ehrenreich, B. Assmus, and B. Schink, "Ferrous Iron Oxidation by Anoxygenic Phototrophic Bacteria," *Nature* 362, 1993, pp. 834–836

Glossary

Abiotic The nonliving component of the environment.

Acid Deposition Rain, snow, fog, or dust that contains higher than normal levels of sulphuric and/or nitric acid. Also called acid rain or acid precipitation.

Aerobic Requiring oxygen.

Agribusiness Large-scale corporate farming.

Agroforestry The practice of interplanting agricultural crops with forest crops.

Albedo The percentage of incident light reflected by a surface.

Anaerobic Not requiring oxygen.

Anthropocentric Considering humans as the most important part of the world and that only the human species has an inherent right to life. All other species exist solely to meet human needs.

Anthropogenic Produced or caused by human activities.

Aquaculture The production of food in a controlled, aquatic environment.

Aquifer Underground, porous, water saturated zone of sand, gravel, or rock yielding large quantities of groundwater.

Arable Land suitable for productive cultivation of crops.

Biocentric Considering that each species has an inherent right to exist and that humans are no more important than any other species.

Biocide Chemicals that are primarily used to kill life (i.e., pesticides, herbicides, fungicides).

Biodiversity See biological diversity.

Biogenic Generated by life.

Biogeochemical Cycle The cycling of nutrients from living components of the environment to nonliving components and then back again (i.e., the carbon cycle and the hydrological cycle).

Biological Control (see IPM). The use of parasites, predators, and diseases to control pest populations.

Biological Diversity The number and variety of species found on Earth, including the genetic diversity among members of a species.

Biomass (1) Total amount of all living material in an ecosystem, usually expressed on a dry weight basis; (2) total amount of all organic matter (including living organisms as well as decomposing organic materials) in an ecosystem; (3) plant and animal matter used as fuel.

Biome A large regional ecosystem having a distinct physiognomy (or appearance); for example, polar tundra, temperate deciduous forests, tropical wet forests.

Bioregion A region having distinctive, natural characteristics, such as plants, animals, landforms, climates, watershed, and soils, transcending any political and/or human-made boundaries.

Biosphere The region of Earth where life is located.

Biosphere Reserve Ecosystems that are to be permanently preserved through a program sponsored by the United Nations.

British Thermal Unit (Btu) The amount of energy required to raise the temperature of 1 pound of water by 1 degree Fahrenheit.

Calorie Amount of energy needed to raise temperature of 1 gram of water 1 degree Celsius.

Carcinogen An agent causing cancer, i.e., asbestos fibers, radon gas, cigarette smoke.

Carrying Capacity The maximum population of a species that a particular ecosystem can accommodate on a continuous and sustainable basis without degradation.

Chlorofluorocarbon (CFC) Organic compounds consisting of carbon, chlorine, and fluorine, used as propellants and as coolants.

Composting A natural waste recycling method using bacteria and fungi to decompose plant and animal matter.

Deforestation Removal of trees in a given area without replanting.

Desalination Removal of salts and minerals from salt water to produce potable/drinkable fresh water.

Desertification Ecosystem degradation, either naturally or by humans, resulting in a 10% or greater decline in biological productivity. In extreme cases desertlike conditions may result.

Dioxin An organic chemical believed to be carcinogenic. It is a contaminant in Agent Orange and is also released as a byproduct from some incineration facilities.

Doubling Time The number of years it takes a population, or some other quantity, to double, assuming a constant growth rate.

Ecosystem An assemblage of interacting species and their associated physical and chemical environments.

Endangered Species A species that is very likely to become extinct in the immediate future.

Epiphyte (Air Plant) Plants that live on other plants without having their roots in contact with the soil.

Estuary A semienclosed body of water where both fresh and salt water mix. Estuaries are critical for the survival of many aquatic animals.

Eukaryote An organism composed of cells containing a distinct nucleus; contrast with prokaryote.

Eutrophication Physical, biological, and chemical processes taking place in a body of fresh water that enrich it with nutrients; known as *cultural eutrophication* when enhanced and accelerated by human impacts on the environment.

Evapotranspiration A combination of natural processes that release water vapor from the soil and plants back into the air.

Extinction When no individuals of a particular species remain alive.

Extractive Reserve An ecosystem reserve in which the sustainable harvest of natural products is allowed.

Geothermal Relating to heat in the Earth's interior created by gravitational collapse during the Earth's formation and by subsequent radioactive decay.

Global Warming An increase in global temperature resulting from an enhanced greenhouse effect when atmospheric concentrations of greenhouse gases (i.e., water vapor, carbon dioxide, methane, nitrous oxide, and ozone) are increased.

Greenhouse Effect The warming of the Earth's surface and lower atmosphere that occurs when infrared radiation emitted from the Earth's surface is absorbed and reradiated by greenhouse gases.

Green Revolution Measures (i.e., fertilizer, pesticides, and new strains of grain) to improve agricultural productivity in less developed countries.

Groundwater Underground water source replenished from precipitation that percolates through the soil into the water system below.

Hectare A unit of measurement equal to 10,000 m² (approximately 2.48 acres).

Hydrologic Cycle Circulation of water from the atmosphere to the Earth's surface via precipitation and the eventual return of that water through evaporation and transpiration to the atmosphere.

Hydroponics Growing plants with their roots in water instead of soil, often with minerals added.

Indigenous Native, innate, inherent, original. Native American Indians are the indigenous peoples of America.

Infant Mortality Rate The number of infants per 1,000 births who die before their first birthday.

Insecticide See biocide.

Integrated Pest Management (IPM) The combined use of biological control, biocides, knowledge of ecosystem function, and certain cultivation techniques to control pests and weeds.

J Shaped Curve A curve formed when geometric or exponential growth of populations is graphed.

Landfill A place where waste is buried. Sanitary landfills are used for ordinary waste and secure landfills are established for hazardous waste.

Laterite An iron-rich subsoil layer occurring in some tropical soils; if exposed and allowed to dry, laterite permanently hardens and forms a pavementlike soil.

Less Developed Country (LDC) A country with low to moderate industrialization and GNP per person; most are found in the Southern Hemisphere.

Malnutrition Inadequate amounts of essential nutrients (such as protein) in one's diet.

Monoculture Cultivation of only one crop for harvest.

Montane Occurring in the mountains.

More Developed Country (MDC) A country with a high degree of industrialization and GNP per person; most are found in the Northern Hemisphere.

Multiple Use Principle of managing land for many uses rather than a single use. It is usually applied to federal lands.

NIMBY (Not In My Back Yard) Syndrome When residents of a particular area do not want a facility like a landfill to be sited in their neighborhood.

Nonpoint Source Source of pollution that cannot be pinpointed.

Organic (1) Carbon-containing chemicals; (2) pertaining to natural methods of crop production; (3) having to do with life.

Overgrazing Consumption of vegetation beyond the carrying capacity of the land, resulting in ecosystem degradation and, in extreme cases, desertification.

Oxisol An order of tropical soils. Usually very infertile, oxisols have a thick subsurface horizon dominated by hydrous oxides of iron and aluminum.

Ozone A molecule containing three oxygen atoms (O_3). A major component of smog, ozone is a known respiratory irritant. Stratospheric ozone absorbs UV-B, which is harmful to life.

Pesticide See biocide.

pH The measure of acidity or alkalinity. The pH scale ranges from 0 to 14, with acidity below 7 and alkalinity above 7.

Photochemical Smog Secondary air pollutants formed from the interactions of sunlight with primary air pollutants (hydrocarbons and nitrogen oxides), often forming a haze that can be harmful to humans and other species.

Photosynthate Products of photosynthesis.

Photovoltaic A process in which the energy in sunlight is absorbed by a semiconductor device (called a photovoltaic, or solar, cell) and is used to directly produce electricity. The cells may be connected to modules to produce the desired voltage and current output.

Point Source Pollution Pollution that can be traced to a specific source.

Polychlorinated Biphenyl (PCB) A group of toxic, synthetic substances used as coolants and plasticizers. Accumulating in fatty tissues, PCBs can be biologically amplified in food chains and can be harmful to humans and other species.

Polyculture The cultivation of multiple crops.

Prokaryote Lacking a membrane-bound nucleus. Found only in the kingdom Monera, prokaryotes include the bacteria and the cyanobacteria.

Protoctista One of the kingdoms of life. It is composed of eukaryotic, single-celled organisms and includes the algae and protozoans.

Radiation (1) The flow of energy across space by electromagnetic waves. The electromagnetic spectrum includes long-wave radiation (radiowaves, microwaves, infrared), visible light, and short-wave radiation (ultraviolet, x-rays, gamma rays). (2) The emission of high velocity particles and electromagnetic radiation from the disintegrating nuclei of radioactive atoms.

Radioactivity The process whereby the disintegrating nuclei of unstable atoms emit alpha and beta particles and gamma rays.

Rain Forest Forests that receive large amounts of rainfall. Although mostly tropical in distribution (see tropical wet forest), a few rain forests can be found in temperate regions having wet oceanic climates.

Reforestation The process of replanting trees in deforested areas.

Renewable A resource such as wood that would be available on a continual basis if used in a sustainable manner. Also, certain sources of energy that are regenerated on a continuing basis from the sun. This includes direct and indirect (wind, hydropower and biomass) forms of solar energy.

Reserves Identified resources that can be economically extracted with current technology.

Salinization Accumulation of salts in soil through irrigation that eventually can cause loss of productivity and death of crops.

Speciation Formation of new species by evolutionary changes in existing species.

Species A group of organisms that interbreed (or have the potential to interbreed) and that are reproductively isolated from other organisms.

Stratosphere The atmospheric layer extending approximately from 10 to 50 km (6 to 31 miles) above the Earth.

Subsistence The minimum level required for sustaining life, such as when a farmer produces only enough food to feed one family.

Surface Water Bodies of fresh water found at the Earth's surface (i.e., lakes and rivers).

Sustainable A process that can be utilized on a continuing basis without degradation to the environment.

Symbionts The partners of a symbiotic relationship.

Symbiosis The living together of two or more species (i.e., corals, lichens, mycorrhiza).

Temperate Having recognizable seasonal changes.

Teratogenic Causing birth defects.

Thermal Inversion/Temperature Inversion When a layer of cool, dense air is trapped beneath a layer of warm, less dense air.

Threatened Species A species that is on the threshold of extinction but is not likely to become extinct in the immediate future.

Tropical Refers to the region within the Tropics of Cancer and Capricorn that experiences little change in seasonal temperature.

Tropical Cloud Forests High elevation tropical forests that receive a significant amount of water from extensive exposure to cloud mists. A special type of cloud forest, elfin woodland, occurs at very high elevations, where the trees are stunted and have a gnarled, twisted shape.

Tropical Seasonal Forests Deciduous tropical forests that experience a pronounced dry season every year. Most of the trees shed their leaves during each dry season. Also called tropical dry forests.

Thornwoods Tropical forests composed of small scrubby trees, many having spines and small evergreen leaves. Usually associated with semi-arid climates. Also called thorn woodlands.

Tropical Wet Forest Evergreen tropical forests that receive abundant amounts of rain, far in excess of evapotranspiration, throughout the year. Also called tropical rain forests.

Troposphere The innermost, or surface, layer of the atmosphere.

Tundra A treeless biome found in the polar regions and also above the treeline at the tops of high mountains.

Ultraviolet Radiation A form of electromagnetic radiation that is capable of damaging biological molecules and thus harming life.

Undernutrition A condition resulting when people chronically have inadequate amounts of food.

Watershed The land area providing water to a particular body of water such as a river or reservoir.

Zero Population Growth (ZPG) A population having no net growth; this occurs when the birth rate equals the death rate (assuming that migration is not a variable).

Index

Acid deposition, 221–225
air quality policies and, 237–240
causes of, 13–14, 76, 218–220
energy efficiency and, 91
impacts of, 13–14, 221–225
Aerosols, 220
control strategies and, 233–234
global climate change and, 229–230
haze and, 228–229
Africa
drought, climate change and, 296–298
Agribusiness
family farms and, 58
human values and, 58
Agriculture
agroforestry techniques in Rwanda
and, 62–63
alternative, 51–52
biotechnology and, 72–74
deforestation and, 154–156, 166–167
desertification and, 178–179, 181–184
ecological soundness in, 54
economic viability in, 54–56
energy and, 40–43
erosion and, 71
hand production in Mexico, 41
humaneness in, 57–58
hydro-development and, 112
industrial, 6–7, 38
land resources and, 39
livestock and, 45–47
methods to help create sustainability,
64
social justice in, 56–57
sustainable, 7, 51–67
sustainable agriculture legislation, 63
sustainable agriculture organizations, 63
technology, natural resources and,
38–50
water resources and, 39–40
worldwide trends in, 38, 51

Agroforestry, 62, 75
and preventing deforestation, 185–
186
Air pollution, 215–241
acid deposition and, 13–14, 221–225
aerosols, particulates and, 220,
233–234
air quality policies and, 236–240
carbon dioxide and, 88, 217, 234–235,
271–273
carbon monoxide and, 217–218,
234–235
effects of, on human health, 91,
222–225
fossil fuel combustion and, 91,
215–221, 231–236, 271–273
global climate change and, 15–16,
87–91, 229–230, 255, 263–264,
269–287, 298–303
haze and, 228–229
hydrocarbons and, 220–221, 234
industrial smog and, 228
input control of, 235–236
nitrogen oxides and, 219–220,
232–233
output control of, 231–235
ozone and, 225–227, 229
photochemical smog and, 225–228
regional, 91
stratospheric ozone depletion and,
14–15, 245–268
sulfur dioxide and, 218–219, 231–
232
Albedo, 270
Animal manure
as organic fertilizer, 43–44
Animals
as food resources, 46–47
Aswan Dam, 115–116
Atmosphere
Gaia and the control of, 17, 336–344

Balbina Dam, 111
Bhopal, India
    background to accident in, 305–307
    consequences of pesticide accident in,
        308–310
    differences and similarities with Cher-
        nobyl, 321–325
    government vulnerability for, 326–327
    lessons learned from, 325–329
    lessons not learned from, 329–330
    pesticide accident in, 16, 308–311
    private corporate vulnerability and,
        325–326
    psycho-social impacts of accident in,
        324–325
    responsibility for accident in, 322–324
    technology transfer issues and, 330–331
Biodiversity
    evolution of, 127–128
Biodiversity crisis, 126–150
    agriculture and, 133–134
    climate change and, 145–146
    ecosystem reserves and, 141–144
    ecosystem restoration and, 144
    forestry and, 134–135
    habitat loss and, 10–11, 131–132
    humans and, 128–129
    hunting and, 131
    introduced species and, 131–132
    net primary production and, 129–134
    pollution and, 132–133, 145–146
    solutions to, 11, 137–144
    species preservation and, 137–141
    UN and, 140, 142–144
    use of nontimber products and, 78–79
Biogas, 76–77
    plants, 8
Biological control of insect pests, 44–45
Biomass, 75–77
Biosphere II, 345–346
Biosphere reserves, 146
Biotechnology
    agriculture and, 72–74
Birds
    ocean pollution and, 208
Brazil
    hydro-development and, 110–111

Capital
    depletion of Earth's natural, 55

Carbon dioxide (CO_2), 217, 271–272
    climate models and, 273–274, 292–296
    control strategies and, 89, 234–235,
        279–280
    global climate change and, 15, 89,
        229, 269–287, 298–299
    ice cores and, 291–292
    increases in atmospheric, 88, 271, 301
Carbon monoxide (CO), 217–218
    control strategies and, 235–236
    global climate change and, 229
Carrying capacity
    food production and, 27, 48
    population growth and, 20–28, 48
Charcoal stoves, 75–76
Chernobyl, former Soviet Union
    causes of accident at, 311–312
    cross-boundary contamination and,
        317–321, 327–328
    differences and similarities with
        Bhopal, 321–325
    economic loss of accident in, 317–319
    government vulnerability and,
        326–327
    impact on nuclear power industry
        and, 328–329
    impacts of accident outside of former
        Soviet Union, 317–321
    impacts of accident within former So-
        viet Union, 312–317
    lessons learned from, 325–329
    lessons not learned from, 329–330
    nuclear power plant accident in, 16,
        310–332
    psycho-social impact and environ-
        mental stigma from, 324–325
    responsibility for accident in, 322–324
Children
    malnutrition and, 26
China
    family planning and, 32–33
    hydro-development and, 112, 122
Chlorofluorocarbons (CFCs)
    eliminating the flow of, 261–263
    global climate change and, 229, 272
    ozone depletion and, 14–15, 230, 245,
        250–251
Clean Air Act, 236, 238–240
Climate, 288–289 (*See* Climate change,
    Greenhouse effect)

desertification and, 180–181
protection strategies, 87–91
warming (*See* Climate change, Greenhouse effect, Ozone layer)
Climate change, 15–16, 269–287, 288–304
  climate models and, 273–274, 293–296
  European community and, 282
  global, 15–16, 229–230, 269–287, 298–303
  Global Climate Convention Treaty and, 282
  greenhouse gases and, 270–273, 298–299
  LDCs and, 282, 284
  Milankovitch theory and, 291–292
  natural, 289–292
  policy solutions to, 279–285
  potential impacts of, 273, 276–279
  United States and, 282–283
Climate models, 273, 292–296
  African drought and, 296–298
  tests of, 299–300
  uncertainties and inaccuracies in, 274, 294–296
Conservation
  land, 71
  water resources and, 71–72
Contraception, 31–32
Convention on Biological Diversity, 144
Convention on International Trade in Endangered Species (CITES), 140
Corn production
  energy inputs and, 40–43
  limiting factors and, 43
  water resources and, 39–40
Crop rotation, 44–45
Crop yields
  energy inputs and, 46–47
Cultural carrying capacity, 23–24
Cyanobacteria, 73

Dam failure, 113
DDT
  impact on birds, 132–133
Debt
  hydro-development and, 120
Debt for Nature Swaps, 169–170
Deforestation
  alternatives to, 159–169

biodiversity and, 157–158
  causes of, 154–166, 166–167
  climate change and, 159, 271–272
  desertification and, 180
  rates of, 151–152
  soil degradation and, 158
  tropical, 74–75
Denmark
  wind technology in, 100–101
Desertification, 12, 174–188
  causes of, 176–180
  climate and, 180–181
  in Africa, 182–185
  in the United States, 181–182
  policies and, 187–188
  preventing, 185–186
  war and, 184–185
Development
  sustainable, 68–85
  technology and, 68–71
  use of nontimber products and, 78–79
Dissolved oxygen
  nutrient loading and, 204–205
Dredging, 197–198
Dust Bowl, 181–182

Ecological reserves
  tropical forests and, 165
Economics
  appropriate technology benefits to, 69–71
  use of nontimber products and, 78–79
Economics and environment
  agriculture and, 54–56
Ecosystem destruction
  hydro-development and, 110–113
Ecotourism
  as alternative to deforestation, 165–166
Electricity
  biomass production of, 76
Endangered Species Act, 137–140
Energy, 86–106 (*See also* Energy resources)
  acid deposition and use of, 13–14, 91, 218–220
  agriculture and, 6, 40–43, 47
  charcoal stoves and, 75–76
  environmental impacts of, 9, 87–91, 221–230, 277–279

growth in LDCs, 75
renewable technologies in LDCs and, 75–76
technology and, 8–9, 99–103
wood stoves and, 8, 75–76
Energy conservation (*See* Energy efficiency)
Energy efficiency, 86–87, 92–106
  agriculture and, 40–43
  buildings and, 97–99, 101
  hydro-development and, 122–123
  impact on economy and, 97–99
  international perspective on, 95–96, 102–104
  MDCs and use of, 8–9
  national economies and, 92–93
  potential for reduced energy use and, 92, 97–99
  reducing greenhouse effect through, 89, 280, 283
  regional air pollution and, 91
  renewable energy use and, 9
  research and development in OECD and, 99–101
  research and development in United States and, 101–102
  technology and, 99–103
  Third World nations and, 9, 96–97
  transportation and, 93–95
Energy resources
  agriculture and, 6, 40–43, 45–48
  air pollution and, 87–91
  biomass and, 87, 89
  coal, 90
  cost of renewables and, 87
  energy expenditures for R&D and, 99–102
  energy policy for, 102–104, 279–284
  fuelwood production and, 75
  hydropower, 10, 90, 107–110
  livestock production and, 45–48
  nuclear (*See also* Nuclear energy), 90, 99, 102
  petroleum, 93–95
  wind, 100
Energy strategy, 86–106
  climate change and, 88–89, 96, 279–284
  market forces and, 103–104

sustainable societies and, 86–87, 102–104
Energy, U.S. Department of
  energy research and development efforts in, 101–102
  global climate change and National Energy Act of 1992, 282
  response to Chernobyl, 329
Environment
  desertification and impacts on, 12, 174–176
  energy use and impacts on, 9–10, 87–91
  hydro-development and impacts on, 10, 110–115
  impacts of Chernobyl on, 312–321
  impacts of development on, 68
  poverty in LDCs and impact on, 81
  protecting with energy efficiency, 97–98
  sustainable technologies and, 70–71, 90
Environmental regulation
  Gaia and, 336–351
Erosion
  agriculture as cause of, 39, 71
Ethanol
  production as fuel, 77
Europe
  impacts from Chernobyl on, 317–321
Evolution, 126–128
External costs of economic goods
  agriculture and, 55–56
Extractive reserves, 77–78, 160–161

Family planning, 31–34
Feedback loops
  effect on climate change, 274–275, 289, 299, 303
  uncertainties in climate change and, 274, 276, 303
Ferret, 130–131
Fertilizers
  alternatives to industrial, 43–44, 73
  industrial agriculture use of, 42–43
Fish
  hydro-development and, 115
  ocean pollution and, 207–208
Food
  size of organic industry and, 63–64

Food resources
    impact on natural resources and, 38–39
    population growth and, 6–7, 24–28, 38, 48
    public health and, 38
    technology and, 38–39, 48
Forest management
    ecotourism and, 77
    natural techniques of, 74–75
Forestry
    LDC population dependence on, 77
    nontimber products of, 78–79
    sustainable, 71, 74–75
Fuelwood, 7–8
    farm production of, 75
Fungi
    soil, 73

Gaia hypothesis, 17, 336–344
    carbon dioxide and, 338–339
    environmental regulation and, 336–344
    global temperature and, 337–339
    humans and, 346–348
    nitrogen cycle and, 339–340
    sulfur and, 346
    superorganism concept and, 344–348
Global climate change, 15–16, 88–89, 229–230, 269–292, 298–303
    biodiversity and, 145–146
    desertification and, 181
Global Climate Convention Treaty, 281–282
Global warming (*See* Greenhouse effect)
Grain production
    population growth and, 27
Green manure, 44
Green Revolution, 6, 38–39, 52
Greenhouse effect, 15–16, 88–91, 229–230, 269–287, 298–299 (*See also* Climate, Climate change)
    adaptation approach to solving, 280
    climate impacts and, 15, 277–279
    climate models and, 273–274, 298–300
    coal use and, 90
    government policies towards, 88, 97–99, 281–284
    mitigation approach to solving, 356–357
    planting forests to reduce, 76, 89, 279
    policy dilemmas and, 15, 281, 284
    policy solutions to, 89–91, 93, 279–285
    potential climate surprises and, 276–277
    potential impacts of, 277–279
    role of CFCs in, 255, 263–264, 272
    scientific studies of enhanced, 273–274, 277, 291–292
    uncertainties in climate models and, 274, 283, 294–296
    uncertainty in climate change impacts, 276–277, 281, 283
    uncertainty in global warming signal and, 274–276, 283, 301–303
    United States energy policy and, 282–283
Greenhouse gases
    carbon dioxide as, 89–90, 270–272, 298–299
    chlorofluorocarbons (CFCs) as, 263–264, 272–273
    methane, 272, 299
    methane from ruminants and, 80–81
    nitrous oxide, 229, 272
    sources and sinks of, 219–220, 270–273

Halons
    effect on ozone layer, 245, 250–251
    eliminating the flow of, 261–263
Haze, 228–229
Herbicides, 44
Hydro-development, 107–125
    as energy source, 107–116
    deforestation and, 166
    economic impacts and, 119–120
    energy use and, 119
    environmental impacts and, 110–115
    health impacts and, 115–117
    social impacts and, 118–119
Hydrocarbons (HCs), 220–221
    control strategies and, 236
    global climate change and, 229
    haze and, 229
    photochemical smog and, 225–228
Hydrologic response, 200

Industrial smog, 228
Integrated Pest Management (IPM), 44–45, 73–74

Introduced species
  ballast water and, 209
  biodiversity and, 131–132
  reforestation and, 163–164
Invertebrates
  ocean pollution and, 206–207
Irrigation
  appropriate technology and, 71–72
  energy use and, 40

James Bay Project, 112–113
Japan
  energy R&D in, 102
Justice
  sustainable agriculture and, 56–57

Land
  desertification impacts on, 12, 174–176
  tropical forests and loss of fertile,
    11–12, 158
Livestock production
  energy inputs to, 45–47
  grain use and, 46–47
  improved conversion efficiency in, 81
  land use and, 46
  methane emissions from, 80–81
  organic techniques for, 62

Malaria, 115
Malnourishment, 26
Mass extinction
  biodiversity and, 127–129
Medicinal plants
  deforestation and, 157, 161–162
Methane, 80–81, 220–221
  global climate change and, 229,
    272–273, 299
Methyl mercury, 114
Mexico
  sustainable agricultural practices in, 61
Milankovitch theory
  climate change and, 291–292
Minimum-till cultivation, 44
Montreal Protocol, 14, 260
Mt. Pinatubo
  global climate change and, 230, 276
  impact on ozone depletion and, 261
Multinationals
  development aid and, 57
Mycorrhizae, 135

National Wildlife Federation
  preservation of tropical forests and,
    167–168
National Wildlife Refuge System, 141
Natural forest management
  as alternative to deforestation,
    162–163
Net Primary Production (NPP)
  biodiversity and, 129–130
  carrying capacity and, 20–23
Nitrogen
  as fertilizer for corn, 42–43
  green manure and, 44
Nitrogen cycle, 339–341
Nitrogen fixation, 73, 134
Nitrogen oxides (NO_x), 219–220
  acid deposition and, 221–225
  control strategies and, 232–233
  photochemical smog and, 225–228
Nitrous oxide (N_2O), 219–220
  as a greenhouse gas, 229, 272
  stratospheric ozone depletion and,
    230, 248
Northern spotted owl, 139–140
No-till cultivation
  energy use and, 44
Nuclear power
  impact of Chernobyl on, 16, 329–330
Nutrients
  in soil, 43–44
  ocean pollution and, 204–206

Ocean pollution, 191–211
  development and, 191–196
  nutrient loading and, 204–206
  pathogens and, 200–201
  sewage and, 201
  toxicants and, 202–204
Oceans
  environmental damage to, 13
OECD (*See* Organization for Economic
    Co-operation and Development)
Oil
  imports and, 93–94
  Persian Gulf war and, 94
  world supplies of, 93–95
Organic farming techniques, 52
  in Iowa, 61–62
Organization for Economic Co-operation
    and Development (OECD)

R&D efforts and expenditures by, 99–101
Ozone, stratospheric, 245–268 (*See* Ozone layer)
Ozone, tropospheric
  global climate change and, 229
  photochemical smog and, 225–228
Ozone depletion (*See* Ozone layer)
Ozone layer
  Antarctic depletion of, 14, 249, 256–257
  Arctic depletion of, 249, 258
  climate change and depletion of, 255, 263–264
  depletion of, 14–15, 245–251
  discovery of hole in, 14, 246, 256–257
  ecosystem impacts and depletion of, 14, 253–255
  human health and depletion of, 14, 251–253
  impacts of depleting, 14, 245, 251–255
  mid-latitude depletion of, 14, 249–250, 258
  natural balance of, 247–248
  protecting the, 14–15, 258–263
  scientific studies and depletion of, 255–258
  space shuttle and depletion of, 255–256
  SST and depletion of, 255
  uncertainties in depletion of, 15, 260–261

Parasites
  pest control and, 44–45
Particulates, 220
  control strategies and, 233–234
  industrial smog and, 288
Persian Gulf
  supplies of oil in, 94
  war of 1991, 94
Pest control
  cultural practices and, 45
Pesticides
  Bhopal disaster and, 305–307
  cultivation methods reduce need for, 45
  ecosystem impacts from, 59
  human health risks from, 56

integrated pest management and, 44–45, 73–74
Petroleum (*See* Oil)
Photochemical smog, 225–228
Phytoplankton
  reduction in, caused by ozone depletion, 254–255
Plants
  resistance to insects and disease, 44–45
Pollution
  air, 13–16, 87–91, 215–244
  energy consumption and, 13–16, 87–91, 217–221, 270–273
  ocean, 13–14, 191–214
  prevention and control using appropriate technology in LDCs, 79–80
Population displacement
  hydro-development and, 112–113
Population growth, 4–6, 19–36
  carrying capacity and, 20–28, 35–36
  deforestation and, 154
  family planning and, 31–34
  health and, 5
  poverty and, 27, 30–31
  resource consumption and, 28–30
  self-worth and, 30
  standard of living and, 28–29, 30–31
  UN and, 33–35
  war and, 26–27, 29–30
Poverty
  environmental degradation and, 29, 81
  grassroot financing to relieve, 69–70
  land reform and, 56
  LDCs and, 69–70
  population growth and, 27, 30–31
Predators
  pest control and, 44–45

Reductionism, 59
Reforestation, 74–75, 163–164
Renewable energy resources (*See* Energy resources, Energy efficiency, Biomass)
Renewable resources
  agriculture and loss of, 48
Resource(s)
  common property management of, 77
  MDCs and overuse of, 4

Rodents
  mycorrhizae and, 134–135
Rubber, 160
Rwanda
  agroforestry techniques in, 62–63

Salinization, 183
Sanitation
  technologies and costs in LDCs and
    MDCs, 80
Schistosomiasis, 115–117
Sedimentation
  hydro-development and, 114–115
Shoreline modification, 198–200
Small-scale hydro, 122
Snail darter, 139
Social carrying capacity, 24
Soil
  loss and impact on tropical forests,
    11–12
  nutrient loss from, 43
Soil conservation
  reducing soil erosion and, 43–44
Soil degradation
  deforestation and, 158
Space Shuttle
  ozone depletion and, 255–256
Standard of living, 28–29
Stratospheric Ozone Depletion (*See also*
    Ozone layer)
  CFCs and, 14–15, 230, 245–247, 250
  human and ecosystem impacts of, 14,
    251–255
  nitrous oxides and, 230, 248
Sulfur dioxide (SO_2), 218–219
  acid deposition and, 91, 221–225
  control strategies and, 231–232
  industrial smog and, 228
  photochemical smog and, 225–228
Supersonic Transport (SST)
  ozone depletion and, 255
Sustainability
  balanced life support systems and hu-
    man, 17
Sustainable agriculture, 7, 51–67, 71–75
Sustainable development, 68–85
  impact of Bhopal and Chernobyl on,
    330–331
  Rwanda and, 62–63

Technology
  agriculture, natural resources and,
    38–50
  appropriate, 7–8, 68–85
  disaster and impacts of, 16, 305–335
  energy use and, 8–9, 40–43, 75–77,
    89–103
  global environmental issues and, 1–18
  hand water pumps and, 80
  improved agricultural, 43–45
  methane reduction from ruminants
    and, 80–81
  nuclear power and, 99, 102, 310–335
  poverty, development and sustain-
    able, 81–82
  reduction of liquid and solid waste,
    80
  transfer implications of Bhopal and
    Chernobyl on, 330–331
Temperature
  global warming and, 270, 273–276,
    301–303
Tissue culture, 73
Transportation
  bicycles and, 77
Tropical Forests, 151–171
  agriculture and, 11–12, 71–75
  deforestation and, 151–171
  soils and, 153
  species diversity and, 11, 152–153

Ultraviolet radiation
  ecosystem impacts of, 253–255
  health impacts of, 251–253
  ozone depletion and, 245, 247–248
  skin cancer caused by, 252–253
  terrestrial life and, 133, 253–255
United Nations
  climate change and, 281–282
  population growth and, 33–35
  role in protecting ozone layer,
    259–260
United Nations (UNESCO)
  international biosphere reserves and,
    142
United Nations Conference on Desertifi-
    cation (UNCOD), 174, 182, 191
United Nations Conference on Environ-
    ment and Development (UNCED)

biodiversity and, 144
population growth and, 34
tropical forests and, 152
United Nations Environmental Program
    (UNEP)
desertification and, 187
endangered species and, 140
stratospheric ozone depletion and,
    259–260
United Nations Fund for Population Ac-
    tivities (UNFPA), 33–35

Vegetarianism
food supply and, 46–47

War
desertification and, 184–185
Waste(s)
biogas production from human and
    agricultural, 76–77
forestry and agricultural, 74
small-scale technologies to reduce, 80
Water (*See* Water resources)
Water conservation
crop choice and, 45

techniques for, 45
Water quality
hydro-development and, 113–114
Water resources
agriculture and, 39–40, 45
groundwater and, 45
population growth and, 21, 26
small-scale industry pollution in LDCs
    of, 79–80
technologies to conserve, 71–72
transpiration and, 39
Water supply
costs in LDCs and MDCs, 80
hand pumps and, 80
Women
income, property and, 56
overpopulation impacts on, 26, 30
Wood stoves
use in LDCs and, 75–76
World Bank
deforestation and, 167–169
hydro-development and, 120–121